COLLINS

Lean Six Sigma for Supply Chain Management

The 10-Step Solution Process

About the Author

James W. Martin is president of Six Sigma Integration, Inc., a Lean Six Sigma consulting firm, located south of Boston. He has served as an instructor at the Providence College Graduate School of Business since 1988. He instructs courses in operations research, operations management, and economic forecasting, as well as related quantitative subjects, and counsels MBA candidates from government organizations and leading corporations. As a Six Sigma consultant and master black belt for eight years, Mr. Martin has worked with organizations in retail sales, residential and commercial service, banking, insurance, financial services, measurement systems, aerospace component manufacturing, electronic manufacturing, controls, building products, industrial equipment, and consumer products. He has trained and mentored more than 1,500 black belts, executives, deployment champions, and green belts in Lean Six Sigma methods including supply chain applications. He holds a M.S. Mechanical Engineering, Northeastern University; M.B.A., Providence College; and B.S. Industrial Engineering, University of Rhode Island. He also holds several patents and has written numerous articles on quality and process improvement. He is a member of the Association for Operations Management (APICS) and has certifications in production and inventory management (CPIM) and integrated resource management (CIRM). He is a member of American Society for Quality (ASQ) and is a certified quality engineer (CQE).

Lean Six Sigma for Supply Chain Management

The 10-Step Solution Process

James W. Martin

New York Chicago San Francisco Lisbon
London Madrid Mexico City Milan New Delhi
San Juan Seoul Singapore Sydney Toronto

The *McGraw·Hill* Companies

McGraw-Hill books are available at special quantity discounts to use as premiums and sales promotions or for use in corporate training programs. For more information, please write to the Director of Special Sales, Professional Publishing, McGraw-Hill, Two Penn Plaza, New York, NY 10121-2298. Or contact your local bookstore.

Lean Six Sigma for Supply Chain Management:
The 10-Step Solution Process

5 6 7 8 9 0 DOC DOC 0 1 9

ISBN-13: 978-0-07-147942-4
ISBN-10:   0-07-147942-2

Sponsoring Editor
Ken McCombs

Editorial Supervisor
Patty Mon

Project Manager
Vasundhara Sawhney

Copy Editor
Margaret Berson

Proofreader
Jayanti Ghosh

Indexer
Stephen Ingle

Production Supervisor
George Anderson

Composition
International Typesetting and Composition

Illustration
International Typesetting and Composition

Art Director, Cover
Brian D. Boucher

Printed and bound by RR Donnelly.

Contents

Foreword

Twelve years ago I began my Lean Six Sigma career as a black belt (BB), then later became a master black belt (MBB) at Honeywell (at that time it was AlliedSignal). My project assignments were in the areas of finance, sales, marketing purchasing, materials planning, and logistics. Then, as I began my second career as an independent consultant, I found the majority of the Lean Six Sigma applications were within the supply chain. But the Lean Six Sigma training was always conducted by generalists from a manufacturing perspective. To this day, basic supply chain theory and methods remain on the "outside" of Lean Six Sigma training. The training focus continues to be from a manufacturing viewpoint, although Lean topics are often thrown into the training to cover some of the supply chain applications. In addition, few master black belt (MBB) instructors are certified by The Association for Operations Management (APICS) or have an understanding of how supply chain systems actually function. I wrote this book to help these MBBs and other "belts," champions, and team members to understand Lean Six Sigma supply chain concepts and applications.

I must thank Ken McCombs, my McGraw-Hill senior editor, for encouraging me to publish this book, as well as the reviewers of the book, who made constructive comments that changed the direction of the book to a more practical and hands-on approach. Along this line of thought, the book has been expanded to include many practical tips for the identification and analysis of Lean Six Sigma projects to improve supply chain performance. In fact, the entire focus of the book is to allow someone with little knowledge of supply chain concepts to quickly "come up to speed" on Lean Six Sigma supply chain concepts. Finally, I want to thank my family and my wife Marsha in particular, clients, and Providence College faculty and graduate students, who have provided the inspiration for this book.

Introduction

Этhis book is the culmination of several years of successful Lean
Six Sigma consulting and training focused on improving supply chain performance. This consulting work was performed
as a series of graduate classes as well as on-site corporate seminars
and operational assessments of major work streams within diverse
supply chains. In this work, cross-functional business unit teams were
brought together to analyze key supply chain metrics and implement
process improvements using a 10-Step Solution Process based on Lean
Six Sigma, operations management, and operations research tools and
methods. These improvement efforts demonstrated the need for a book
showing Lean Six Sigma "belts" how to analyze supply chain work
streams, and inventory systems in particular, to identify and implement
Lean Six Sigma improvement projects. This book brings together the
specific tools, methods and techniques that are relevant to supply chain
analysis and optimization together into an organized format in one place
for the first time to improve supply chain operational performance.

The goals of this book are to help the reader understand how the
major work streams within their supply chain work and make process
improvements within the context of Lean and Six Sigma methods as
described in the 10-Step Solution Process. However, bringing diverse topics and metrics together to understand their inter-relationships and impact
on supply chain performance requires quantitative tools and methods not
found in Lean and Six Sigma training. Some of these required tools and
methods are considered basic in supply chain analysis and can be found

within Operations Management (OM). Others are associated with Lean methods, Six Sigma or demand management; however, their full analytical potential has not been realized because their users are not familiar with supply chain work streams, how the major supply chain systems work, inventory basics or system modeling. Finally, many supply chain problems are best solved using operations research (OR) tools and methods to understand how major components of the system dynamically change. The correct mix of analytical tools and methods for a particular project should be based on the nature of the root cause analysis. We want to avoid a situation in which the analysis is being forced in the wrong direction by using an ineffective set of tools resulting in a solution which is not optimal. Understanding all aspects of the major work steams found in a supply chain will enable improvement teams to develop supply chain (or inventory) models to systematically improve key financial and operational metrics for their organization in the most effective manner.

The book has several other goals. The first is to expose people who are new to the field of supply chain management or Lean Six Sigma principles to practical supply-chain terminology and methods. The second goal, related to the first, is to provide the reader with commonly accepted tools and techniques useful to manage demand, lead-time, and related activities to accelerate the deployment of Lean Six Sigma projects. The third goal, for those more advanced in supply chain concepts, is to develop simple models of major work streams using an inventory model as an applied example to show how its analysis can be used to understand relationships between process inputs, that is, "X's" such as lead-time, demand and service levels, and their impact on key process outputs, that is, "Y's" such as cash flow, profitability, inventory investment, and inventory turns. This will help us understand the famous Lean Six Sigma relationship $Y = f(X)$ as it applies specifically to inventory investment. The fourth goal is to facilitate development of Lean Six Sigma projects to systematically improve key supply chain metrics. These metrics are listed in Figure 1-2 and defined in Appendix I. To enable the fourth goal, almost every chapter contains examples of Lean Six Sigma project applications across the major supply chain functions. The fifth goal, once we have a firm understanding of current operational performance baselines, is development of feasible improvement targets or internal benchmarks. Finally, the sixth goal of the book is to introduce standard project management methods to properly execute the Lean Six Sigma improvement projects. These project management techniques will be described in the coming chapters as they pertain to Lean Six Sigma supply chain improvement projects.

Understanding customer requirements through voice-of-the customer (VOC) translation is a major focus of a Lean Six Sigma initiative. These VOC requirements are categorized into high-level categories of quality, cycle time, and cost, which form the basis for the concept of "customer value." Working backward from specific customer value metrics identified through marketing research activities, Lean Six Sigma improvement projects are identified and translated into internal organizational requirements using the "critical-to-quality (CTQ)" concept. The objective of the CTQ concept is translating VOC requirements from a strategic to tactical level throughout an organization to identify operational performance gaps and the projects necessary to close these gaps. The objective of the improvement projects is to move the process on target (accuracy) with minimum variation (precision). Effective project execution is accomplished through methodical process analysis and identification of the root causes for poor operational performance. The 10-Step Solution Process, shown in Table 1-3 and Figure 1-8, consists of a series of ten steps that, when executed, will move the process toward its design intent (entitlement). These ten steps are correlated to the Lean Six Sigma "define-measure-analyze-improve-control (DMAIC)" process, but have been modified slightly to provide more focus on those improvement actions which are most applicable to supply chain improvement, that is, less statistical analysis and more emphasis on supply chain tools and methods found in the field of operations management and operations research.

The advantage of using the structured 10-Step Solution Process is that clear improvement goals and objectives can be created by the organization. This methodical approach ensures effective execution of strategy. It also ensures that tactical objectives are consistently met. This situation enhances the organization's capability to achieve and sustain supply chain performance breakthroughs. The success of the 10-Step Solution Process is directly related to its practicality. There are management reviews of each project to ensure that the Lean Six Sigma project remains business focused and on schedule. The basic philosophy underlying the problem-solving methodology is that variations in process inputs cause variations in outputs and lowers operational capability and customer satisfaction. Understanding relationships between process inputs and outputs facilitates improvements in quality, capability, and customer satisfaction. This is often abbreviated as the DMAIC equation, that is, $Y = f(X)$. These concepts are used as the initial justification for the Lean Six Sigma project through performance analysis that measures how well the process meets customer requirements. This is called a "capability analysis." A process that exhibits a small amount of variation will produce fewer defects than one that has significant variation. These Lean Six Sigma concepts are used to characterize the ability of a supply chain to meet its customer requirements.

Another important characteristic of the 10-Step Solution Process is the use of Lean Six Sigma methods to identify the root cause for process breakdowns and development of an optimum solution to eliminate them. After an optimum solution has been identified to eliminate the root causes of the process breakdown, control strategies are required to fully integrate and sustain these improvements. Although there is a hierarchy of control strategies that can be used to ensure that process improvements are sustained over time, specific combinations of control tools vary by project and are based on the project's root cause analysis and specific countermeasures necessary to eliminate the root causes related to poor process performance. Another, critical aspect of the process control strategy is transition of the project to the process owner and local work team. An effective project transition implies that the people in whose work area process improvements have been made have the necessary training and tools to ensure that process controls remain effective over time. This situation implies that both the process owner and the impacted local work team agree on the basic steps necessary to transfer the improvement project. Integration of numerous Lean Six Sigma projects across the supply chain ensures continuous process improvement over time and higher customer satisfaction.

In addition to improving key supply chain metrics, inventory analysis and control are important competencies which will help a supply chain identify operational breakdowns throughout its supply chain. This is because inventory investment and its turns ratio are barometers of how well the supply chain utilizes its assets. Efficient asset utilization is also a characteristic of a "Lean" supply chain. Since inventory is usually a chronic problem within most supply chains, it is used as an example, throughout the book, of how to tie together the diverse concepts of Lean Six Sigma, supply chain management, and strategic alignment of improvement projects to generate significant business value for senior management. Using the Lean Six Sigma philosophy of "acting-on-fact," we want to avoid situations in which inventory levels are arbitrarily reduced to satisfy year-end financial goals and objectives without regard for long-term customer service levels or operational efficiency. But arbitrarily maintaining higher inventory investment is not an effective long-term solution. In fact, lower inventory levels are always preferred, but only if the operational system can satisfy external customer demand at the required target per-unit service level based on an item's lead-time and capacity constraints. Instead, investment should be calculated optimally, by item and location. Also, development of realistic internal investment benchmarks, as opposed to arbitrary external benchmarks, should be a major organizational objective. Thus, inventory analysis and investment are a common theme throughout this book to show how Lean Six Sigma methods can be used to attack this chronic asset utilization problem, but many other supply chain applications are also discussed throughout the book. The focus of Lean Six Sigma improvement teams should be development of internal competencies to provide a strategic framework from which to fully integrate and sustain supply chain improvements into the local work team's daily work activities.

The information presented in Chapter 1 shows why it is important to identify key financial and operational metrics prior to creating Lean Six Sigma improvement projects. This ensures business alignment. Twenty important operational and financial metrics are presented in Chapter 1. Twelve of these metrics are important to Lean Six Sigma projects at a tactical level, but there are many other types of performance metrics, which vary by the specific supply chain design. Chapter 1 also presents a long list of lower-level operational metrics common to operations within human resources, distribution, materials planning, and other major supply chain functions. An important theme of Chapter 1 is that Lean Six Sigma improvement projects should link to the voice-of-the customer (VOC) to ensure that customers remain satisfied with your organization's products and services. Once the improvement team identifies the critical VOC requirements, balanced metric scorecards are created to baseline current performance levels against the VOC, the 12 key metrics and higher level financial metrics presented in Figure 1-2. This allows the team to systematically track process improvements across several supply chain functions and projects to ensure that everything remains integrated from a senior management viewpoint. Also, since process inputs and outputs are inter-related, it is important to evaluate them as integrated work streams. Integral to project identification is development of system models. System models are useful in understanding how the process works and which key process input variables ("X's") impact operational performance. Another useful tool to translate senior management's goals and objectives into tactical improvement projects are the input/output matrices shown in Figure 1-4. A project charter and effective team management are also required to successfully execute the team's project.

In Chapter 2, we begin to focus on basic supply chain management techniques and methods. An important discussion shows why proper execution of the sales and operating plan (S&OP) requires all components of the supply chain to perform according to plan. The discussion in Chapter 2 shows how the S&OP team facilitates the communication process across the supply chain, since process breakdowns often occur because of lack of communication between organizational functions. The S&OP team is also a very good source of Lean Six Sigma project ideas since they are strategically aligned with senior management and see major process breakdowns as well as their impact on the supply chain. An important topic of Chapter 2 is the importance of lead-time in driving supply chain costs including inventory investment. Inventory investment and its turns ratio are two of several key supply chain metrics measuring the efficiency of supply chain asset utilization, that is, the "Leanness" of the supply chain. Chapter 2 shows that lead-time reduction is one of the major ways to reduce inventory investment to "Lean-out" the supply chain using tools such as elimination of process steps and operations, bottleneck management, application of mixed-model scheduling, application of transfer batch methods, and deployment of Lean, just-in-time (JIT).

The subject of demand management is discussed in Chapter 3. Effective demand management is important to ensure that products and services are available as promised to the customer. In this context, the sales and operations planning (S&OP) team is essential to the identification of Lean and Six Sigma improvement projects that will improve demand management and be aligned to business goals and objectives. An important discussion in Chapter 3 is why violation of the organization's stated lead-times, that is, time fences, causes significant operational problems that may necessitate creation of Lean and Six Sigma improvement projects. These projects identify and eliminate issues across supply chain functions contributing to poorly estimated demand. Finally, it is shown that Lean Six Sigma belts can help forecasting analysts develop and interpret advanced forecasting models to more effectively manage products having unusual demand patterns. Effective demand management will also improve new product forecasts. Poor forecasts of new product sales are a major contributor of excess and obsolete inventory.

Lead-time reduction strategies are further discussed in Chapter 4. In Chapter 4 it is shown that creating a value stream map (VSM) of a major process work stream and breaking its operations down into smaller time elements is a powerful way to identify areas that can be improved within the process. It is also shown that effective lead-time reduction requires identification of a system's critical path. In addition to calculating a system's critical path, its takt time must be calculated to ensure that the process will produce according to actual customer demand.

Takt-time calculations will show the Lean Six Sigma improvement team the resources necessary to maintain material or information flow through the supply chain's major work streams. The VSM and takt-time analyses help the team to identify Lean Six Sigma projects associated with process waste. Chapter 4 also provides the reader with several powerful tools to reduce lead-time to improve supply chain performance.

In order to understand the process improvement potential within a supply chain, belts must know the "nuts and bolts" of its operation. Chapter 5 shows why understanding your organization's master production schedule (MPS) and material requirements planning (MRPII) systems will allow your Lean Six Sigma belts to identify many areas for improvement projects. These improvement projects attack chronic problems at receiving, within production activity control (PAC), within purchasing, and in areas related to inventory investment. Many of these chronic problems are associated with MPS and MRPII system issues. It is later shown that belts also need to learn operations management and operation research tools and methods in order to properly understand how a supply chain changes dynamically over time. The discussion in Chapter 5 also discusses the comparative advantages and disadvantages of make-to-stock, assemble-to-order, and make-to-order systems. This will allow your Lean Six Sigma improvement team to identify projects to improve your supply chain performance. Chapter 5 also discusses why analyzing system capacity, especially at bottlenecks and system-constrained resources, is important to allow an organization to increase material and information throughput rates using pull scheduling systems to reduce inventory, improve quality, and reduce operational costs.

Since inventory is a major barometer of supply chain financial performance, Chapter 6 discusses inventory basics from several viewpoints. First it shows why Lean Six Sigma belts should calculate optimum inventory levels, using an inventory model, to identify where to strategically deploy Lean Six Sigma projects to reduce inventory investment and improve other supply chain metrics. The model is built using lead-time, demand variation, service levels, and other relevant information as major factors to determine inventory investment levels. Second, it identifies the various types of Lean Six Sigma projects that are associated with inventory problems. Third, Chapter 6 shows why the identification and elimination of excess and obsolete inventory is an important step to improve supply chain operational and financial performance.

Chapter 7 introduces the concepts of "Lean supply chain" and third-party logistics. The initial discussion in Chapter 7 is how to define a Lean supply chain. It is shown that asset utilization efficiency is one of several key metrics measuring the degree of supply chain "Leanness." The discussion continues into why Lean Six Sigma deployments are more complex in these environments. Another focus of Chapter 7 is the identification of Lean Six Sigma projects within major supply chain functions. These functions include order import/export, carrier management, claims management, warehousing operation, and fleet management, as well as others. Several examples of Lean Six Sigma project opportunities are provided in Table 7-1. The chapter also discusses third-party logistics and why it is important to integrate third-party logistic providers into a Lean supply chain using Lean Six Sigma methods. Specific Lean Six Sigma applications in retail food and commodity supply chains are also briefly discussed in Chapter 7.

Chapter 8 shows how Lean Six Sigma tools and methods are used to identify relationships between key process inputs and outputs to develop the Lean Six Sigma $Y = f(X)$ relationship. The discussion begins with measurement system analysis (MSA) to determine the accuracy and precision of the MSA. It continues with a discussion of estimating process capability of key process output variables (KPOVs). Once process capability has been established for the KPOVs, data collection and analysis is conducted within the process to understand the relationships between key process input variables (KPIVs) and KPOVs. This analysis is designed to answer critical questions related to the project's problem statement and objective. In addition to classical Lean Six Sigma tools, other analytical tools such as queuing analysis, linear programming, scheduling algorithms, simulation, and financial modeling may be necessary to identify the root causes of the process breakdown. Although these advanced analytical methods are not usually part of traditional Lean Six Sigma training, some supply chain projects may require their usage. These are discussed at the end of Chapter 8.

In Chapter 9, after the Lean Six Sigma improvement team analyzes the collected data and finds the root causes of the original process breakdown, it evaluates alternative solutions using a cost-benefit analysis. This ensures that the organization attains the maximum business benefit from the Lean Six Sigma project with a minimum level of risk. Once an optimum solution to the problem has been identified, the improvement team evaluates the solution under controlled experimental conditions. This is called conducting a "pilot" investigation. The pilot investigation shows the team how to set the levels of the KPIVs to put the KPOV on target with minimum variation. Using a pilot study minimizes project implementation risk. It is also important in gaining support from the process owner and local work team since they can see the impact of the process changes under controlled conditions. Chapter 9 also shows why it is important to integrate process changes into current control systems such as the International Standards Organization (ISO) to ensure that the process improvements are incorporated into daily work. Also, that they are effectively communicated to the organization.

Our objective in Chapter 10 is to develop a simple inventory model to show some of the concepts related to modeling major supply chain work streams. Lean Six Sigma projects are identified using this model. In Chapter 10, a simple Microsoft Excel model is used to show relationships between KPIV and KPOV variables as well as their impact on inventory investment and its turns ratio. Using the inventory model, we will be able to develop internal benchmarks for some of the financial and operational metrics shown in Figures 1-1 and 1-2. Also, we will also be able to identify Lean Six Sigma projects associated with high inventory investment as well as some of the root causes for long lead-times and demand variation. This methodical approach will also ensure that the correct operational and financial data will be available to build the model. If necessary, Lean Six Sigma improvement projects can be applied to areas within the process where data is inaccurate or not available and there are operational breakdowns. Using an optimized model will also enable the team to identify and implement "quick hit" opportunities through simple rebalancing of the inventory by item and location as well as obvious process issues. It will also be shown that further reductions in investment, over the initial balance quantities, will require identification and execution of Lean Six Sigma improvement projects to eliminate the root causes of the chronic process problems.

In the Conclusion, the 10-Step Solution Process is reviewed one last time. At the end of the Conclusion, the next steps in the continuing improvement process are discussed including design-for-Six-Sigma (DFSS) concepts, organizational change, and Lean Six Sigma maturity. Finally, the appendices contain information useful to understand the supply chain metrics discussed in this book as well as key observations from each chapter. The glossary defines common terms used in the book.

The advantage of using the structured 10-Step Solution Process is that clear improvement goals and objectives can be created by the organization to ensure effective execution of strategy. It also ensures that tactical objectives are consistently met through the deployment of Lean Six Sigma projects. To facilitate this discussion, the ten chapters of this book discuss supply chain improvement from an elementary to an advanced level. For green and black belts not familiar with supply chain concepts, the first few chapters briefly describe basic supply chain structure, demand management, lead-time analysis, materials planning, and basic inventory topics and models. Subsequent chapters describe data analysis from a Lean Six Sigma, operations research, and inventory modeling perspective. The final chapters discuss project control strategies based on proven Lean and DMAIC methodologies. This format allows the experienced reader to skip chapters containing familiar information to concentrate on more interesting chapters. As an example, if the reader has experience in supply chain methodology, materials planning, forecasting, and inventory analysis, then Chapters 2, 3, 4, 5, and 6 can be skipped to concentrate on voice-of-the customer (VOC), project management, inventory modeling, and analysis, and then improvement and control of the process changes. To show the linkage between Lean Six Sigma concepts, tools and methods, Chapter 10 contains more than 25 Lean Six Sigma examples covering many aspects of supply chain operations for those not experienced in applying Lean or DMAIC methodologies within supply chain environments.

Chapter 1

Using Lean Six Sigma Methods to Identify and Manage Supply Chain Projects

Key Objectives

After reading this chapter, you should understand how to identify and execute supply chain projects to ensure operational linkage with senior management's high-level goals and objectives and understand the following concepts:

1. Why it is important to identify financial and operational metrics, to ensure business alignment, prior to creating Lean Six Sigma improvement projects.
2. Why projects must link to the voice-of-the customer (VOC).
3. Why it is important to establish metric scorecards prior to deploying projects.
4. Why system models of major work streams are useful in understanding their impact on supply chain performance.
5. Why input/output matrices are useful in translating senior management's strategic goals and objectives into Lean Six Sigma projects.
6. Why it is important to create a list of questions relevant to the project's problem statement and objective prior to collecting and analyzing data.
7. Why a project requires a project charter and effective project management in order to be successful.

Executive Overview

In this chapter, the initial metric discussion is based on improvement of one or more of the 12 key supply chain metrics shown in Figures 1.1 and 1.2. An additional 8 supply chain metrics focused on financial performance are also listed in Figures 1.2. The 12 metrics consist of financial metrics including inventory investment, profit and loss expense (P/L), and excess and obsolete inventory investment, as well as operational metrics such as inventory turns, on-time supplier delivery, forecasting accuracy, lead time, unplanned orders, scheduling changes, overdue backlogs, data accuracy, and material availability. The 12 key supply chain metrics are operational, they can be used as the focus of your Lean Six Sigma improvement projects. In addition to the 12 key metrics, Lean Six Sigma projects can also be identified and deployed using these last 8 metrics as a baseline. These metrics include customer service–level targets, *net operating profit after taxes (NOPAT), asset efficiency (turns), fixed asset efficiency (turns), account receivables efficiency (turns), gross profit margin, return-on-assets (ROA),* and *gross-margin-return-on-assets (GMROI).* Appendix I contains basic definitions and additional information for all 20 metrics. In addition to the 20 metrics, there are other operational metrics that could serve as the basis for Lean Six Sigma improvement projects. These are listed later in the chapter in Tables 1.5, 1.6, and 1.7.

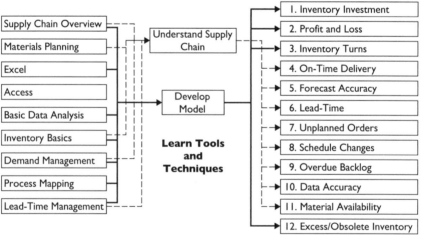

Figure 1.1 Twelve Key Metrics

Metric Scorecard	Current	Quarter 1	Quarter 2	Quarter 3	Quarter 4	YTD Percentage Improvement
Twelve Key Metrics						
1. Inventory Investment						
2. Profit/Loss						
3. Inventory Efficiency (Turns)						
4. On-Time Supplier Delivery						
5. Forecast Accuracy						
6. Lead-Time						
7. Unplanned Orders						
8. Schedule Changes						
9. Overdue Backlogs						
10. Data Accuracy						
11. Material Availability						
12. Excess & Inventory						
Additional Metrics						
13. Customer Service Target						
14. NOPAT						
15. Asset Efficiency						
16. Fixed Asset Efficiency						
17. Receivables Efficiency						
18. Profit Margin						
19. ROA						
20. GMROI						
Average Improvement						

Figure 1.2 Tracking Twelve Key Metrics Against Baselines and Targets

The Voice-of-the-Customer (VOC)

Every Lean Six Sigma project touches the external customer relative to time, cost, or quality, if only indirectly. As the improvement team begins to identify and scope project opportunities, they must always consider the impact of the project including its eventual solution on the external customer. As an example, inventory exists to meet customer demand in environments having limited manufacturing capacity; in this context, Lean Six Sigma inventory projects must balance lead-time, demand, and customer service–level targets to ensure that delivery promises to customers are kept. However, the only way to ensure meeting customer delivery promises is to understand what your customer wants, what you promised to provide to the customer, and how well your system actually meets customer requirements over time, that is, your process capability.

However, understanding customer requirements is a complex process. It starts with the concept of customer value. At a high level, value can be broken into components of convenience and price. When an organization provides goods and services to external customers, it should ask, "How convenient is it for the customer to access, understand, and use our product or service?" Convenience can be broken into the sub-components of time and benefits. Finally, customer benefits can be broken down further into components of relative importance and usefulness to the external customer in addition to how well the product or service actually performs in practice. Every organization should at least consider the concept of value when marketing its products and services. When we take the VOC value concept to an operational level, it is important to understand that the external customer's value expectation is what our operational systems should deliver every day. The customer is unhappy when there are breakdowns in operational performance. The goal of every organization should be to design operational systems that will deliver customer value according to internal *voice-of-the business (VOB)* and external VOC expectations. If the systems do not perform according to expectations, then Lean Six Sigma or other types of process improvement projects are necessary to improve operational performance.

The reason for improving supply chain processes and delivering customer value is that organizations having a "best-in-class" supply chain can operate at lower relative costs versus their competitors, but also perform at levels exceeding customer expectations. These best-in-class supply chains have major competitive advantages over their competitors. Taking the concept of customer value another step, we need to ask who our customers are. And how well do we really know them? The answers to these questions should reflect the strategic goals and objectives of an organization using a market segmentation viewpoint. Identifying customer values as well as their specific needs requires understanding who, why, where, when, and how customers use our products and services. This approach will identify not only internal process breakdowns but also opportunities to create Lean Six Sigma improvement projects to satisfy customer needs.

What is the voice-of-the-customer (VOC)? It is a set of tools, methods, and techniques that allow the Lean Six Sigma improvement team to methodically collect and analyze customer needs and how customers value those needs. Customer needs can be broken into basic, performance, and excitement needs using a Kano model. Basic needs are unspoken and expected by the customer. If these needs are not met by a supplier there will be extreme customer dissatisfaction. An example would be buying an automobile but finding it will not start. On the other hand, performance needs differentiate one supplier from another relative to price, functionality, or the ability of the supplier to provide the product or service in less time than a competitor. An example would be buying a car and finding it has a longer warranty or better handling performance than a competitive vehicle for the same price. Excitement needs are not normally known by a customer in advance, but they clearly separate a product or service from current designs. An example would be finding a warranty has been lengthened from 60,000 to 100,000 miles without an increase in purchase price, or if maintenance for the first 10,000 miles of vehicle usage is provided free by the seller. Understanding customer needs as well as how they value these needs shows the Lean Six Sigma improvement team what is most important to the external customer. This information can be used to identify and deploy Lean Six Sigma improvement projects.

Quantifying the VOC begins by breaking down the customers into their market segments including segments based on relevant demographic criteria. Common demographic criteria include geographic location, products purchased, type of customer, and size of customer, as well as other criteria. As an example, we could segment customers by how, where, and when the customer uses the product or service. This marketing information can help focus our attention on critical customer requirements, which can be translated into Lean Six Sigma projects. After market segmentation, improvement teams obtain information from the customer in a structured way using a variety of methods. These methods include analysis of customer complaints, operational reports, and competitive intelligence, as well as customer surveys, interviews, and on-site customer visits. This allows the team to understand the major drivers of customer satisfaction by market segment. These major drivers of customer satisfaction, called *critical-to-quality (CTQs)*, are later analyzed and broken into the big "Ys" representing customer requirements at an internal operational metric level, that is, the project *key process output variable (KPOV)* level. In the last step of translating the VOC, the team uses *quality function deployment (QFD)* and other information management tools and methods to compare customer requirements, in the form of KPOVs, against current internal systems (or process work streams) designed to satisfy those requirements. Performance gaps serve as the basis for Lean Six Sigma projects. "Belts" are assigned to the Lean Six Sigma projects to close performance gaps. We will now discuss each of these concepts in more detail.

Market Segmentation

Understanding our customers is not an easy task. We often inadvertently substitute our voice for theirs. This common practice causes miscommunication and disappointment for both the customer and supplier. Many Lean Six Sigma projects are based on the process breakdowns occurring due to inadequate VOC information. As an example, failure of a supplier to deliver on time to the customer is a common problem in supply chains. However, the concept of "on-time delivery" needs to be carefully defined between all parties in the supply chain. It is also important to understand all elements of the delivery interaction between the supplier and customer. In particular, lead-time must be carefully defined prior to setting the on-time delivery target. A good question to ask is, "Given the order arrives on time and in good condition, what else does the customer require from this delivery transaction?" Other relevant questions are, "What are the customer's unspoken needs and requirements?" "What does the customer expect as minimum performance?" "What would differentiate us from our competitors?" "What would really excite the customer and ensure that we remain the supplier of choice?" Obtaining this VOC information is a complex task using a variety of marketing research tools and methods.

Marketing research attempts to understand customer needs and requirements as well as their relative importance to the customer. Important in these evaluations are customer perceptions regarding supplier value relative to price, time, utility, functionality, and relative importance to the external customer. Using the delivery example again, "How does the customer value the delivery transaction? Is low-cost delivery the major criterion determining customer value? Or is it timeliness or flexibility of delivery time? Does the supplier do value-adding tasks associated with the delivery such as invoicing, self-scheduling, or unloading the trailer? How well does the supplier function or perform over time? How consistent is the supplier's delivery performance?" Understanding how the customer values and perceives operational performance relative to these criteria can help differentiate the supplier from competitors. Also, closing any identified performance gaps using Lean Six Sigma projects will ensure that the supplier is preferred by the customer. It is important to properly segment customers to understand their value needs since different market segments place a different emphasis on each value element. Suppliers should also understand that value perception varies by market segment.

Customer segmentation is a very useful process since the better we understand our customers, the better we can provide products and services that can be efficiently and profitably used by our customers. Understanding how our goods and services are valued by our customers as well as their customers, who are further down the supply chain, can show us how to improve their value elements. As an example, imagine a Lean Six Sigma black belt riding in a delivery truck and observing the entire delivery transaction. These observations would include careful recording of all work tasks relative to material and information flow. Using this hands-on approach, the black belt would identify basic transaction elements critical to the customer's standard operations as well as ways in which to enhance the customer experience and differentiate the supplier from its competitors. This process would identify basic, performance, and excitement needs as well as relevant value elements. Using United Parcel Service (UPS) as an example, UPS is exceptional in providing standardized customer service. I receive many UPS shipments at my office, which is in my home, and I am never disappointed with the service from UPS. I also receive Federal Express shipments, which are also delivered in a very standardized manner with very good service. In my view both suppliers are equivalent in their service. However, neither organization really keeps me informed of their service offerings. I guess I should read the fine print. It would be great if they would periodically tell me the current service offering; for example, will they pick up at my house if I call them so I won't have to travel to drop the package off at a remote location? Of course I could check the service offerings myself, but I never really seem to have the time to do this. Either organization could greatly expand their market share and ensure customer loyalty if they just went one step further through market segmentation to understand the customer's Kano needs as well as their value elements.

In manufacturing this concept can be taken one step further. When was the last time a supplier actually came out to their customer's distribution center or job site to understand how their products and services were actually used by their customer or customer's customers? Although the supplier might have many good ideas on how to improve their operational performance, much can be learned directly from the customer. Understanding how customers use our products and services is a powerful way to understand what is valued by the customer to improve performance. As an example, if a supplier visited a customer's distribution center, they may find that the packaging of their products does not provide sufficient protection from moisture or normal handling within their customer's distribution center. This situation may be causing high levels of repacking of the product by the customer. The customer may even complain periodically of product damage. The customer may also learn to live with the damage problem if the supplier does not effectively respond to complaints. As another example, orders may be arriving with their loads shifted, damaged, or not loaded in proper sequence resulting in higher customer labor expense and lower satisfaction. Lean Six Sigma improvement projects can be created to close performance gaps or modify products and services to increase one or more value elements from the customer perspective. This will lower supply chain costs and increase customer satisfaction.

Obtaining VOC Information

There are many methods available to obtain information from the customer. Each method has its strengths and weaknesses relative to its quality of information, cost, and ease of obtaining the information. The important concept is that the supplier should have a structured plan to allow systematic understanding of the VOC information. This plan should include one or more standard methods useful in obtaining the VOC information. One important way in which VOC information enters an organization is in the form of complaints about current products and services and requests for modified or new products and services. These are obvious areas in which to satisfy the customer. But, what are not normally communicated to the supplier are the internal operational breakdowns, which are not measured by either the customer or supplier, that is, the "hidden factory." Also, the customer's customer may have needs and requirements that neither the customer nor suppliers fully understand. To obtain VOC information from the customer over and above that which is currently available requires using standardized methodology to gather the VOC information.

Organizations routinely use surveys, interviews, marketing research of industry trends, and focus groups, as well as on-site visits, to obtain VOC information directly from the external customer. Surveys involve sending a list of questions to specific market segments in either written or electronic format. Surveys are relatively easy to develop, but they must be carefully designed to ensure that the information that is ultimately collected is accurate and useful for understanding the VOC. The advantage of surveys is that relative to other ways to obtain VOC information they are relatively inexpensive, but since they often have a very high non-response rate, the cost per response might actually be of the same order of magnitude as other methods used to obtain the VOC information. Surveys also generate large amounts of quantitative data, which can be analyzed with statistical confidence. However, in addition to the large percentage of non-responses, customers may also misinterpret questions and not fully understand all aspects of the survey process. This situation could result in inaccurate and biased responses, which will result in inaccurate VOC information. For this reason survey conclusions must always be validated on a limited sample prior to acceptance as the VOC.

Customer interviews are another way to obtain VOC information. An advantage of using customer interviews is that they allow follow-up questioning by the interviewer. Interviews can be conducted by telephone or in person. Both methods have advantages and disadvantages. As an example, telephone interviews are less expensive, but the supplier does not have the customer's full attention nor can the supplier ascertain the customer's body language during the interviewing process. In-person interviews are more expensive, but the supplier can better interact with the customer and analyze body language. Proper preparation, including important questions that need to be answered at the interview, is critical to success since the customer's time is valuable. While the major benefit of interviews is the quality of information (if properly conducted), the disadvantage is the small sample size.

In addition to customer surveys and interviews, external marketing research can be conducted on the customer and its competitors as well as relevant industry trends. This external marketing information can be very useful in identifying new products and services or showing operational weaknesses in the supply chain, which are common within the industry. Lean Six Sigma projects can be developed to provide a competitive advantage for the supplier in these situations. Marketing research also involves focus group interviews. Focus group interviews are structured and facilitated exchange of information between the supplier and targeted customers by market segment. The advantage of focus groups is the quality of information obtained relative to use of products and services by customers. The disadvantage of focus groups is their expense and limited sample size. An extension of in-person interviews is walking through a customer's operations to see how your product or service is used on a daily basis by the customer. This is an extremely useful way to understand the VOC at its most basic level.

Analyzing VOC Information

Since VOC information is obtained using several different methods and formats, the collected information needs to be organized and analyzed by the improvement team and ultimately translated into major customer requirements. These major requirements are called *key process output variables (KPOVs)*. The team organizes all available information into categories to begin the task of identifying major customer requirements. These categories vary by industry and organization, but at a high level they include the value elements associated with time, price, function, and utility to the customer. The team should organize all the customer statements concerning value elements by similar categories. Each major statement would be decomposed into operationally oriented statements, which more nearly represent internal supplier specifications. These operationally oriented statements form the basis for defining and developing specifications for the KPOVs. As an example, suppose the Lean Six Sigma improvement team has collected VOC information for a delivery process for a particular market segment and relative to its value elements. Customers may have said they want deliveries to arrive on time plus or minus one hour; the product should be loaded according to industry standards, including having the loads stepped down to prevent shifting; and the loads should be protected from moisture by placing plastic wrap at the end of the trailer. Other comments may be that heavier product must be on the bottom of each pallet with lighter products on top of the pallet to prevent product damage. Additional comments might be that the pallets should have standardized dimensions and be properly labeled to allow easy storage in the distribution center. Perhaps there is other information relative to the driver bringing the paperwork into the distribution center in a timely manner or helping unload the trailer according to customer requirements. There would probably be many other customer comments obtained from the various methods used to obtain the VOC. These comments might be obvious to the supplier, that is, "basic needs," or may be surprises that offer opportunities to enhance customer satisfaction and differentiate the supplier from its competitors, that is, "excitement needs."

Identifying Performance Gaps and Projects

The quantified customer requirements, that is, KPOV variables, are compared against baselines representing current system performance to identify and prioritize Lean Six Sigma projects. As an example, if customers specify that deliveries must arrive on time plus or minus one hour, but our current performance reports show an arrival variance of plus or minus three hours, this performance gap could serve as a potential Lean Six Sigma project. The business benefits gained from closing this performance gap are not only increased customer satisfaction, but also lower supplier delivery costs when non-value-adding time elements are eliminated from the process. The identification of performance gaps should be conducted in a systematic way to ensure that Lean Six Sigma improvement projects are properly prioritized prior to their deployment. There are many ways in which prioritization can be done including using quality function deployment (QFD), Hoshin planning, or the current strategic planning process. Regardless of the prioritization method, the result should be a prioritized project listing based on performance gaps. Resources would then be assigned to the project to ensure that the performance gaps are closed according to schedule.

The 10-Step Solution Process

The team's balanced scorecard is shown in Figure 1.2. The balanced scorecard ensures strategic alignment of the 12 key supply metrics through actionable projects at an operational level. This methodology ensures that the project team remains aligned with the organization's strategic goals and objectives as projects are deployed across various supply chain functions. The metric scorecard is very important because it is used to track, at a high level, process improvements against their original baseline levels. Using the methods described in this book, internal benchmark targets for each metric are calculated based on analysis of the specific work stream. As performance gaps are identified for each metric, improvement projects are created to close those gaps. This methodical approach has been shown to significantly improve operational effectiveness. Basic modeling concepts are summarized in Chapter 9 along with an applied example that shows how to build and analyze a simple inventory model. This applied example is based on a make-to-stock inventory system, but other types of models can be created by the improvement team based on specific system constraints and assumptions. Using the model, Lean and Six Sigma improvement projects are identified and deployed to systematically improve supply chain performance using the inventory model as a guide.

The central idea of the balanced metric scorecard is to form interfunctional teams around one or more key metrics and develop Lean Six Sigma improvement projects to improve the baseline levels of the financial and operational metrics. The interfunctional team, that is, the Lean Six Sigma improvement team, begins a detailed analysis related to their metric(s). This requires reviewing the historical performance of the process and its associated output metrics to identify one or more areas that will improve process performance. After this initial team meeting, the improvement team begins to build its business case for the project.

Effective project management and execution of Lean Six Sigma improvement projects will generate significant business benefits for the organization. Tables 1.1 and 1.2 show, at a high level, some of the key concepts of the 10-Step Solution Process. In particular, Table 1.2 is used to break the original problem statement down into applied projects. The 10-Step Solution Process is the basis for improving the 12 key metrics and is used to systematically improve process performance in conjunction with the team's balanced scorecard shown in Figure 1.2. The team uses these concepts to collect the data required to answer the questions associated with the original project's problem statement and objective. This may be a difficult process since the required data may be scattered across several databases or software platforms. As part of the data collection effort, the team should ensure that the required data can be brought together into one place for the analysis. Once the data has been brought together, the team begins to build a simple work stream model reflecting the questions and information the team needs to improve the

1. Develop a list of questions related to the project's problem statement that must be answered to complete project's objectives.
2. Ensure required data is available to complete the analyses.
3. Build a simple supply chain/work stream/ inventory model. Conduct sensitivity analysis of inputs and outputs.
4. Analyze the model relative to the project's objective. Record other business improvement opportunities for subsequent projects.
5. Identify the root causes for the problem using Lean Six Sigma methods.
6. Ensure that countermeasures are fact-based and tied to the root cause analysis.
7. Eliminate the underlying root causes adversely impacting the key metrics.
8. Complete the "*target*" and "*baseline*" portions of the metric score card.
9. Develop long-range plans to sustain improvements over time.

Table 1.1 Supply Chain Improvement Program

Measurable Improvement	Enabler	Productivity Opportunities	Productivity Link
Increased RTY	Six Sigma experimental design	Reduced rework	Reduced labor hours and material
Improved schedule attainment	Lean, TPM, Supply chain and DOE	Reduce over-scheduling of material and labor	Reduce labor, material, premium freight, etc.

Table 1.2 Operational Linkage

work stream operations. This model enables the team to optimize the work stream operations. A sensitivity analysis can also be conducted by varying the model's inputs according to the relationship $Y = f(X)$. This relationship is either known, as in the case of inventory modeling and other supply chain systems, or must be developed by the Lean Six Sigma improvement team. Based on the sensitivity analysis of the model, one or more Lean Six Sigma projects can be created to improve operational performance. As the team begins to focus on one or more related process breakdowns, the root causes responsible for the poor operational performance are identified and eliminated from the process. Based on analysis of the work stream model, the team makes process changes to improve operational performance. These process changes improve one or more of the KPOV variables identified on the balanced metric scorecard or which

1. Align project with business goals
2. Ensure buy-in from process owner, finance, and others
3. Communicate project results
4. Prove causal effect $Y = f(x)$
5. Improve measurement systems
6. Develop detailed improvement plan
7. Integrate countermeasures to their root cause analysis
8. Standardized procedures
9. Implement training and audits
10. Apply control strategies

Table 1.3 10-Step Solution Process

are specific to the project. Transfer of the modified process back to the process owner and local work team is the last major action taken by the improvement team. All recommended process controls must be easy to implement and maintain over time by the process owner and local work team. As the first project is transferred to the process owner and local work team, the improvement team begins work on other work streams. This methodology systematically improves process performance over time and is reflected as improvements relative to the key metric baselines in the balanced metric scorecard.

Building and analyzing supply chain models often is a significant communication breakthrough because people can gain agreement on the operational reasons for the process problem. This moves an organization toward a solution at a faster learning rate because people can agree on the underlying reasons for the problem because of fact-based analyses. A model also facilitates an interdisciplinary team approach to the identification and solution of chronic supply chain problems. The key chapters necessary to develop a work stream model (with an emphasis on inventory applications) are Chapters 6, 8, 9, and 10. Chapters 2, 3, 4, 5, and 7 present basic supply chain concepts. These latter chapters are important for those not familiar with basic supply chain concepts, to properly identify the improvement projects. Relative to the inventory modeling discussion in later chapters, it is assumed the reader has an elementary understanding of Microsoft Excel.

Understanding How Your Process Works

Deployment of the program to improve a major work stream begins with creation of a system model map as shown in Figure 1.3. A system model implies that the team understands the major inputs "X's" and outputs "Y's" of the work stream as well as their inter-relationships. This is the famous $Y = f(X)$ concept in Lean Six Sigma. Identifying the inputs and outputs of the system begins with a team brainstorming session. In the brainstorming session, the Lean Six Sigma improvement team identifies all the possible inputs, that is, "X's" and their associated outputs, that is, "Y's." Using these inputs and outputs, the team creates the process map. The map describes spatial and dynamic relationships between all process activities. As an example, in inventory analysis, inventory level (or quantity) is a KPOV. It can be shown to depend on all four of the inputs shown in Figure 1.3, that is, lead-time, demand variation, service

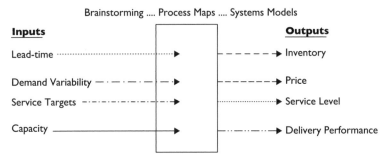

Figure 1.3 System Model Map

targets, and capacity. Depending on the supply chain design, there may also be other inputs not listed in Figure 1.3 impacting service level. Using the Lean Six Sigma concepts we can say "Inventory Level = f(lead-time, demand variation, service targets, and capacity)." Delivery performance is another output variable that may be impacted by lead-time, demand variation, service targets, and system capacity as well as other *key process input variables (KPIV)* variables.

The inputs and outputs of the high-level system map shown in Figure 1.3 are used to develop input/output matrices to quantify metric linkages at functional interfaces throughout the work stream as shown in Figure 1.4. Figure 1.4 shows how mapping strategic goals and objectives into tactical improvement projects is done using the input/output matrix. Mapping is a critical task of the Lean Six Sigma improvement team because there must be one-to-one correspondence between high-level strategic metrics and lower-level operational metrics driving their performance. The input/output matrix forces the Lean Six Sigma team to align

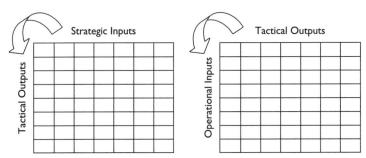

Figure 1.4 Input/Output Matrices

improvement projects with business goals and objectives. The mapping process also facilitates the work of several independent teams since they can each work a part of the larger supply chain problem without interfering with each other's work or "double-counting" business benefits. This is a common deployment practice since several concurrent projects may be required to improve a KPOV. As an example, if on-time delivery is adversely impacting inventory investment, several smaller projects may have to be executed to ensure that the on-time delivery metric achieves its target performance level. One project may focus on supplier quality issues. Another project may focus on internal demand variation issues. Still another project may focus on inventory accuracy issues. Together they improve the on-time delivery metric from an external customer's viewpoint.

Strategic Project Execution in the 10-Step Solution Process (Steps 1, 2, and 3)

To ensure strategic and tactical alignment through the project identification and prioritization process, metric linkage should encompass the entire supply chain including external suppliers and customers. This deployment strategy ensures project alignment of high-level organizational strategic goals and objectives throughout the supply chain. As an example, Table 1.2 shows how lower-level operational metrics are linked to higher-level strategic goals and objectives through enabler initiatives. Lean Six Sigma is an enabler initiative having complementary toolkits, that is, Lean and Six Sigma toolkits. Projects strategically linked in this manner will generate significant business benefits for your organization. As projects are deployed, the Lean Six Sigma improvement team should work with the process owner as well as the organization's financial representative to ensure that project benefits will link to financial statements as the operational metrics improve. This will ensure a direct productivity impact from the improved process.

The specific type of financial linkage depends on the type of project as well as the root cause analysis. As an example, the reasons for poor schedule attainment may vary due to several operational factors. For this reason, the project benefits associated with improving schedule adherence will vary based on the specific root causes impacting the process. Also, the specific Lean Six Sigma improvement tools necessary to understand the root causes of the problem will vary depending on the root cause analysis. As an example, if the project focus becomes one of process simplification, waste elimination, or work standardization,

Lean methods would be more appropriate to the process analysis and final solution rather than a reliance on complicated statistical tools and methods. On the other hand, if the data analysis required advanced statistical analysis or designed experiments, then the Six Sigma tools would be more appropriate. In either case, the "correct enabler initiative" should be used to ensure operational improvements link to the organization's financial statements. Ideally, linkage will impact the profit and loss (P/L) statement to generate incremental revenue, higher margin on revenue, or reduced cost. Alternatively, a positive impact to the balance sheet will increase cash flow and reduce expenses on the P/L statement.

Table 1.1 provides a good example of how to transform the concepts contained in Figures 1.1 and 1.2 into an organized improvement program that is easy to understand and communicate to the organization. Relative to inventory investment, analysis of the work stream model will show how to balance the inventory quantities at an item and location level relative to lead-time and demand. This balance analysis will identify "quick hit" opportunities through simple rebalancing of the inventory population item-by-item. This inventory balance analysis also identifies obsolete and excess inventory by stocking location. Additional operational improvements can be gained through effective execution of Lean Six Sigma projects. These projects are developed in areas offering the highest opportunity to reduce inventory investment. However, if significant obstacles to reduction of excess and obsolete inventory are encountered by the team, development of longer-term solutions may be required to reduce inventory investment. An example would be inventory investment levels which are difficult to reduce because of several years of on-hand supply, that is, excess inventory. Or the item might be obsolete due to design changes. In other cases, inventory levels may be higher than optimum due to long lead-times, large lot sizes, or other operational conditions. In these situations, Lean Six Sigma projects are useful to effectively reduce inventory investment.

Strategic Project Selection

Projects should be strategically selected based on their potential business benefits. Referring to Figure 1.5, we can see there are many ways to identify business benefits. Some are obvious and can be found on financial statements or management reports. Others require an operational analysis. The balance of project opportunities arises from customer, employee, and regulatory requirements. The first project identification

Figure 1.5 Project Selection Versus Process Improvement Strategy

strategy is analysis of performance gaps within an organization's strategic plan. Since organizations have different strategic goals and objectives, their performance gaps may be different. As an example, every organization targets improvements in sales, cash flow, and profitability. But some organizations also target improvements in inventory investment, asset utilization, or increases in the rate of new product introduction, as well as strategic gaps identified by senior management. These gaps are senior management's high-level goals and objectives against which specific improvement projects have not been identified to close the performance gap, that is, "unidentified task." The second project identification category is analysis of budget variances. Budget variances are easily seen on management reports but must be analyzed carefully. As an example, a variance could exist because the original budget target was incorrectly set or something unusual occurred over the time period of analysis. However, in the absence of these situations, budget variance analysis is a very good way to identify improvement projects. Cost-avoidance is the third category of project opportunity. Cost-avoidance projects cannot be quantified using the current financial statements since the performance problem has not occurred. However, it may be well known that unless a project is undertaken to avoid the future cost, it will in fact occur in the immediate future. For this reason cost-avoidance projects, if properly defined, can form the basis for very good improvement projects. A common example of a cost-avoidance project is the case where regulatory pressure requires a technological change. Perhaps the organization uses a key material that will not be allowed in the future. Failure to replace this material in the near term might result in a heavy fine to the organization in the long term. Benchmarking is the fourth category of project opportunity. Benchmarking compares operational performance of one system against another. The concept is that the poorer-performing system could be improved by using operational methods similar to the better-performing system. Benchmarking can be very useful if the systems have been designed on the same basis. As an example, two distribution centers may have significantly different inventory-turns-ratio performance. But, prior to assuming that the distribution center with a lower turns ratio could actually perform like the distribution center with the higher turns ratio, questions must be asked to ensure that both centers have similar customers, products, suppliers, and inventory systems. If they are similar then lessons may be learned from the better-performing distribution center. If the distribution centers are not really comparable, the project to improve the inventory turns ratio might in fact be based on false benchmarking assumptions. In the process assessment category, a team would "walk the process" to study how it actually works. One of the most useful tools used to conduct a process assessment is a *value stream map (VSM)*. A VSM is a detailed and quantified process map discussed in Chapter 4. The VSM describes every operation within a work stream including their inter-relationships. It also quantifies the flow of material and information through the system. Process assessments are useful because they identify waste reduction opportunities not easily seen on management reports. These are called the "hidden factory" because they are not measured or known to exist. The customer, regulatory, and health-and-safety issue categories are also good areas in which to identify improvement projects. Financial benefits should be clearly estimated for any of the projects created using one of the eight project identification strategies shown in Figure 1.5.

Once a project opportunity has been identified at a high level, it is necessary to begin the project deployment process. There are many ways to do this, but the basic questions that should be asked are, "Which improvement strategy should be used to solve the problem and generate the business benefit? If the solution to the problem is known and can be easily handled by professionals within the organization, then why overly complicate the project by bringing in Lean Six Sigma belts? Also, if the project can be solved by capital expenditures or reengineering, then why go through root cause analysis using Lean Six Sigma tools and methods?" In other words, Lean Six Sigma projects should have characteristics that make deployment of teams and root cause analysis beneficial to the organization. This leads to the discussion of what makes a good Lean Six Sigma project.

Key considerations in deciding if a project should be deployed as a Lean Six Sigma project include estimated business benefits, the implementation time frame, resource requirements, and project risk. Project benefits can be broken down from a financial viewpoint into four categories. These are impact to the profit and loss (P/L) statement, cash flow improvements, significant cost-avoidance opportunities, and soft savings (benefits that are difficult to quantify). At a high level, operational benefits are seen as reductions in time, material, and similar expenses; increases in revenue and margin; and cost avoidances, as well as improvements in quality, customer satisfaction, and employee satisfaction. At a lower level, operational benefits can be broken down in hundreds of specific operational metrics an organization uses every day to control its systems. Timeframe of implementation refers to the fact that projects should be executed within a time period useful to the organization, which is typically within twelve months. The better the project focus (objectives), the shorter will be the required time to execute the project. Good project management will keep the project on schedule. The advantage of most Lean and Six Sigma projects is that they do not require large capital outlays since they are improving a process back up to its original design intent (entitlement level). However, anticipated resource requirements in terms of team members and budgets for data collection and root cause analysis are important considerations prior to project deployment. Relative to risk level, the projects should not depend on other projects, new technology, or conflicting organization goals and objectives (strategic and political conflicts). As an example, the project should not be deployed if the process will be eliminated in the intermediate future, if it requires technology not in existence, or if it interfaces with other processes that cannot be included within the project scope.

The justification of the business case should include the context for the problem because building the business case for the Lean Six Sigma project is important to obtain senior management and process owner support. Important considerations when building the business case include showing how eliminating the process problem will benefit senior management. In other words, the project's problem statement should be aligned with senior management's high-level business goals and objectives. The business case should also identify gaps in current performance relative to organizational goals and objectives as well as the time period of problem occurrence. The Lean Six Sigma team can write the project's objective only after defining the overall business problem from a strategic perspective. The project objective should state the project's success criteria, the timeline for completion of the project, and the specific operational KPOV variables to be improved, including their current baselines and improvement targets. These operational KPOVs should be consistent and well defined at every level of the project. The business case for the project should be documented in the form of a written project charter.

Lean Six Sigma Operational Assessments

Conducting operational assessments is a structured process of estimating potential business benefits from a Lean Six Sigma deployment. Assessments that are properly conducted will firmly establish the Lean Six Sigma deployment and ensure a high *return-on-investment (ROI)*. The major goals of an operational assessment are to train senior management and champions to identify Lean Six Sigma projects within their operations. This identification process is different from normal ways of project identification, as shown in Figure 1.5. The assessment process starts with senior management training sessions. In these training sessions, senior management's high-level goals and objectives are systematically broken down into high-level areas or opportunity by major strategic goal and objective. These areas of opportunity are called project clusters. The goal is to map project clusters into project charters through champion training and operational assessments. In addition to identifying projects, the executives and champions are shown how to customize the Lean Six Sigma program to their organizational culture to accelerate the deployment schedule.

Executive Training Sessions

An operational assessment must have support from senior management. The best way to gain senior management's support is to show them the success characteristics of the Lean Six Sigma initiative. A second objective is to show how the program can be modified to improve their organization's operational effectiveness. In this executive training session, real-world examples are drawn from the organization's work streams to show how the program could be deployed within their organization. This also includes information related to the types of Lean Six Sigma projects that would be deployed for this specific organization, the profiles of the people who would be involved in the deployment including the belts, champions, support personnel, and team members, and the training curriculum, as well as typical business benefits obtained from similar deployments. The major goal of the executive training session is to gain senior management's commitment to proceed to project champion training and on an operational assessment to identify Lean Six Sigma project opportunities within the business units.

The first topic of executive training covers the proposed deployment structure for the organization. The deployment structure includes the organizational framework describing both key participants from the organization as well as those from the consulting organization. This discussion includes all roles and responsibilities. The deployment should be controlled by a steering committee consisting of executives from all major functional areas within the organization as well as the consulting organization. In Chapter 2, we will suggest that later in the deployment, the sales and operations planning team should control the steering committee since they are responsible for all major supply chain functions. Also, the *sales and operating planning (S&OP)* team has a vested interest to optimally deploy the Lean Six Sigma initiative since they control the resources necessary to work the projects and receive direct business benefits from the Lean Six Sigma projects. The executive steering committee is assisted by functional support people and other key deployment personnel. Functional support includes people from information technology (IT) who extract data files and information relevant to analysis. Having IT support will accelerate the time to close the Lean Six Sigma projects. In the improve and control phases of the project, IT support people will also be useful in making minor to moderate IT system changes to help implement process improvements and controls. Finance is another important support function. During the operational assessment process, finance must review every proposed project charter to verify financial assumptions are reasonable. Also, finance will help estimate return on investment (ROI) for the program. In the improve and control phases of the project, finance will be required to sign off on the actual project benefits and work with the process owner to ensure that the financial benefits accrue over time according to the deployment plan. Other important support people include the corporate communication department (CCD). The CCD works with the deployment team to ensure that all communications concerning the Lean Six Sigma deployment are consistent across the organization with respect to both the Lean Six Sigma message and the organization's strategic goals and objectives. Human Resources (HR) is the fourth major support function. HR helps move people into their new roles and explains their responsibilities or directly hires new people to take on these Lean Six Sigma deployment roles and responsibilities.

Major roles and responsibilities within the Lean Six Sigma deployment include deployment leaders who direct the deployment within business units and report to the executive steering committee. Other important roles include the master black belts who lead the assessments and train Lean Six Sigma belts. Champions work with the belts to identify significant business opportunities against which are aligned senior management's goals and objectives. Champions are the business people within the deployment and the belts are the technical people. Process owners help support the Lean Six Sigma projects with their local resources, which include the project team members. Also, the process owner will eventually help implement process improvements within their local work areas. The belts lead the Lean Six Sigma projects and help the local work team collect and analyze data. Toward the end of the project, they will also work with the local work team to implement process improvements within the modified process. Black belts typically work across business functions while green belts typically work within a specific business function.

The Lean Six Sigma initiative has been very successful for several reasons. Critical success factors include ensuring support from senior management, that is, top-down deployment. Senior management's support will guarantee resource commitments and other support from the organization. Another critical success factor is basing the deployment on the right people. The right people are people who have historically performed above average during the years prior to the deployment. If an organization cannot spare a large percentage of their top performers, then they might be better off planning a smaller deployment. The best people will get the project completed on schedule and in a very professional manner. Poor performers must be managed to complete their projects. This is very difficult since projects may be deployed in diverse areas in which the poor performers have not worked, resulting in an inability to accurately measure their performance. For poor performers this is an ideal situation, but it can slow the deployment down to a snail's pace. Another important success factor in Lean Six Sigma deployments is selecting the correct projects. The Lean Six Sigma projects should be selected when there are no known solutions and the root cause analysis has a high probability of using the training tools. If the preceding success factors exist, then obtaining quick wins should be relatively easy for the deployment. Quick wins are very important to ensure the deployment's success. Success is important both to maintain the target ROI and to show the organization that the Lean Six Sigma concepts apply to their organization. The final success factor is effective communication of the deployment process. This communication includes information on roles and responsibilities, and the types of projects that will be worked as well as the anticipated impact on the organization. Success stories will greatly accelerate the Lean Six Sigma deployment. As the deployment matures, each organization must customize its deployment characteristics to fit its strategic and tactical goals and objectives.

The executive workshop should also expose executives to key Lean Six Sigma tools and concepts, but only at a very high level to show them the types of projects and problems that would be attacked within their functional areas. The goals at this point of the executive training session should be to show the executives key aspects of Lean Six Sigma projects and how to recognize them within their functional areas. Part of this technical review should include a group breakout in which each executive and their key support staffs begin to scope out areas of Lean Six Sigma project opportunities using their current goals and objectives as basis for the project identification process. The goal of this breakout exercise is to identify operational performance gaps, at a high level, against which the executives do not currently have projects. In other words, the Lean Six Sigma project identification process considers operational and financial goals and objectives the executives cannot improve in the foreseeable future. The executives should also be introduced to the next step in the deployment, which will include champion training followed by an operational assessment within their areas of responsibility.

Champion Training

Following the executive training session, executives choose champions within their functional areas, that is, "direct reports." These lower-level executives are typically middle managers. The purpose of champion training is to expose champions to more deployment details. These details include information regarding the technical tools and methods of Lean Six Sigma, the roles and responsibilities associated with the deployment, project selection, how to write project charters, and finally how to conduct operational assessments to create project charters. These project charters will eventually be assigned to Lean Six Sigma belts. Other aspects of champion training include information concerning the deployment structure. The discussion of deployment structure includes the executive steering committee functions, the typical roles and responsibilities associated with the deployment, and how projects will be selected during the assessment.

As the champion training continues through the week, assessment details are discussed with the champions. The champions then create an assessment plan for their functional areas of responsibility. As the assessment proceeds, the champions will be working with the consultants and local black belt candidates to create project charters based on preliminary operational and financial analysis of their operations. These project charters will be reviewed by local management and finance. Once each black belt has two to four project charters, each one having the savings objective specified by senior management, the Lean Six Sigma training will begin in earnest. The project charters' savings are rolled up to a business unit level and the deployment ROI becomes firm. During the course of project execution, although a specific project may be over or under its original savings target, the overall business benefits from the entire project mix should be on target to deliver the required ROI for the organization. At the conclusion of the champion training session, the champions should present their operational assessment plans to the group prior to departing the training session for the operational assessment.

Conducting Operational Assessments

Operational assessments allow an organization to customize the Lean Six Sigma deployment according to the business opportunities within each of their business units. The Lean Six Sigma black belts are selected to work improvement projects that will improve operational or financial performance. Operational assessments are a structured way to investigate and identify a Lean Six Sigma project within a business unit. The project charters become the basis of the Lean Six Sigma deployment within each business unit.

A major purpose of operational assessments is to gain support from local management. After all, the deployment will rely on their resources. Most likely these resources have already been assigned to other projects within the business unit. In the initial meeting with local executives, the Lean Six Sigma deployment team must convince them that they will be an integral part of the operational assessment. This includes evaluation of potential projects. At the initial meeting with these local executives, the deployment team should review key aspects of the deployment, but at a high level. In particular, it is important to review the overall deployment strategy that impacts them at a local level. This includes training dates, financial projections, what will be taught, and how the project identification process will work within their facilities. At the end of the assessment the proposed Lean Six Sigma projects identified within their business units would be the last discussion prior to leaving their facility.

The operational assessment consists of several major parts. These include interviews with functional managers, analysis of financial and operations reports, the creation of value stream maps (discussed in Chapter 4) to identify operational issues related to quality, long cycle times, high costs, and high inventory levels as well as the data related to gaps in operational and financial performance. At the end of the assessment, and before the team leaves the facility, project charters must be created by the assessment team. These charters can be modified later as additional data becomes available to the assessment team. But the assessment team should not delay creating the project charters until a later date since it will be more difficult to obtain the necessary functional manager review and sign-offs of the project benefits. If the assessment team cannot create project charters to show performance gaps and potential business benefits by project, then the operational assessment has not been successful. The assessment team should remain at the location until its work has been completed according to plan. The failure to successfully complete the operational assessment will mean the Lean Six Sigma black belts will struggle to validate their project charters during training. This situation will delay their project execution by 30 to 90 days.

Local Leadership Interviews

As the assessment team begins the interviewing process, it is important to review with each local executive the operational breakdowns that prevent them from achieving their goals and objectives. The assessment team could also ask the leader to recommend specific areas or work streams for investigation. Part of this interviewing process should include a review of the local executive's operational and financial performance reports to evaluate current performance baselines and metrics used by this organizational function. At the end of the interviewing process, the assessment team should arrange to work with the executive's local work team to value-stream-map major work streams within the functional area to look for process breakdowns not currently captured by management reports. This information should be checked against the areas of opportunity identified during the executive training session to ensure consistency across the organization. These functional interviews are also important to obtain resources to continue the assessment within each functional area.

Opportunities for Lean Six Sigma projects will emerge as the assessment team identifies critical processes that have performance gaps. Data may be available to immediately analyze performance gaps and create Lean Six Sigma project charters. Alternatively, the assessment team should create data collection plans to collect and analyze data that will be sufficient to make the business case for projects. It is important the assessment team collect data sufficient to identify and establish the business case for the project during the assessment; otherwise, the Lean Six Sigma black belts will have to do this work during the training cycle. At the end of the assessment, the assessment team creates a deployment plan for the local business unit. This deployment plan includes the initial round of projects to be deployed as well as the second and third round of projects. Also, an initial ROI can be estimated for local leadership to gain their support for the deployment process. At the end of the operational assessment, the team creates a report that embodies all the information gained from the assessment. The last action item is a meeting with the local leadership team to close the loop on the assessment process.

Lean Six Sigma Black Belt Training

If the assessment team and deployment leaders did their jobs well, the Lean Six Sigma black belts will have well-defined project charters. This will put them on a firm basis relative to data collection and analysis. The students will then be able to focus on learning the tools and methods of Lean Six Sigma rather than having to define their projects. The Lean Six Sigma training program is normally made up of training cycles consisting of one week of classroom training followed by three weeks of project work. However, there are many variations on this theme. If the Lean Six Sigma black belts have a very good technical background, the classroom training can be reduced to just two weeks. Or if the organization needs sequenced training, the training cycle might begin with one week of Lean training followed by one week of green belt training and finally by two to three weeks of black belt training. Overall, I believe the sequenced training cycle is the most effective since the students can apply their newly learned tools to three projects in total rather than just one project. Also, the students learn to simplify a process using Lean tools and methods prior to applying the statistically oriented Six Sigma tools and methods. It is also important the training emphasize the correct mixture of analytical tools since "one size does not fit all." In particular, supply chain projects should really have more of an operations-management or operations-research blend of analytical tools. This concept will be discussed in Chapter 8.

Lean Six Sigma black belt training consists of a combination of training modules designed to bring the black belt through the *define, measure, analyze, improve,* and *control* phases of the DMAIC process. In the define phase, the black belt is trained to identify project opportunities using the voice-of-the-customer (VOC) and voice-of-the-business

(VOB) information to create or better scope their project charters. The project identification process will be discussed later in this chapter. In the measure phase of the training cycle, the Lean Six Sigma black belt is trained to conduct *measurement system analyses (MSA)* to ensure that the key process output variable (KPOV) can be measured accurately and precisely and determine the process capability of the KPOV. Capability analysis is discussed in Chapter 8, but it consists of comparing the VOC relative to the VOB (or comparing customer requirements to process variation). The final major activity of the measure phase is to ensure that the team has a list of potential input variables, which may or may not be impacting the KPOV. In the analysis phase of the training cycle the Lean Six Sigma black belt is trained to analyze the collected data to identify those input variables having the greatest impact on the KPOV. During the analysis phase the Lean Six Sigma black belt could use a variety of tools and methods to identify key process input variables (KPIVs) and their combined impact on the KPOV. Finally the black belt and improvement team change the process to improve the operational metrics, that is, KPOV, in the improve and control phases.

Lean Six Sigma Green Belt Training

A common problem in green belt training is "watering down" the Lean Six Sigma tools and concepts rather than designing the training modules for the types of people and projects that will be part of the green belt deployment. The major problem with green belt training is it tends to be overly complicated, with too much emphasis on black-belt-level statistical tools and methods. There may also be a lack of emphasis on completing applied projects having significant business benefits. Green belt training is very important to the organization, and the students should be supported with good training materials and held accountable for business results in the same manner as the Lean Six Sigma black belts are held accountable. Also, the training should adequately prepare green belts for the next step in their career development, that is, black belt training.

Design-for-Six-Sigma Training (DFSS)

Whereas the Lean Six Sigma projects follow the DMAIC methodology to help eliminate breakdowns in the current process, Design-for-Six-Sigma (DFSS) black belts create new processes (or products) to meet VOC requirements. This work requires specialized tools and methods that are the basis for Design-for-Six-Sigma. The DFSS methodology consists of several phases each having inputs and outputs designed to create solutions to meet customer requirements. DFSS is discussed in the Conclusion chapter under "Next Steps." It should be noted that current DFSS tools and methods require major modifications for supply chain applications because they were developed based on the manufacturing concepts, that is, product design. Also, the specific combination of tools and methods necessary to design supply chain systems will also vary by supply chain type. For this reason DFSS must be modified for different industries and even functions within an industry. As they say, "One size does not fit all." Or as people often remark, "Everything isn't a nail even though you know how to use a hammer."

Building the Project Charter

The project charter is the communication vehicle for the project. It takes several forms depending on the organization and industry, but it has several common elements. These are "stating the organization's business problem"; "stating the project objective, that is, part of the higher-level problem statement"; "the required resources to execute the project"; and "the financial and other benefits anticipated from project completion." A typical project charter is shown in Figure 1.6 on the next page. Reviewing Figure 1.6, we see the project charter is broken into administrative, project-definition elements, operational metrics, resource requirements, including team members, and a financial summarization section, which calculates net business benefits. The project charter is updated as the Lean Six Sigma improvement team works through the root cause analysis.

Project Problem Statement

The problem statement should be a complete description of the business problem stated in such a way that the entire organization can understand the importance of deploying the project. It should include current performance baselines related to financial and operational metrics including quantitative background information. This background information describes where the problem occurs, how it is measured, who is impacted by the problem, and by how much they are impacted. Other critical information should include all major internal and external impacts on the business due to the existence of the chronic problem including those affecting the customer. The project's metrics should clearly link to

Administrative Information

Project Name: Date:

Business Unit:

Champion: Process Owner:

Project Timeline:

Project Definition Information

Problem Statement:

Project Objective:

Project Metric (s):

Resource Requirements

Team Members:

Project/Team Leader:

Financial Information	Margin Revenue	Cost Reduction	Cost Avoidance	Other
One-Time Cost Savings				
Project Costs				
Total Benefit				

Figure 1.6 Typical Project Charter

high-level business goals and objectives. It is also important the problem statement contain no solutions or other information that may bias data collection and analysis efforts.

Project Objective

The project objective should be a part of the problem statement. It should be linearly linked to the overall problem and be in the same financial and operational metric format. As an example, if the higher-level problem is stated in terms of "excess inventory investment" and "low inventory turns," then the project objective should have at least these same metrics. However, as the root cause analysis proceeds, the team may be required to bring additional operational metrics into the project charter. As an example, through root cause analysis, the team may find that lead-time for a certain product category or supplier has a major impact on inventory investment and turns. The project objective might then be focused on lead-time reduction for this product category. Lead-time would then become another operational metric measured by the Lean Six Sigma improvement team. In addition, the team must state the relationship of all project metrics. As an example, the team should be able to make a statement like: "If lead-time is reduced by 50% for product category *xyz*, then "inventory turns at a product level will increase by 20%, and at a business unit level they will increase by 10% to reduce inventory investment by 10%." In this manner, the project objective changes as the team works through the root cause, but senior management always sees the overall business impact of the project.

Required Resources

The resources for the project will vary depending on the type of project. In Lean Six Sigma projects the goal is not to make major changes in the process that would require large capital outlays or significant resource commitments by the organization. This is not to say some high-leverage projects should not be resourced using capital, but only that this situation should be the exception, not the rule, in Lean Six Sigma deployments. The three categories requiring resources are the people taking part in the project, paying for data collection and analysis, and the implementation of the process improvements. The people making up the project team, as well as support personnel from information technology, finance and other professionals within the organization are often very busy. Their time is valuable. As a result, the project should only use the necessary level of human resources required to meet its objectives. This doesn't mean the project should not be properly resourced, but only that thought must be put into who should be on the team as well as what they will do when on the team. Resources required for data collection and analysis are another important consideration. Some data collection efforts require that surveys be conducted or data be purchased. Examples are data required to obtain customer buying preferences, laboratory testing, and payments to consultants or other professionals for their time. Finally, improvements will cost money. Examples include making minor changes to software code to mistake-proof data entry, modifications to the employee training program, or minor process changes. Resources should be managed well since they are expensive.

Project Financial Justification

Project financial benefits can be categorized into at least four major areas. The first is direct cost savings visible on the profit and loss (P/L) statement. The second is incremental margin on current or improved sales. The third is lower carrying costs associated with working capital. The fourth category is significant cost avoidances that can be clearly documented to the organization. In addition to these four categories, Lean Six Sigma projects should improve or at least not degrade customer metrics such as on-time delivery or other service levels. It is important to correctly classify the projects in terms of their financial benefits as well as month-to-month timing, so senior management can manage the Lean Six Sigma deployment to the business unit's financial goals and objectives. Financial analysts should also be an integral part of the project team. All anticipated financial information should be captured within the improvement team's project charter.

The P/L lists all major cost categories that provide direct opportunities to identify and deploy Lean Six Sigma improvement projects. Obvious project opportunities include reductions in the cost of scrap, rework, overtime expenses, warranty expense, premium freight, interest expense, direct labor, materials, and transportation, as well as several other categories. Using Lean Six Sigma methods, numerous projects have been identified to reduce these P/L costs as well as others. In fact, this heavy emphasis on cost reductions has been one criticism of the program. While this may be true, cost savings should always be a business priority. But, to prevent an unbalanced deployment strategy, a mixture of Lean Six Sigma improvement projects should be created from each of the four major business benefit categories. Also, every project should improve or at least not deteriorate customer satisfaction. The actual percentage of projects within a particular category will vary by organization depending on its goals and objectives.

P/L financial benefits associated with revenue increases include sales percentages over and above those already in the annual operation plan or increases to gross margin. When identifying projects in these areas, it is important to very clearly establish current performance baselines as well as anticipated operational improvements by the sales function. This is to ensure that the Lean Six Sigma team is recognized for their contribution since the organization cannot allow revenue increases to be double-counted by both groups. Having said this, there are many examples where revenue has been increased through the use of Lean Six Sigma teams. In these examples, the team usually translated the VOC requirements from the customer back into their internal process into KPOVs and internal specifications. Then a project charter was created by the team to improve the KPOVs using root cause analysis of the problem. As an example, perhaps sales were being lost due to long cycle times or poor quality. The Lean Six Sigma improvement team may have created a value stream map to identify wasted time within the process. Another example where sales revenue can be increased is by improvements in gross margin. Perhaps the organization can analyze current product or service offerings and modify them to both increase sales and obtain incrementally higher gross margins. Or perhaps current gross margins have been artificially reduced due to product returns and allowances. Lean Six Sigma projects can be deployed within these areas to make the necessary improvements. Finally, through VOC analysis, the organization might identify new marketing opportunities that would increase sales revenue.

The third category of business benefits includes cash flow improvement projects. These projects include reductions in account receivables and asset conversions including inventory as well as asset divestiture and facility outsourcing. The P/L impact from these Lean Six Sigma projects depends on the prevailing interest rates, but the cash flows do reflect the value of the conversions as well as their interest expense. In these types of projects, the Lean Six Sigma improvement team attempts to free up available capacity to remove entire elements of the asset base. As an example, if productivity and throughput are increased, then perhaps fewer manufacturing lines or facilities will be required to satisfy customer demand. Or if inventory turns are increased, less distribution center floor space will be required. In fact, if enough floor space is eliminated there may be no further need for a satellite warehouse. Other Lean Six Sigma projects could evaluate the need to own assets rather than lease them to free up invested capital. There are many other examples where Lean Six Sigma projects have significantly increased cash flow.

The last category includes business benefits obtained through cost avoidance or other customer and stakeholder satisfaction improvements. There are many examples where a cost-avoidance project may be a much higher priority than a cost-saving project. But most cost avoidance projects must be very carefully defined and their assumptions must be realistically verified by finance. Examples where Lean Six Sigma improvement projects have been successfully deployed to avoid cost include material substitutions in which, without the removal or addition of certain materials, the organization would be forced to cease operations or use much more expensive materials. Health and safety issues are other areas where cost-avoidance projects have been successfully deployed to prevent anticipated injury or death to employees. As a final example, a customer may threaten to take business away if a chronic problem is not eliminated. This is a classic example where Lean Six Sigma improvement projects are useful to the organization. The business benefits of cost avoidance projects to the organization can be enormous, but they will not be captured on the current year's financial statements.

Project Planning

After creation of the project charter and approval by management, the project plan must be developed by the team. Development of the project plan requires support from the process owner and their work team. Using the project charter as a focal point, the improvement team refines the original project scope. Refinement of the project scope is accomplished by review of the system map shown in to Figure 1.3. In many organizations, this system map is called a *SIPOC*. The SIPOC is a high-level, quantified system map that shows process input/output relationships. The term SIPOC means "**S**upplier"; "**I**nputs to the process including materials, labor and information; "**P**rocess," which converts inputs into outputs; "**O**utputs" including material, labor or information; and the "**C**ustomer," who receives the output from the process. Eventually, as the team continues the analysis and refinement of the project objective, more detailed process maps are created to reflect the process analyses. To develop the project plan, the improvement team breaks the project into a series of well-defined activities or tasks. These activities are spatially and temporally related to each other, that is, series or parallel tasks separated by a time sequence. Every operational element has an expected duration or completion time, that is, how long it takes to complete the activity. Once an activity has been clearly defined relative to its sequence and duration, the Lean Six Sigma improvement team estimates the overall project resource requirements and control strategies. Effective project management and control is facilitated by development of key project milestones. The 10-Step Solution Process provides key project milestones to control Lean Six Sigma improvement projects. These milestones guide the team through the analysis, root cause identification, and solution process as well as eventual transition to the process owner and local work team. Establishment of effective process controls, reporting systems, and audits based on the root cause analysis are important to the project transitioning process.

One of the key principles of process improvement (and consulting) is to understand what the focus of the improvement activities needs to be in the first place. The best way to get started is to bring the team together and brainstorm questions that must be answered by the improvement team's analysis. Management must have direct input into this process since their questions are important to the success of the Lean Six Sigma project. One of the most important tasks to ensure project focus and resource optimization is to list the relevant project questions in advance prior to doing any data collection. The specific questions and their associated data fields will vary by industry and organization. This project definition process can be very difficult because organizations seldom have the data necessary to answer the important questions. In fact, this is why the questions have never been satisfactorily answered in the first place. The typical situation is that required data fields, if they exist at all, are spread over several databases. But, to answer the specific questions necessary to complete the project, these required data fields must be brought together in one place to build the work stream model.

Using inventory investment and its turns ratio as an example, the simple question "What should inventory turns be by item and location?" requires accurate estimates of lead-time and expected demand as well as demand variation by item and location since these are the necessary inputs into the inventory model. However, there are real issues in accurately estimating lead-time and demand by item and location. The data may not exist. Other relevant data elements often include *cost-of-goods sold (COGS)*, service-level targets, and lot size requirements. There may be other required data elements depending on the Lean Six Sigma project. But effective inventory management requires answering these types of questions during the analysis. Table 1.4 provides a partial listing of questions relevant to inventory analysis in most organizations. The first question on the list is applicable to most industries holding inventory in raw material, *work-in-process (WIP)*, or finished-goods form. The second question on the list also is applicable to most organizations using a forecasting system. The advantage of listing questions prior to data collection and analysis is that the necessary data fields can be collected and brought together in one place to build a quantified model of work streams directly associated with the list of critical questions. This ensures project alignment with business goals and objectives as well as operational alignment with the data directly linked to the business metrics that must be improved by the team. Quantitative analysis of the inventory model will ensure that the correct questions are answered by the analysis because data collection activities were aligned upfront

1. What should inventory targets (turns) be?
2. How do demand variation and lead-time impact inventory?
3. What is the inventory investment baseline?
4. How can it be improved?
5. Which suppliers contribute to excessive inventory?
6. How do lot size and long lead-times impact inventory?
7. What are the levels of excess and obsolete inventory?
8. Should low-volume items be centralized or decentralized?
9. What are the forecast accuracy baselines?
10. What is the impact of unplanned customer orders on inventory?
11. How many unplanned schedule changes occur?
12. What is the overdue order backlog?
13. What is the material availability status at production start?
14. What is accuracy of Material Requirements Planning (MRP)?

Table 1.4 Relevant Questions

with project objectives. These concepts will be more fully discussed in Chapters 6, 8, and 10 in the context of inventory models. But the concepts also apply to the analysis of other supply chain work streams. Asking the right questions upfront will go a long way towards a project solution. As they say, "A problem well defined is half solved."

Project Risk Assessment

As the Lean Six Sigma project is being defined, the improvement team must analyze external and internal project risks including their impact on projected business benefits and resource requirements. Project risks can arise for many reasons. These reasons might include project cost overruns, shortages of resources, technological problems, changes in organizational strategy, changes in organizational structure, and changes in customer requirements. Project cost overruns can occur due to many factors including incorrect specifications and poor utilization of materials and labor. Shortages of resources could occur due to loss of key suppliers or unanticipated demand for the resource due to external situations. Technological problems can occur due to unforeseen problems with machinery, test equipment, or materials.

The result will be lower-than-expected process yields. These issues are normally associated with the design of new products and processes or *research and development (R&D)* projects. Risks also occur if an organization changes its strategic focus (that is, closes facilities or shuts down manufacturing lines); the projects associated with these processes will be at risk. If the organizational structure changes, projects that are currently supported by certain process owners and champions may be left without organizational support or resources. Finally, if the project touches external customers and their requirements change, the premise of the project must be modified to ensure its success. Although project risks exist, they can be managed by the improvement team and contingency plans made to eliminate or minimize their impact on the Lean Six Sigma project.

Project Management

Effective project management is critical to the improvement program. In particular, it is important to build the business case for the project and establish its success criteria. Key project management tasks include developing the overall project plan with team members, allocating resources to project work activities, implementing project control procedures, and ensuring an effective transition of the modified process from the improvement team back to the process owner and local work team. These key project management elements are necessary to ensure efficient resource planning and project control to meet cost, quality, schedule, and improvement objectives.

A project brings people from different organizational functions together for a specific purpose, that is, to solve a problem for the organization. A project consists of people, materials, information, machines and other resources that are used to accomplish the project's goals and objectives. More specifically, a project is a network of activities that are related to each other. These activities have a time duration, which varies and consumes resources. Some activities begin before other activities and some are parallel with each other. The management of a particular project consists of developing the project goals and objectives with the project team, breaking these goals and objectives into key milestones or events, and finally breaking down all the activities comprising a milestone into work tasks. This breakdown process is called the creating the *work breakdown structure (WBS)*. Several projects combine in a Lean Six Sigma deployment. The Lean Six Sigma deployment consists of numerous inter-related projects which in aggregate help execute the organization's strategic plan.

Fundamentals of Project Management

The project management consists of ensuring that the project scope is correct, the work is done correctly, and the project remains on budget and on time relative to resource utilization. Project scope defines where in the work stream the project begins and ends. A key tool to scope a project is the high-level process map called the SIPOC (Supplier-Inputs-Process-Outputs-Customer). Project scope is communicated differently in different organizations. It is called the project charter objective, statement of work (SOW), internal performance specifications, or related terms depending on the context. The quality of the project management process relates to how well the team executes their deliverables and the overall project results. It is important the team periodically measures its effectiveness relative to quality of project execution including the cost associated with the original budgeted targets for the project. The project should remain within budget if the project fundamentals are correct and all schedules related to project milestones are completed on time.

Specific project management methods vary depending on the size and complexity of the project as well as its business value to the organization. Large, complex, and important projects require rigorous management of the project while small and less complex projects may require only basic project management techniques. The basic elements of project management are defining the required project tasks using the work breakdown structure (WBS), planning how the tasks will be done bringing the team together, measuring and controlling all the inter-related project tasks and ensuring the project closes on time and within budget.

Selecting the Team

Lean Six Sigma improvement teams require effective project management techniques. Because the team members are assigned on a temporary basis to the team and may have other assignments, it is important to stay focused on the project's goals and objectives, that is, problem statement and objective. Also, the team members will probably be from several different organizational functions such as logistics, manufacturing, marketing, sales, and purchasing. However, the core portion of the team should reflect the process under investigation as defined by the SIPOC. Core team members should include people who are part of the process under investigation as well as those who either supply or receive materials, labor, or information at the process input and output boundaries. In addition the team may require technical people to help in the collection and analysis of data. Once the improvement team is formed, it will progress through several phases ranging from disagreement to consensus. The team is more apt to reach some form of consensus as data is collected and analyzed in an unbiased manner. A fact-based approach will often indicate a clear course of action for the team.

In the initial formation of the team, the team leader should bring the team together to review the original project objectives as represented by the project charter's problem statement and objective. This review should include a thorough discussion from all team member perspectives. Throughout this review process, the team must stay focused on the management questions that must be answered by the analysis. In this process, a diverse team is important and proper facilitation of the team is critical to avoid either groupthink or derailment of the team by forceful personalities.

Work Breakdown

Work breakdown is an essential part of project planning. It involves taking the project deliverables and systematically breaking these deliverables into major milestones. The milestones are then broken into specific work activities and finally work tasks. Deliverables include the specific goals and objectives of the project. In the 10-Step Solution Process, the milestones represent each of the ten steps of the solution process. This is shown in Figure 1.7. The Lean Six Sigma improvement team breaks down the ten steps into their work activities and finally breaks down each work activity into its associated work tasks. As the work activities within a given step are completed, the improvement team moves to the next project milestone or step of the 10-Step Solution Process. Work tasks include major portions of work having starting and ending dates as well as required resources. The work tasks are the specific items that must be completed by a small group of people or a specific person.

Depending on project complexity, there may be easier ways to display project status than others. However, certain combinations of communication will usually be required to communicate to the improvement team, the process owner, and local work team as well as senior management. Complicated projects should be tracked using project management software such as Microsoft Project. Simpler projects can probably be managed using Microsoft Excel lists of each work activity with their associated work tasks including beginning and ending dates for each task. The advantages of using project management software is that project status is displayed in Gantt chart format with activities listed by row and time duration by column with bars to indicate the relationship of one activity to another. An example of a Gantt chart is shown in Figure 1.7. Another advantage of project management software is that when there are scheduling changes casued by late work or resource scarcity, the software can immediately re-establish the work schedule and required resources (crashing the schedule).

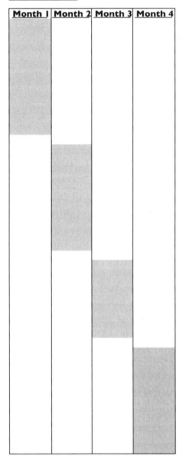

Key Project Milestones: **Date Completed**

	Month 1	Month 2	Month 3	Month 4

1. Project aligned with business goals
  a. Identify project opportunity
  b. Estimate business benefits
  c. Select team and complete project charter

2. Buy-in from process owner, finance, and others
  a. Develop project plan with process owners
  b. Develop as-is process map and baselines
  c. Refine project charter as required

3. Communicate project results
  a. Develop communication strategy

4. Prove causal effect Y = f(X)
  a. Identify key metrics (outputs)/improvement targets
  b. Validate key metric measurement systems
  c. Develop data collection strategies on inputs/outputs
  d. Collect data and build models
  e. Analyze models and identify root causes
  f. Test the optimum solution (pilot)
  g. In parallel, create a list of other projects and charters
  h. Provide charters to champions
  i. Help champions integrate charters
    into Lean Six Sigma deployment

5. Measurement systems improvement
  a. Update measurement systems on key inputs/outputs
    as well as other identified MSA issues

6. Detailed improvement plan
  a. Analyze pilot/develop list of countermeasures by root cause
    to ensure their elimination
  b. Refine project timeline

7. Countermeasures tied to root cause analysis
  a. Develop action plans to eliminate root causes

8. Standardized procedures
  a. Standardize revised procedures

9. Implement training and audit
  a. Implement training and audits as required

10. Apply control strategies
  a. Implement and integrate all other control actions

Figure 1.7 Gantt Chart

Communicating Project Status

Communication of improvement activities should be coordinated with the organization's communication department and deployment leaders. It is important the organization hears a simple, clear, and consistent message from all the improvement teams. Communication should be standardized to the highest degree to ensure that everyone easily understands the deployment information including project status. Lean Six Sigma initiatives communicate project status using easy-to-understand visual displays of data including its analysis and conclusions. This process is done using a combination of project status charts and presentation templates. There are many forms of project status charts, but the most common is the Gantt chart shown in Figure 1.7, which lists every activity in sequence and its duration by time period. Using a Gantt chart the team can measure the project's progress against its original project timeline. The second key communication vehicle is the presentation of the team's project status to others within the organization. The most common presentation format is a PowerPoint presentation, but every organization has its preferred form of communication. It is important to provide the various Lean Six Sigma improvement teams with a common presentation template so that every project can be easily compared to others. Figure 1.8 has a list of presentation deliverables

10-Step Solution Process		DMAIC Project Presentation
D	1. Project aligned with business goals	1. Team picture
		2. Problem statement
	2. Buy-in from process owner, finance, and others	3. Project objective
		4. Process baseline
		5. Process map
M	3. Communicate project results	6. Brainstorming the inputs (X's)
		7. Measurement on the output (Y)
	4. Prove causal effect Y = f(X)	8. Capability analysis on the output (Y)
		9. Business benefit verification
A	5. Measurement Systems Improvement	10. Elimination of many trivial inputs (X's)
		11. Selection of key inputs (X's)/build model
	6. Detailed improvement plan	12. Final solution
		13. Integrated control plan on the inputs (X's)
I	7. Countermeasures tied to root cause analysis	14. FMEA
		15. Instructions and training
	8. Standardized procedures	16. Verified business benefits
		17. Next steps
C	9. Implement training and audit	18. Translation opportunities
		19. Lessons learned
	10. Apply control strategies	

Figure 1.8 Communicating Project Status Using DMAIC

commonly used in Lean Six Sigma deployments. The list of deliverables corresponds to the 10-Step Solution Process as well as the Six Sigma define, measure, analyze, improve, and control phases (DMAIC).

Identifying Lean Six Sigma Projects

It has become widely recognized that Lean Six Sigma methods are useful throughout an organization's many supply chain functions. Tables 1.5, 1.6, and 1.7 contain generic lists of operational metrics that have served as the basis for Lean Six Sigma improvement projects. Other metrics could be also added to this list depending on organizational requirements. This is especially true if an organization is not measuring important metrics because of resource limitations or ignorance. Over the next several chapters we will provide numerous examples showing how Lean Six Sigma methods can be applied in practice by your organization. In the interim, it would be a good idea to use the metric lists contained in Tables 1.5, 1.6, and 1.7 to begin the project identification process within your organization.

Finance	Billing
■ Accounts payable cycle time	■ Billing errors
■ Variance to budget	■ Mailing expense
■ Margin improvement	**Call Center**
■ Overtime expense	■ Average handling time
■ Account receivable cycle time	■ Call transfers
Quality Assurance	■ Cost per Abandoned Call
■ Defects	**Purchasing**
■ Customer complaints	■ Inventory investment and turns
■ Claims	■ Year-over-year cost reduction
■ Rework	■ Number of suppliers
■ Scrap	■ Cost per invoice
■ Warranty	■ Purchasing errors

Table 1.5 Typical Project Examples Part A

Administration	Marketing and Sales
■ Utilities expense	■ Sales dollars per salesperson
■ Insurance costs per employee	■ Turnover per customer
■ Facility costs per employee	■ Marketing cost per dollar sales
■ Material and supplies expense	■ Percentage of sales closed
HR	■ Quote to actual project cost
■ HR staff per total employees	■ Order changes
■ Absenteeism rate	■ Sales policies and procedures
■ Training hours per employee	**Operations**
■ Employee cost to hire and retain	■ Lead-times
■ Health costs per employee	■ Late orders
■ Lost time accidents	■ Average cycle time per order
■ Disability costs	■ Scrap
■ HS&E	■ Rework
	■ Emergency maintenance

Table 1.6 Typical Project Examples Part B

R&D	Distribution
■ Man-hours per project	■ Number of shipments
■ New products per year	■ Freight charges (inbound and outbound)
■ Current projects	■ Inventory investment and turns
■ Project cycle time	■ Excess and obsolete inventory
■ Projects waiting	■ Shortages
■ Engineering changes per year	■ Premium freight costs
IT	■ Retuned product
■ System capacity	■ Product transfer between facilities
■ System crashes	■ On-time delivery

Table 1.7 Typical Project Examples Part C

Summary

The 10-Step Solution Process leads the reader through several key project management tasks including project definition and alignment. The result is a Lean Six Sigma project supported by the process owner, work team, finance, and senior management. This fact-based approach also facilitates communication of project milestones to the organization. As the team progressively works through creation of a process model of the work stream including its root cause analysis, the underlying relationships between KPIV and KPOV variables become apparent. In the Lean Six Sigma world, this is called development of the $Y = f(x)$ relationship. This methodology allows development of detailed improvement plans based on countermeasures tied to the project's root cause analysis. To effectively transition the project back to the process owner and local work team, the improvement team develops a control strategy for the modified process. These concepts will be discussed in the coming chapters.

Chapter 2

Deploying Lean Six Sigma Projects Using Lean Tools

Key Objectives

After reading this chapter, you should understand how Lean tools and methods are useful to simplify processes and reduce process waste, and understand the following concepts:

1. Why the sales and operating planning team (S&OP) should direct the deployment of Lean Six Sigma projects across the supply chain.
2. Why Lean methods are useful in identifying and eliminating process breakdowns.
3. The types of Lean tools and methods as well as their applications.

Lean Six Sigma Project Alignment

Deployment and strategic alignment of high-level goals and objectives is qualitatively shown in Figure 2.1. Here, strategic planning begins with *shareholder return-on-equity (ROE)* targets. At the next level down, ROE targets are delayered or broken down into lower level financial metrics, that is, sales, cash flow, and income targets. These financial targets are also delayered to link to specific operational metrics with firm performance targets. As an example, cash flow can be increased by using Lean Six Sigma projects to attack the root causes of high inventory investment

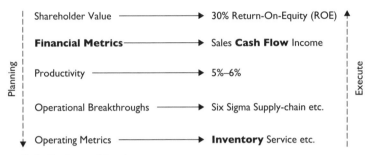

Figure 2.1 Business Metrics

which ties up cash decreasing its flow. In this example, the higher level financial metric cash flow is driven by lower level operational metrics such as inventory investment. In turn, inventory investment or its turn's ratio is driven by lead-time and demand variation. It is important that higher level financial metrics link at every organizational level to ensure alignment of goals and objectives behind those of senior management. In particular, return-on-equity (ROE) is an important high-level metric (although it can be distorted by inventory valuation and other expensing policies of an organization). ROE is a ratio calculated by dividing net income by equity capital. ROE is important to senior management since it is a measure of the return shareholders receive by investing in the organization. It is important to measure an organization's effectiveness using these financial metrics as well as others since commercial organizations exist to return profit to their investors. As an example, if you placed $100 in a bank account having minimum risk, you would expect an annualized return commensurate with this risk level. However, if you invest in more risky investments like corporate stock, it makes sense to expect a higher annualized return on investment. To summarize the strategic flow down process, once the ROE targets (and other high-level financial targets) have been set by senior management, the organization breaks these down into the sales, cash flow, and income targets necessary to achieve the ROE objective. This information is used in part to develop year-over-year productivity ratios. The productivity ratio, in its simplest form, is a ratio of the sum of the productive outputs divided by the sum of the inputs necessary to achieve the outputs. The ratio should be larger than one. In fact, well-run organizations routinely achieve productivity levels in the 4% to 8% range adjusted for inflation, interest and foreign exchange rates.

The organization has many options available to achieve its sales, cash flow, and income targets. As an example, an organization could buy capital equipment or implement operational initiatives such as Lean Six Sigma to improve its operations and increase productivity by reducing the quantity of inputs to the process. The objective is to improve operational metrics to achieve higher-level financial goals and objectives while balancing the inter-relationships between the metrics. As an example, an inventory reduction project could increase cash flow and improve income, but inventory shortages may adversely impact sales, cash flow, and income. This implies that Lean Six Sigma projects and their improvements must be evaluated using more than one metric to ensure a balanced deployment.

To ensure alignment with available resources, high-level strategic goals and objectives must be integrated into the improvement projects. Using this concept, major areas of business opportunity called project clusters are identified at a high level by the Lean Six Sigma improvement team. During champion training, very focused project charters will be developed within the project clusters. This systematic approach to project deployment is important to ensure linkage of operational improvements to customer and business objectives as well as to ensure resource optimization across many projects. Over time, projects that have been properly aligned and deployed will generate project benefits including lower operational costs, reduced cycle time and higher customer satisfaction.

S&OP Control of Lean Six Sigma Projects

Lean Six Sigma improvement projects focus on the process breakdowns that create performance gaps. In supply chains, these breakdowns are usually related lead-time or demand management issues. This is why understanding the components of lead-time as well as the tools and methods useful to reduce lead-time is important. Similarly, understanding demand variation and its components as well as the tools and methods that can better manage demand is also very important. In the identification and deployment of Lean Six Sigma projects, joint project coordination between one or more functions is required because these improvement projects often touch several organizational functions. One effective solution to the challenge of joint project coordination is to use the sales and operations planning (S&OP) team as a conduit through which Lean Six Sigma improvement projects are created and executed by the organization.

Using the S&OP team as a conduit for Lean Six Sigma project deployment is an effective strategy because overly complex organizational structures, typified by functional silos (shown in Chapter 7, Figure 7.2), require a cross-functional team to coordinate and prioritize resources including capacity to satisfy customer demand. Lack of coordination will result in poor execution of tactical plans and a failure to achieve long-term strategic supply-chain goals and objectives. Using the S&OP team, an organization can systematically improve supply chain performance over time through the effective identification and execution of Lean Six Sigma projects. A cross-functional team approach is also required because each organizational stakeholder has a limited view of the entire supply chain. In contrast, the purpose of the S&OP meeting is to consider inter-related constraints impacting demand, capacity, logistics, purchasing, and inventory planning as well as other organizational functions. Disparate viewpoints are eventually reconciled by the team to achieve the organization's operational objectives. Figure 2.2 shows the complex issues the S&OP team must consider. These include cash flow, budget to plan, material availability, on-time material delivery, and execution of manufacturing and shipment schedules as well as other issues.

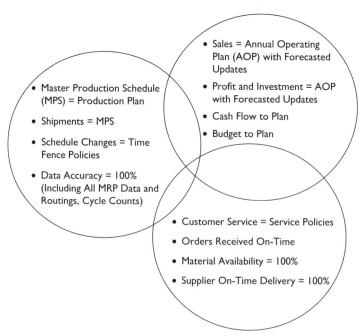

Figure 2.2 Execution of the Plan

The annual sales operating plan (SOP) is derived, in part, from the sales targets necessary to achieve the organization's higher-level financial goals and objectives. The SOP includes sales, profit, investment, and cash flow targets. The sales portion of the plan feeds the *master production schedule (MPS)* (discussed in Chapter 3). The MPS aggregates demand coming into the organization in order to set manufacturing schedules by product and location. Once the manufacturing schedule is set, the organization should not violate its time fence policies, that is, not break into its manufacturing schedule within the published lead-time policy (shown in Chapter 3, Figure 3.6).

Also important to plan execution is the availability of correct supply chain information. This information includes material requirements planning (MRPII) constants, forecasting parameters, asset estimates, and so on. Finally, all orders from suppliers to the organization as well as from the organization to its customers should be on time. Material availability should be 100%. Breakdowns in any of these areas may be good opportunities to create Lean Six Sigma improvement projects to improve process performance. The S&OP meeting forms the basis of effective supply chain management practice and will be discussed further in Chapter 3.

Lean Basics

Table 2.1 shows typical improvements observed across numerous Lean projects relative to time compression, higher quality, and cost reduction. Improvements in these operational metrics are the reason for the strong interest in integrating Lean concepts into the supply chain. Similar benefits have been documented in service, administrative, and manufacturing organizations and functions across many industries over the past 30 years. However, effective project execution is required to ensure success of Lean initiatives since lean capability is built slowly over time through a sequential deployment of key Lean tool sets. An ad hoc approach to Lean deployment is a major reason for its failure in some organizations. Organizational misalignment is a second major reason for deployment failure. Organizational misalignment leaves isolated and poorly integrated improvement projects scattered across the supply chain. It is difficult to accurately estimate business benefits in these situations. In contrast, effective deployment of Lean methods across the supply chain requires an integrated set of activities designed to put in place key Lean tool sets and to execute well-defined Lean Six Sigma projects. If integrated properly, successfully executed Lean Six Sigma projects will build momentum and effectively change an organizational culture over time. This will create a truly Lean supply chain (discussed further in Chapter 7).

Category	Improvement
Process Development	25%–75%
Labor	15%–50%
Floor Space	25%–50%
Errors	25%–90%
Excess Capacity	25%–75%
Throughput Time	25%–95%
Delivery Time	25%–80%

Table 2.1 Typical Lean Benefits

It takes hard work to develop and deploy a Lean system. There can be no shortcuts. But the effective application of Lean tools and methods may not yield significant process improvements and business benefits. Most studies of Lean deployments show reductions in cycle time, increased labor utilization, reductions in required floor space, error reductions, an increase in system capacity, and improvements in on-time delivery. However, as mentioned earlier, some organizations achieve these results on an isolated basis. This is because full implementation of Lean methods requires implantation of precursor systems each having a unique set of tools and methods, which build on each other over time. These precursors include implementation of performance measurements, operational stabilization using *just-in-time (JIT)* techniques, implementation of a total preventive maintenance program, institution of a *single-minute exchange of die (SMED)* program, implementation of standardized work, mistake proofing, maintenance of high quality levels, and creation of a visual workplace. The precursors also include ensuring that raw material and *work-in-process (WIP)* containers are sized and designed based on internal lead-time and demand and that every function throughout the organization has similar performance metrics. These key Lean system components are listed in Table 2.2 without any specified order of importance.

Two components in particular must be in place to ensure a good foundation for a Lean program. These are performance-based metrics and operational standardization to minimize process variation. Unfortunately, not every organization has the discipline to implement these systems due to a combination of poor education, incorrect measurements, and improper resource alignment. We will discuss each of the system components

■ System Performance Measurements
■ JIT/Stable System
■ Total Productive Maintenance (TPM)
■ Single-Minute Exchange of Die (SMED)
■ Standardized Work
■ Mistake Proofing
■ High Quality
■ Visual Workplace (5S's)
■ Container Size (Design)
■ Supplier Agreements

Table 2.2 Key Components of a Lean System

listed in Table 2.2 in detail over the next several pages. The purpose of this discussion is to present basic Lean system concepts to provide the Lean Six Sigma improvement team with ideas to identify and implement Lean Six Sigma projects where Lean tools are the major focus of the process improvements.

Performance Measurements

Performance metrics are used to monitor the process over time. Some of the more common performance metrics typical of a Lean system are listed in Table 2.3. Notice most of the metrics are time-based. Improvements in time-based metrics will significantly reduce lead-time and increase system flexibility relative to external customer demand. Customer satisfaction, customer on-time delivery, and inventory investment are three additional metrics that are necessary to implement Lean systems.

■ Customer on-time delivery/Inventory investment
■ Throughput of materials and information (in units)
■ Quality of work/Floor space utilized
■ Value-added time/Total time/Machine up-time (available time)
■ Supplier on-time delivery
■ Machine changeover (especially at bottleneck resource)

Table 2.3 Common Performance Metrics

"Customer on-time delivery" measures whether an item was delivered to our customer's facility on-time, in the exact quantity ordered and defect-free. These three conditions must exist for the order to be considered to have arrived on time. As an example, receiving the correct quantity but having 5% to 10% of the shipment be outside the customer specification means the order did not arrive on time. The same concept applies to quantity; if only 90 out of 100 items were received, then the order did not arrive on time. Also, delivery lead-time must be specified very carefully by both the customer and supplier. As an example, if the delivery lead-time is specified by the customer in terms of work days, that is, five days per week, but by the supplier as calendar days, that is, seven days per week, then the order will not arrive on time relative to customer expectations. "Inventory investment" must also be carefully defined by the organization. As an example, the amount invested in inventory needs to be broken into current versus legacy inventory as well as by location and type. The task of implementing process improvements will be more difficult for legacy inventory. Inventory investment should also be converted to an inventory turns metric because products with higher sales will normally have higher inventory investment, all other things being equal, unless their lead-times are shorter than those of the other inventory categories. In fact, teaching the importance of calculating optimum inventory investment quantities is one of the major topics of this book since inventory investment is a major barometer of supply chain operational effectiveness.

"Throughput of materials and information" is a critical Lean metric showing how efficiently a system converts inputs into outputs. The shorter the conversion cycle time, the faster an organization can recover its investment in materials and labor. The general concept of constraint management is the throughput of the entire system is controlled by the operation having the lowest throughput rate. In capacity constrained systems, Lean Six Sigma improvement projects should be focused on the system's constraints to increase throughput rates to meet customer demand. Constraint management will be discussed in more detail in Chapter 4.

"Quality of work" is another useful metric. As a process is simplified using Lean methods, mistakes are eliminated and prevented, resulting in higher system quality. Higher system quality reduces cycle time and cost.

"Floor space utilized" is a good measure of how well a process has been simplified and designed to use less floor space. When floor space is reduced other products can be manufactured using the freed-up space. Also, the travel distance between operations is reduced saving time to transfer materials between these operations. This is because a simpler process having fewer operations and rework loops will require less floor space than the same process having a greater number of work operations. Also, a "U-shaped work cell" layout will take less floor space and provide more operational flexibility than a straight line design. This is because workers can travel easily between operations that are located across from each other within a U-shaped work cell; but, in a straight line design the workers cannot easily move between operations. U-shaped work cells allow the number of workers to be easily increased or decreased if they are multi-skilled relative to the operations within the U-shaped work cell.

The percentage of "value-added time versus total time" measures the percentage of operations which must be part of the overall system because they are valued by the external customer. Systems characterized as high value-adding have minimized unnecessary movements, inspection, rework, scrap, and other operations that waste time, material, and labor. Value stream maps are very useful in mapping the flow of value throughout a process to identify those operations which can be eliminated versus those which must be maintained as part of the system. Value stream maps are also useful in analyzing system constraints, inventory investment and quality levels. Value stream maps will be discussed in Chapter 4.

"Machine up-time" is a measure of how well the preventive maintenance systems work. Well-maintained machines will not break down as frequently as those which are not well maintained to disrupt manufacturing operations. Lean systems require stable operations. An important characteristic of stable operations is availability of machines to maintain the required production rate or takt time. Takt time calculation will be discussed in more detail in Chapter 4.

"Supplier on-time delivery" is similar to "customer on-time delivery" except that it is a measure of how well suppliers ship to our facilities. On-time supplier delivery is necessary to maintain stable manufacturing operations otherwise excessive amounts of raw or work-in-process (WIP) inventory is required to protect manufacturing from variations of material deliveries.

"Machine changeover time" is a critical metric that measures how efficiently we can change over from one product to another. The more frequently we can economically set up to manufacture a product, the shorter will be the resultant lead-time to build products. This will result in lower raw material and work-in-process (WIP) inventory investment and enhanced customer satisfaction.

JIT/Stable System

Just-in-time (JIT) is one critical Lean system component or pillar holding up the Lean system. Major characteristics of JIT are continuous flow of product and information through the system and demand pull through the system starting from the external customer, resulting in a relatively level production schedule that matches external customer demand patterns. In a JIT system, continuous flow is matched to external customer demand through the takt time as well as implementation of mixed-model production schedule. *Takt time* is the rate (time per unit) at which the system must produce a product to meet external customer demand. Takt time and lead-time reduction will be discussed further in Chapter 4. However, JIT scheduling and continuous flow cannot exist in an environment of poor quality and operations that are not standardized. As an example, stable operations cannot exist if quality is poor since work must be done more than once. On the other hand, high quality cannot exist without standardized operational tasks. Operational stability is the cumulative result derived from the stabilization of labor, machines, and the system control elements. JIT also depends on the Lean system components shown in Table 2.2 and Figure 2.3.

Total Productive Maintenance

The concept behind *total productive maintenance (TPM)* is to proactively eliminate all machine failures related to maintenance breakdowns. TPM consists of four major methods. These are preventive maintenance (PM), corrective maintenance, eliminating maintenance where possible, and

Figure 2.3 Three Pillars of a Lean System

emergency maintenance. Preventive maintenance is a combination of methods and techniques that attempt to keep machines running without breakdown to ensure continuous flow of materials or information through the system. PM systems are characterized by maintenance schedules, standardized materials, procedures, and training to perform maintenance. The concept behind corrective maintenance is to have workers routinely record daily maintenance procedures and identify any observed maintenance problems. Using this corrective maintenance history, work groups are encouraged to identify ways to make the PM program more efficient and reliable over time by the local work team using continuous improvement methods. In the third maintenance category, there will be higher equipment reliability if portions of the required maintenance can be minimized or eliminated through design changes of the machinery. Emergency maintenance is associated with situations where machines break down unexpectedly. These situations cause higher cost and other operational problems.

There are many contributing factors to maintenance problems and machine breakdown. Some of these factors are related to how clean and standardized the work areas are kept by the local work team. Others are related to management issues such as the standardization of maintenance operations and the proper design of machinery. If the local work area is dirty with many oil and lubricant leaks, or if the equipment is not kept clean, it will be very difficult to identify maintenance issues as they arise. As a result, in a TPM environment, all machinery is cleaned and inspected on a daily basis by the people using the machinery. This helps identify obvious maintenance problems including minor leaks and loose components on the machine as well as other problems that might contribute to an unexpected machine breakdowns. To aid machinery inspection, the local work team develops inspection and maintenance standards that help identify critical maintenance tasks.

Lean Six Sigma projects are deployed in maintenance areas for a variety of reasons. Process yields may be low due to machines breaking down or not being able to keep critical tolerances due to wear. Lead-times may have increased resulting in schedule interruptions or higher inventory investment. Some of the root causes found in Lean Six Sigma projects can be traced back to improper maintenance practices. The required process improvements are usually implementation of TPM in the local work areas. However, an organization does not have to wait for an immediate problem to develop in local work areas before deploying TPM. TPM methods have been shown to maintain continuous material flow though the system by improving system up-time, reliability, and operational standardization. The TPM concept can also be extended to the human side of the system in the form of ergonomics. In ergonomically optimized environments, the work is structured so that people will perform at standard rates without adverse physical effects, that is, injuries to the workers.

SMED

Single-Minute Exchange of Die or SMED is a combination of techniques designed to ensure that process changeovers from one type of product to another are made quickly to maintain the takt time (discussed in Chapter 4). Integral to SMED is the reduction and stabilization of lead-time associated with machine or job changeovers. Common SMED techniques are listed in Table 2.4. The basic SMED concept is to reduce machine or job setup time to allow more setups per unit time. The more setups that can be done per unit time, the shorter will be the overall lead-time for all products. Shorter lead-times translate directly into less inventory investment for an organization. As an example, suppose the setup costs for a product require an *economic order quantity (EOQ)* of one week of production. But, after a SMED study, the setup costs can be reduced to allow a smaller EOQ. This will result in more jobs run with smaller lot sizes and less overall inventory.

The SMED team studies every operation related to setting up the job including the tools and methods used in the setup process. The first step of the analysis is to identify all work tasks comprising the setup process. This is done using a floor plan of the work cell area. This floor plan includes the setup work tasks as well as their time duration and spatial relationships. Also included are the locations of materials and tools. Setup procedures and inspection tasks are also brought into the analysis by the local work team. Nonessential tasks are immediately eliminated from the setup process. The remaining setup tasks are separated into internal and external setup tasks.

- Identify individual elements of setup
- Separate internal from external setup
- Move elements to external setup
- Simplify elements
- Eliminate "adjustments"

Table 2.4 SMED Elements

External tasks can be performed offline and should have no impact on the actual lead-time necessary to set up the machine or job. However, the lead-time associated with internal setup tasks should be carefully analyzed since this time component is, in fact, the lead-time necessary to change over from one job to another. Internal setup tasks should be moved to external tasks or simplified using standardized tools, templates, and methods, as well as through application of mistake-proofing devices to reduce setup rework. The goal is to eliminate manual adjustments to the setup tools. Removing complicated and time-consuming work from the setup process will allow quick changeovers from one job to another. This will reduce the lead-time necessary to manufacture products and lower their inventory investment levels.

Standardized Work

Standardization of work provides several important benefits for the organization. These include lower scrap and rework associated with mistakes made because work is done in different ways resulting in higher variation. Training of new employees and maintenance of International Standards Organization (ISO) certification is also easier in environments using strategies to standardize work. Since the concept of standardized work seems so easy. why is it so difficult to implement and maintain in practice? The reason is that an operation must be carefully studied and broken into its elemental work tasks in order to find the "best" way to do the job and develop the documentation necessary to ensure that everyone understands how to correctly complete each work task associated with the job. To effectively implement standardized work throughout the facility, it is important to gain support from senior management. Effective communication is also important to show people the importance of standardization as well as how to implement standardized work techniques and methods. Improvements to work tasks can be easily identified, implemented and documented by the local work team.

Standardized work is based on a combination of best way to do the work, relevant inspection standards as well as floor layouts and diagrams describing how to do the work in the most efficient manner. The normal sequence of activities used to develop standardized work practices includes analysis of the work tasks, development of standardized times to complete these work tasks, and a listing of all tools and methods necessary to ensure that the work is correctly completed by the worker. When the listing is in completed form, the worker will have a sheet of paper showing each work task, how long it should take to complete each task, including material handling and walking distance to load and unload the work, and the cycle time allocated for the work as well as the allowed work-in-process (WIP) inventory to be kept in the work area. There will also be a diagram showing the flow of material within or between machines.

Mistake Proofing

Mistake proofing involves a combination of techniques and methods designed to prevent errors or detect them before they can cause operational problems. Mistake-proofing techniques help standardize the process to ensure continuous flow of material and information through the system. Mistakes cause error conditions and their resultant defects for a variety of reasons. These reasons include variations of incoming material, incorrect work and inspection instructions, poor worker training, maintenance problems, and worker mistakes in perception or judgment. Without proper application of mistake-proofing strategies to product and process designs, errors and their defects will continually occur in the process. The result will be degradation of process performance.

The best way to eliminate an error is to ensure that it will never occur. Examples include designing part-mating surfaces so there is only one way they can be assembled by a worker. There are many ways to ensure 100% mating of part surfaces with the design of asymmetrical part features being the most common approach. Other common mistake-proofing strategies we encounter everyday include tamper-proof bottles to prevent accidental access to drugs by children, and seat belts to prevent injury and death. Another common example of mistake proofing we encounter every day is cross-checking data entry by automated computerized systems to ensure the information is correct, for example, the correct credit card number and address have been entered by the user. Spell checking by a computer is another common example of mistake proofing.

However, if errors should occur even at very low rates, the next best option is to detect the error condition before a defect is made by the operation. In these situations, there may be leading indicators of machine failure. Leading indicators include sensors placed on the machine to detect error conditions. The final strategy to minimize errors and mistakes is to immediately detect the defects when they are created to prevent them from moving down into the process and out to the external customer. Mistake proofing is also a critical control tool used in the improve phases of Lean Six Sigma projects to ensure that process improvements will be reliably sustained over time.

High Quality

High quality is required to ensure that work does not have to be redone. Rework results in wasted time and material. In Lean, the goal is to achieve zero defects through changes in product and process designs and the use of mistake-proofing strategies. But, at a higher level, quality improvements must be strategically deployed through the quality organization using enabler initiatives such as continuous improvement or Lean Six Sigma breakthrough methods. There are many inter-relationships and similarities between the two improvement methods. "Breakthrough" is associated with the Lean Six Sigma method. "Continuous improvement" is associated with the total quality management (or improvement) method. These quality improvement programs differ in the rate and magnitude of their resultant process improvements. As a general guideline, the Lean Six Sigma method makes focused process improvements across more than one organizational function within a very short period of time. In contrast, continuous improvement teams make moderate process improvements within a function over an extended time period. The common elements of either approach are analytical tools ranging from simple to complex and executed through process improvement teams. The resultant process improvements are obtained through standardization of work methods, mistake-proofing work, and modifications in product and process design. Most organizations use an International Standards Organization (ISO) system to integrate the improvements.

The best way to improve quality is up-front by developing simplified and well-tested products. Products that have been designed and tested using clearly understood customer requirements, that is, voice-of-the-customer (VOC), and built using standardized components, methods, and testing strategies will have higher quality than products that have not been designed in a similar manner. After the VOC has been clearly understood, the product is developed using an evaluation strategy focused on a clear understanding of how system elements impact product characteristics and on setting the optimum levels of these system elements to ensure specification targets are achieved with minimum variation under conditions of expected usage. The final product design should use the minimum number of standardized components according to *design for manufacturing (DFM)* principles. This will ensure that the product can be manufactured at a high quality level. There are many tools and techniques that will aid in the product design process.

One key tool used to ensure high quality is the *Design Failure Mode and Effects Analysis (DFMEA)* document. A generic version of an FMEA (which is also used to track process improvements) is shown in Chapter 9, in Figure 9.5. The DFMEA shows how the design could fail, that is, its failure modes and the causes of each failure mode. In addition, the DFMEA shows the severity of the failure on the external customer, the probability of this failure actually occurring, and the probability of detecting the failure internally, with current internal controls, prior to reaching the external customer. Given that a failure may occur, the design team attempts to put appropriate control mechanisms in place to either reduce the failure probability or increase the manufacturing's ability to detect the failure occurrence. In parallel, manufacturing engineers create the *Process Failure Mode and Effects Analysis (PFMEA)* using the DFMEA. The PFMEA document shows how manufacturing may fail to achieve the product's design specifications or other VOC requirements by evaluating the severity of not meeting design specifications, the probability of the occurrence for a failure cause, and the detection probability of the failure cause using current manufacturing controls. The manufacturing team uses the PFMEA information to develop more efficient manufacturing controls to reduce the occurrence probability of failure causes and increase their detection probability. Quality assurance uses this information to build the customer's quality control plan. The quality control plan ensures that the customer receives the product as originally requested prior its product development phase.

When there are breakdowns in the ability to meet all customer requirements, Lean Six Sigma projects can be deployed to analyze and eliminate the root causes of the process problem. Lean tools and methods can be applied if the root causes are associated with process waste or long lead-times. The various types of process waste are discussed later in this chapter in the section "Rapid Improvement Using Kaizen Events." More complicated problems require use of Six Sigma tools and methods. Six Sigma is a structured problem-solving methodology using advanced statistical tools and methods to understand relationships between input and output variables. The input/output relationship is commonly expressed as $Y = f(x)$. The Six Sigma method is characterized by five phases. These five phases are *define, measure, analyze, improve,* and *control (DMAIC)*. In the define phase, the problem is defined from the customer and business viewpoint and broken down into one or more projects to be worked by green belts or black belts. The belts take the project charter and work with process improvement teams to collect and analyze data relevant to the project's objective. After data analysis, the team should have a small list of input variables (X's), which most likely impact the output variables of interest (Y's). Using this information, the improvement team develops solutions to the problem by evaluating the impact of changing levels of each input variable and their impact on the output variable using experimental design methods. This optimization process is called the improve phase. After the best solution has been found, the team evaluates it under controlled conditions in the actual process environment with the local work team. This is called a "pilot" study of the solution. After the pilot study, the team works with the process owner and local work team to implement the required process changes. Controls are incorporated into the process solution to evaluate the modified process and ensure that improvements remain effective over time.

Visual Workplace

The concept behind the visual workplace is that "process problems and status can be recognized at a glance." This saves time and expense since anyone can quickly identify areas requiring immediate attention. But, there are several stages (improvement levels) through which an organization must pass in order to successfully implement the "visual workplace." 5-S is the first phase of the visual workplace. Table 2.5 lists basic elements of a system called "5-S" wherein each of the five elements is described by Japanese words beginning with the letter "S." The five elements of 5-S consist of sorting out things that should be in the work area from those which should not be present, straightening out what is left in the work area, sweeping or cleaning the work area so tools can be found easily, standardization of all work methods, and sustaining the improvements over time.

5-S must be implemented prior to deployment of other improvement initiatives because improvements cannot be sustained within processes lacking operational standardization. Advantages of 5-S include more local work team involvement, better control of the workplace including its tools and equipment, and safer working environment. 5-S also results in shorter lead-times and lower overall inventory investment. The

- Organize workplace (Sort)
- Order everything (Straighten)
- Clean (Sweep)
- Standardize operations
- Sustain improvements

Table 2.5 5-S Elements

second stage of the visual workplace is sharing information between work groups using visual displays. The final stage of 5-S is establishment of visual controls throughout the work place. These visual controls range from simple alarms when the process has changed to actual prevention of errors and defects through extensive mistake proofing.

After the improvement team applies 5-S to the process, process status can be visually displayed in real time. This allows the local work team to see key metrics hour-to-hour. Over time, by expanding 5-S and visual displays, the work team can move toward visual control of the process, that is, a "visual workplace." In a visual workplace, the local work team can see process status at a glance. This reduces the complexity of control because abnormal process conditions will be easily visible. The result is lower cost, shorter lead-time, and higher quality levels. Examples of key visual control elements include "Kanbans," "Andons," and visual display boards. Kanbans are used to control the production and movement of materials by requiring that specific inventory quantities be placed in standardized containers. Andons signal abnormal process conditions. The more advanced versions of the Andon concept contain mistake-proofing systems. These systems trigger alarms or stop the process if it deviates from target levels. Visual display boards show the performance of key process metrics versus their targets. These display boards show information related to the status of quality, people, and machines within the work area. Work areas are marked for multiples of Kanban quantities to control inventory.

Container Size (Packaging)

Container size is determined based on the calculated kanban lot size. This concept will be discussed more fully in Chapter 4. Kanban lot size is based on lead-time and expected demand over this lead-time with a safety factor used to cover minor uncertainties in lead-time or demand. Reductions in the kanban lot size and hence the container size and work-in-process inventory (WIP) levels will result from reductions in lead-time or demand variations.

Supplier Agreements

The full benefits of a Lean supply chain can only be realized if suppliers and customers are fully integrated into the system using shared performance metrics agreed to by contract. Every step, function, or organization of the supply chain should have consistent performance measurements to ensure attainment of system-wide goals and objectives. Two performance metrics of particular importance are accurate estimates of demand and lead-time. These interorganizational metrics must be properly defined to ensure smooth functioning of the supply chain. Without proper demand and lead-time definitions, significant problems will occur at the functional interfaces.

Lead-time and demand issues negatively impact operational performance including inventory investment and customer service levels. As an example, if all parties agree to a 30-day (calendar) lead-time with expected demand of 5,000 units, but the required minimum lot size is 10,000 units (a 60-day supply), the effective average inventory gets pushed up from 2,500 to 5,000 units (assuming a linear inventory depletion rate). Inventory can also be increased due to poor quality. As an example, if 5,000 units are promised within 30 days, but a second order must be placed because of poor quality, lead-time has been increased over the initial lead-time agreement. Safety stock is required in this latter case to protect against poor quality product. It will be shown in Chapter 6 that inventory is driven by both the magnitude as well as variation of lead-time and demand.

Rapid Improvement Using Kaizen Events

A Kaizen event is used to quickly eliminate obvious waste from a process. The Kaizen event is facilitated by people experienced with the methodology. This is a "hands-on" methodology requiring that facilitators have shop floor experience conducting previous Kaizen events. The Kaizen event should be supported by local management as well as union officials if the workforce is unionized. The local work team plans and executes the Kaizen event through a sequenced set of activities which include onsite training of lean principles, analysis of the operations, identification of required process changes, and actually changing the process according to the Kaizen plan. The Kaizen event usually takes between three to five work days from analysis to actual process improvements.

The event begins when a work area has been selected for improvement. The selection of the work area is usually made through prioritization of business benefits by local management or is the result of a root cause analysis from a Lean Six Sigma project. In the latter situation, the root causes of the process breakdown may have been attributed to factors contributing to process waste.

The team is trained to identify and brainstorm ways to eliminate the several categories of process waste. One form of process waste is due to overproduction. Overproduction causes an increase in inventory levels and destabilizes work schedules since too much capacity is assigned to some products at the expense of others. Defective products are another form of process waste since every unit that is defective must be replaced or reworked to meet customer order requirements. Process waste associated with excess motion includes walking or handling. Excess motion (time variation) destabilizes system takt time, resulting in higher per-unit direct labor costs as well as many other problems. Another type of process waste is unnecessary delays. Process delays result in work batching and extra handling of material. Delays also hide quality problems since materials are not immediately inspected after they have been manufactured. Finally, there is process waste when materials are transported distances greater than necessary or moved several times. Not only are these process waste sources immediate targets for Kaizen events, they also are often the root causes of many Lean Six Sigma projects.

Lean E-Supply Integration

In mature supply chains, procurement activities are integrated across all business functions to reduce lead-time and minimize demand variation. In these systems, supply chain integration is achieved using standardized IT platforms to electronically buy and sell goods and services and exchange information. In this sense, e-supply is an evolutionary process building on previous IT system architectures. This evolutionary process involves four major activities. These are the identification and selection of value-added processes throughout the supply chain, application of integration technology to major portions of the supply chain, optimum asset allocations across the supply chain, and rules for sharing the asset base and status information. Identification of value-adding processes is a direct application of the Lean methodology. Value stream maps are very useful in identification of processes that should remain part of the supply chain, because they add value from an external customer viewpoint.

The capture of real-time demand and order status information throughout the supply chain is accomplished through technology integration. In these highly integrated systems, the goal of optimum asset allocation requires that assets be held in strategic positions across the supply chain to minimize total system cost. An example is the optimum allocation of inventory investment at critical locations throughout the supply chain based on lead-time and expected customer demand patterns. In addition, behavioral rules must be predetermined in advance by contract between all parties to ensure system integration and goal alignment.

■ Cost reductions greater than 10%
■ Reduced WIP greater than 20%
■ Reduction in cash-to-cash cycle time greater than 50%
■ Reduction in sourcing lead-time greater than 90%
■ Production cycle time reduced by greater than 70%

Table 2.6 E-Supply Benefits

Properly deployed by all parties, e-supply systems can yield significant operational improvements and business benefits. Table 2.6 shows typical e-supply chain benefits found in many applications. Over the past several years, e-supply implementations have significantly reduced inventory investment and improved asset utilization and cash flow.

Providing information in real time to all participants within the supply chain drastically reduces the system's lead-time The productivity benefits derived from lower lead-times are tremendous since products are built in less time allowing the money invested in inventory is recouped faster by the organization. This results in shorter cash-to-cash cycles and higher cash flow. The cycle time to source materials is drastically reduced over non-e-supply systems because system status information is available to all participants in real time,

E-Supply Applications

E-supply systems allow online purchases of raw materials and maintenance, repair & operating supplies (MR&O) as well as their efficient disposal if these items are no longer required for usage. Examples of these e-supply systems include trade association and eBay auctions which allow competitive bidding for the items. In addition to purchased materials and supplies, labor can easily be purchased from various online outsourcing agencies. Examples include monster.com, careerbuilder.com, and similar websites. Almost anything can be outsourced including accounting, engineering, manufacturing, logistics, and asset management of buildings, equipment, and inventory.

E-supply systems also eliminate unnecessary tasks and the lead-times associated with ordering materials and information. Process breakdowns are sometimes eliminated through process automation using e-supply systems. Or in some Lean Six Sigma projects, breakdowns within the current e-supply system become the focus of a Lean Six Sigma improvement team. The first step necessary to develop an e-supply system is to partner with a key customer or supplier to improve major processes. The second step is to systematically integrate IT systems between the organizations to create real-time access to order status and asset availability.

Let's look at an example for purchase of an automobile shown in Table 2.7 to see how advanced these systems have become using e-purchasing technology. Historically, the process for purchasing an automobile involved many steps resulting in long lead-times and high purchasing cost due to several intermediaries within the process. The purchasing process began with advertisements to the consumer and the consumer obtaining information, usually in person, relative to various automobiles of interest from a dealer. This process required several visits to dealers, which consisted of obtaining basic information about the vehicle as well as its availability and purchasing cost. The customer placed the order with the dealer once a purchase decision was made. If automobile inventory was not available at the time of ordering, the vehicle order status was periodically checked by the dealership until the vehicle was shipped from the manufacturing facility. The vehicle was invoiced to the customer at time of shipment. In this process, the consumer often dealt with just one dealer over an extended period of time. In contrast, the Internet allows the consumer to check availability and vehicle prices within any desired geographical radius of the customer. This real-time information shows all vehicle inventories by model within

Process Step	Old Way	New Way
1. Obtain information	Go to dealership	Use Internet
2. Check availability/ price	Read newspaper	Use Internet
3. Order	Go back to dealer	Use Internet, electronic data interchange (EDI)
4. Check order status	Phone dealer	Use Internet
5. Ship order	Ship from manufacturing	Third party
6. Invoice	Fill out forms	Use Internet, EDI
7. Pay invoice	Give check, credit card/loan	Use Internet, EDI

Table 2.7 Ordering an Automobile

the geographical area of interest. In addition, the availability of real-time information increases dealer competition, resulting in lower overall vehicle purchase price. In fact, in this system, visits to the dealer can be completely eliminated because orders are placed online by the customer. Eventually, these online systems will eliminate entire components of this supply chain.

Not every organization can implement an e-supply system across the entire supply chain. It takes expertise in many aspects of e-supply to build IT platforms for the myriad operations typically found in modern supply chains. However, almost every organization can convert portions of their supply chain into Internet-based systems. The first step is to create an overall strategic plan and a high-level model across the supply chain. This is coordinated by the largest organization within the supply chain. The second step is to select partners to begin to build the required technology platforms by focusing on one key subsystem within the supply chain at a time. These partners will need to have the required resources and to execute well-defined projects within the converted system to build operational capability over time. The third phase is analysis and improvement of these modified technology platforms over time through the identification and execution of applied projects. As projects are executed and operational capability increases, "lessons learned" are leveraged to all participants throughout the supply chain.

Order Entry Example

As an example, consider the hypothetical manual order entry process shown in Figure 2.4. In this scenario, customers call the customer service department to place orders. Prior to any process improvements, the

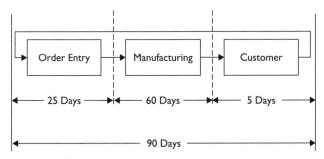

Figure 2.4 Order Entry—Part I

customer service department manually records every order. Booked orders are released to manufacturing in batch mode (system accumulation of several orders before release). After batched order release, the system schedules, manufactures, and ships product to the external customer. Using value stream maps (quantified process maps using customer metrics); operations within each process step are analyzed to determine their lead-time components, quality levels, and utilization rates. The purpose of this analysis is to identify operations (time components) that are not adding value to the process. Value in this sense implies that the customer would be willing to pay for the operation if they were aware of its existence. The value content could be material, labor, or information. The goal of the analysis is to ensure uninterrupted value flows throughout the process, that is, no discontinuities are caused by batching of work. To achieve this process flow, the complexity of the process must be significantly reduced leaving only those operations that add value, or in a worst-case scenario, only those non-value-adding operations that must be maintained because of technical constraints, that is, *business value-adding (BVA)* operations. BVA operations will be discussed further in Chapter 4 along with the key Lean tools listed in Table 2.8.

We can see how the implementation of an e-supply system might proceed using a simple example consisting of receiving, manufacturing, and shipping an order to a customer. Figure 2.4 shows the process broken down into three subsystems, which can be further broken down into detailed time components as shown in Figure 2.5. Figure 2.4 shows that the overall lead-time through the system between placement and receipt of an order by the customer is 90 days. A detailed analysis would be required to identify the greatest opportunities to reduce lead-time.

- Eliminate steps
- Eliminate operations
- Manage bottlenecks to maximize flow
- Minimize setups by using a mixed model
- Transfer batch versus process batch
- Apply Lean, JIT, and Quick Response Manufacturing (QRM) techniques

Table 2.8 Basic Lean Tools

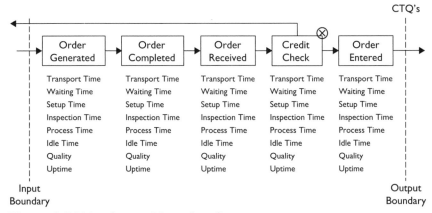

CTQ's

Order Generated	Order Completed	Order Received	Credit Check	Order Entered
Transport Time	Transport Time	Transport Time	Transport Time	Transport Time
Waiting Time	Waiting Time	Waiting Time	Waiting Time	Waiting Time
Setup Time	Setup Time	Setup Time	Setup Time	Setup Time
Inspection Time	Inspection Time	Inspection Time	Inspection Time	Inspection Time
Process Time	Process Time	Process Time	Process Time	Process Time
Idle Time	Idle Time	Idle Time	Idle Time	Idle Time
Quality	Quality	Quality	Quality	Quality
Uptime	Uptime	Uptime	Uptime	Uptime

Input
Boundary

Output
Boundary

Figure 2.5 Value Stream Map—Part 2

Although only time is quantified in the current example, this high-level map should also be quantified relative to cost and quality (defect) levels. The next step would be to analyze each subsystem to determine which operations add value and what are the areas of greatest processing waste. An operational analysis is required to identify improvement opportunities because a subsystem having a smaller overall lead may have more waste than a larger subsystem. As an example, although the manufacturing subsystem has the longer lead-time, perhaps most of the operations in this subsystem are value adding? Before deciding where the greatest process improvement opportunities might be, the team should analyze all operational time elements. In other words, although the magnitude of lead-time is an important consideration, so also is the opportunity gap between the current performance level and a reasonable lead-time target. A reasonable target would consist of value-added and business-value-added time, that is, the current lead minus obvious wasted time. Chapter 4 discusses this concept in more detail.

As an example, after breaking the order entry subsystem into operational elements, there appear to be several manual operations that could be eliminated through automation. The process begins when the order entry person receives the order from the customer over the telephone. The order form is filled out by the order entry person and sent to the credit person who checks the customer's credit status. If the credit policy has been met, then the order is entered into the order entry system and released to manufacturing. Using a high-level

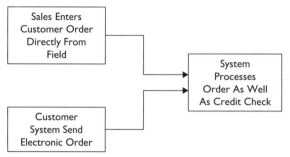

Figure 2.6 Order Entry—Part 3

value stream map (value stream mapping is discussed in more detail in Chapter 4) as shown in Figure 2.6, we see that work tasks categorized within each operation can be categorized by the type of time they consume. As an example, transporting material or information from one location to another is a non-value-added time element. A second-non-value-added time element occurs if the order waits for any reason. Setting up portions of an order prior to completing it is a non-value-adding time element. Inspecting any part of the order, for any reason, is a non-value-adding time element. If the process has defects, these add to the overall time of the process. If the process is stopped for any reason, this adds time to the ordering process. Finally, any other wasted time is categorized as idle time. Identification of wasted time elements is important to reduce lead-time. Lower lead-times reduce inventory investment.

If improvements to the order entry example were carried to their logical conclusion, the ordering process would be fully automated to eliminate all manual operations, resulting in elimination of all process discontinuities. The automated system would handle all aspects of the ordering process to effectively reduce lead-time from the current 25 days to just minutes. The resultant ordering system, which is shown in Figure 2.6, is a very simplified process characterized by a very short lead-time. Additional improvements can be made through outsourcing or implementation of electronic data interchange (EDI). If the process cannot be fully automated, then perhaps major time components can be eliminated from the process. These basic Lean tools and methods are listed in Table 2.8. Their impact on lead-time reduction will be discussed further in Chapter 4.

Summary

Supply chain effectiveness is critical to profitability since manufacturing and distribution costs represent 50% to 70% of an organization's budget, A major supply chain cost is inventory investment which often represents 10% to 20% of sales. The efficient utilization of supply chain assets, and in particular, inventory investment, is a primary operational and financial concern of every manufacturing organization. The key to effectively improving supply chain performance is the identification and execution of Lean Six Sigma projects to reduce variation in both lead-time and demand while maintaining customer satisfaction. Major enablers to supply chain improvement are information technology, partnerships with customers and suppliers, and application of Lean methods. Lean methods have been the major focus of this chapter since many Lean Six Sigma projects use Lean tools and methods both to identify the root causes of process breakdowns and to sustain process improvements.

Chapter 3

Demand Management Impact on Lean Six Sigma Projects

Key Objectives

After reading this chapter, you should understand how demand management tools and methods are useful to accurately estimate demand, and understand the following concepts:

1. Why the sales and operations planning (S&OP) team is essential to the identification of Lean and Six Sigma improvement projects that are aligned to business goals and objectives.
2. Why violation of the organization's stated lead-times, that is, time fences, causes significant operational problems that necessitate the creation of Lean Six Sigma project teams to work across supply chain functions.
3. Why Lean Six Sigma black belts can help forecasting analysts develop advanced models to more effectively manage products with unusual demand patterns and prevent poor forecasts of new product sales, which are major contributors of excess and obsolete inventory.

Demand Management Overview

Demand estimation, including quantitative forecasting models, directly impacts supply chain performance and inventory investment. Demand management is an important topic for efficient supply chain management since inaccurate estimates result in too much or too little inventory in the system. Inventory investment is a barometer measuring supply chain effectiveness. Inventory investment levels or turns can also be used to identify where process breakdowns are occurring in the supply chain to create Lean Six Sigma projects. Poor demand management impacts business performance from two viewpoints. High demand forecasts that are not realized result in too much inventory and increased operational cost, and if the inventory quantities are high, such forecasts contribute to excess or obsolete inventory. Later, when excess and obsolete inventory are written off the balance sheet below their original standard cost, the organization incurs a financial loss. At the other extreme, if demand forecasts are too low, lack of inventory may result in lost sales and higher operational costs. Higher costs result from the manufacturing schedule changes and order expediting that may be necessary to maintain customer delivery schedules due to lack of finished goods inventory. Also, demand estimates are required to estimate future capacity and develop contingency plans to ensure timely shipment of products to the external customer.

Demand can be more accurately estimated and managed by the organization through operational integration using real-time information. As an example, in advanced retail organizations, aggregate demand for an item is accumulated as individual items are scanned at the checkout counter when purchased by the customer. This purchase (demand) information can be shared with suppliers throughout the supply chain on a daily basis through shared information technology (IT) infrastructure. Sharing demand and supply-chain status information across the supply chain reduces demand variation and lead-time. This allows the entire supply chain to build more accurate manufacturing schedules based on actual external customer demand patterns. This situation is in contrast to those supply chains relying on combinations of people and technology to estimate demand. In these complex and highly manual systems, lead-times and demand variation are normally high resulting in inventory investment levels exceeding optimum.

Sales & Operations Planning (S&OP)

The basic functions of the S&OP team are shown in Figure 3.1. Communication between all organizational functions in the supply chain is critical to ensure effective coordination throughout the supply chain and lowest overall system cost. The S&OP meeting and team are also critical to facilitation and communication between diverse

- Include top management

- Cross functional process

- Thorough preparation

- Decisions relative to demand/ supply balance

- Focus on product family level

- Use facilitation methods

Figure 3.1 S&OP Key Elements

organizational functions. The S&OP meetings and team are also excellent conduits through which to manage the Lean Six Sigma deployment across the supply chain. One major objective of the S&OP meeting is to provide a consensus forecast through reconciliation of the strategic revenue forecast and current business conditions. After the S&OP review of current supply and demand constraints, the system forecast is updated to reflect the consensus forecast. It is then fed into the master production schedule (MPS).

The MPS aggregates demand information across the organization. It also feeds the aggregated demand information to the master requirements system (MRPII). The MRPII system sets component manufacturing schedules and planned purchases based on several MRPII parameter settings, of which lead-time and lot size are critical to inventory management. The MPS and MRPII systems will be discussed later in this chapter.

Effective S&OP meetings improve supply chain performance and customer service levels and reduce inventory investment. Key elements of the sales and operating planning meeting (S&OP) include planning for new products, demand, finance, and supply. New product planning includes products that are either completely new or extensions of current product families. Since new products are designed and developed as part of joint task teams, everyone on the S&OP meeting should be familiar with them. One common problem with planning new products is that demand estimates may be overly optimistic since there is often pressure to overstate the demand forecast to meet financial goals and objectives. However, the advantage of the S&OP meeting is that all functions can openly discuss the feasibility of the new product forecast.

This is important because if new product demand has been incorrectly estimated, the result will be either lost sales or excess and obsolete inventories. Since the organization does not want to hold large amounts of excess and obsolete inventory, the tendency will be to challenge the marketing forecast. On the other hand, demand planning for extensions of current products is a straightforward process. This is because the organization should already be familiar with historical sales for similar products. In fact, customers may have even asked for the product extension. In addition to new product demand planning, demand forecasts need to be estimated to plan and allocate capacity across the supply chain. Demand planning for new and current products will be discussed in more detail later in this chapter.

The financial planning aspects of the S&OP meeting ensure that forecasts of revenue, cost, and cash flow are linked to operational planning objectives. As an example, finance is concerned that sales forecasts are reasonable and adequate to meet revenue projections and gross operating margin targets. Finance also wants to be sure there is adequate system capacity, by time period and location, to manufacture the forecasted product mix and meet ending inventory targets. In this process, sales and inventory planning forecasts are used to determine what will be manufactured by location as well as determining all sources of supply necessary to build the product to ensure that financial targets will be met on schedule.

Several important tools and methods are important to ensure that the S&OP process works smoothly given that it is composed of several organizational functions whose goals and objectives are not always aligned with each other. First, the people composing the S&OP meeting must be able to make decisions for the organization relative to supply and demand as well as inventory investment. This decision-making authority typically resides at the directorial or vice-presidential level in most organizations. The reason for this decision-making requirement is that decisions must be made by the S&OP team based on available facts, and the decision often cannot be delayed several days or weeks. A second requirement necessary to ensure effective S&OP meetings is that everyone should arrive at the meeting prepared with up-to-date facts and figures relative to their organizational responsibilities. If necessary, the people attending the meeting can also bring along key support people in case unexpected questions should arise during the meeting. The purpose of the meeting is to make decisions relative to demand and supply, so these topics should be on the agenda for every meeting. As an example, relative to demand, it is important for everyone to be made aware of the expected additions or loss of major customers since these situations impact the ability of the organization to either supply materials or adjust inventory levels. Relative to supply, it is important for everyone to know if there may be work stoppages or lost capacity of any kind during the time periods under consideration. As an example, if significant changes in capacity or sales are expected, either due to seasonal effects or changes and losses of major customers, adjustments in inventory investment may be required in the future. At S&OP meetings, the discussion is at the product family level by location across the supply chain. This is because the purpose of the meeting is to adjust capacity and inventory to meet expected demand. Throughout the S&OP process it is important to ensure that the meetings are properly facilitated so that information flows freely across organizational functions. It is also important that the meetings are run efficiently without unnecessary rework so that goals and objectives are consistently met on schedule over time.

Demand Aggregation

Demand is accumulated from all sources in the organization into the master production schedule (MPS) module. A high-level schematic of the overall process is shown in Figure 3.2. The demand streams include firm orders that have been agreed to in advance by contract as well as forecasts. Firm orders are normally associated with major customers. The demand should be quantified by item and location.

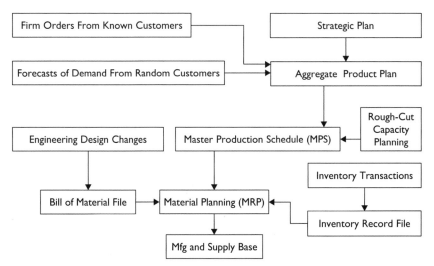

Figure 3.2 System Overview

Some organizations only manufacture to firm orders while others only forecast items. If forecasts are made, then they need to be reviewed by the forecasting department to ensure that the underlying forecasting model has been built using optimized forecasting parameters. It is also important to ensure that the forecast is made using the relevant historical baselines. As an example, if a product has been manufactured for a long period of time, then its demand may be decreasing or lumpy. Using a forecasting historical baseline for several prior years will probably overstate the forecast. On the other hand, new products need to be forecast very carefully since their growth patterns may not be clearly understood by the forecasting analysts. The MPS aggregates the various demand streams into item quantities, which are allocated to future time periods. This is the *aggregate product plan (APP)* for the organization. The APP should agree with the strategic plan since it is the actualization of previous planning by senior management.

Firm order demand is known in advance because it is specified by contract or agreement. Forecasts of independent demand by item and location are made using one of several forecasting methods ranging from judgmental forecasts (nonquantitative) to quantitative methods such as time series models. In addition, promotional and seasonal demand patterns are factored into these aggregate demand estimates. Demand estimates for item quantities by location are aggregated upward to the product family level and finally to one number for the organization. As part of the aggregation process, unit demand is converted into monetary units using standard costing information. Unit demand at the product level is used to develop the aggregate production plan by facility for every product line. The forecasting time horizon is determined by cumulative component and manufacturing lead-time based on the manufacturing build sequence. The aggregate production plan is executed by the material requirements planning (MRPII) and *production activity control (PAC)* systems. Breakdowns in the MPS, MPRPII, or PAC systems can serve as the basis for Lean Six Sigma projects. Alternatively, the presence of high investment levels may eventually be traced back to one or more of these demand and planning systems by the Lean Six Sigma improvement team.

Forecasted Demand

Referring to Figure 3.3, item demand can be independent or dependent. Independent demand items (also called "end items") are directly tied

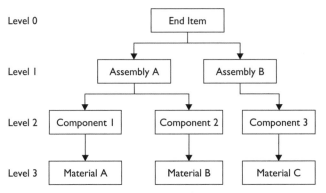

Figure 3.3 BOM Hierarchy

to external customer demand patterns. Dependent demand items are directly tied to independent demand items through the bill of material (BOM) hierarchy or structure. Some dependent demand items are also sold directly to external customers as service parts. In these cases, the service parts are also independent demand items since their demand varies according to customer demand patterns. For this reason calculating dependent demand is a straightforward computational task performed by the MRPII module using demand calculated by the MPS and information from the BOM and on-hand inventory modules. Engineering design changes determine this hierarchy and specifically the product build sequence by component. Scheduling problems will occur if, after having executed portions of the original MRPII production plan, the MPS changes due to demand fluctuations.

The BOM shown in Figure 3.3 is used by the master requirement module to link end-item demand for a product with all the materials that make up the product. These materials and other components are called dependent demand items. Dependent demand items are represented by the BOM in a hierarchical structure with lower-level components building into subassemblies and finally into the finished product. Also included in the BOM are packaging and related materials that will be included in the final customer product. The more complicated the product, the longer the BOM component listing. This has the result of increasing transaction and ordering costs for the product as well as inventory levels at every stage of manufacturing. Adding to the problem of complex BOMs is the fact that they may contain errors resulting in additional ordering costs, excess inventory, and rework. Lean Six Sigma projects attacking certain operational problems associated with lack of materials, scheduling changes, rework, and scrap are sometimes traced back to incorrect BOM component specifications.

Forecasting System

Figure 3.4 shows the major elements of a forecasting system. These major elements are new demand sources, forecasting system controls, model parameters, historical demand, and the forecast-modeling software. It is important to identify critical functions and control points of the forecasting system. This will allow the organization to systematically improve forecasting accuracy over time based on root cause analysis of system breakdowns. In addition to placing certain items on firm order book, that is, contract, to remove their demand variation from the system, some items may be placed on reorder points within the MRPII module, that is, "min/max systems." These items would be characterized as having a combination of low demand variation and cost. In a min/max system, orders are placed at reorder points determined by item and location using historical usage and lead-time offsets. The advantage of using a min/max system is that replenishment of items can be reliably estimated by the MRPII system without forecast analyst intervention. This will allow the forecasting analyst to focus attention on items having repeatable demand patterns, large volume, and high standard cost, that is, the important items for which accurate quantitative models must be developed by the analyst. Min/max systems will be discussed in more detail in Chapter 6.

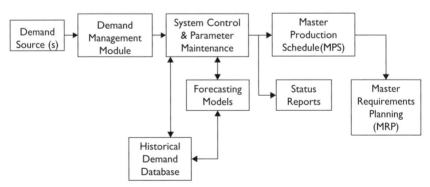

Figure 3.4 Forecasting System

Forecasting System Objectives

The objectives of a forecasting system are to integrate sales, operations, purchasing and finance; to provide forecast visibility for all users; to provide best mathematical forecasts for the organization (allow aggregate forecasts and hierarchical reporting by the system); to gather demand data across all geographical regions of the enterprise; and to build forecasts by item and location. Another objective of the forecasting system is to provide information necessary for continuous improvement of the forecasting process. Best-in-class forecasting systems have these characteristics. The result is that the integration of sales, operations, purchasing, and finance is accomplished using the "one-number forecast." This system integration allows organizations to reconcile forecast estimates across the supply chain and its many organizational functions to ensure that everyone is working to the same manufacturing schedule for every item and geographical location throughout the supply chain. The forecasting system also increases forecast visibility by providing users with screens and hard-copy reports. Integral to this entire process is the sales and operations planning (S&OP) team.

On a monthly basis, trained forecast analysts update the system forecast item-by-item using historical demand patterns and the most appropriate forecast model. Effective modeling of actual demand patterns requires using a combination of one to three components. The level component is specified by the parameter "alpha." It models the average demand of the time series. The trend component is specified by the parameter "beta." It models the upward or downward trend of the time series. The seasonal component is specified by the parameter "gamma." It models seasonal demand. In addition to development of an optimized forecast, the forecasting team is responsible for related system elements including their design. These elements include process and logic flowcharts, user screens and reports, maintenance of miscellaneous databases, cross-functional team support, and guidance for installation of the forecasting system modules, training of users, and monitoring of system performance.

The forecasting system consists of people, information, and information technology systems that manage demand forecast for every item by its location over the forecasting time horizon. Forecasting analysts are normally assigned product groups to manage. These product groups contain a mixture of products each having a different demand pattern. As an example, some high-volume products may be mature and have a very stable demand pattern that is easy for the system to forecast. But another low-volume product may be in the decline phase of its life cycle. This product may have a very lumpy demand pattern that is impossible to forecast. In the latter case, the analyst must carefully monitor the product's demand, without a forecasting model, to ensure product availability.

The forecasting analysts choose the forecasting parameters that build the forecasting time series or demand pattern for the product. These parameters model the average level of demand, a trend (if present), and seasonal patterns. Another important input for the forecasting model is the length of the historical record required to build the time series. Products having demand patterns that have significantly changed over the past several months should be modeled using the more recent historical demand data. Also, products with seasonal demand patterns require at least three years of month-over-month unit demand history by item and location to determine the monthly seasonal indices required to build the seasonal model. These forecast-modeling activities take place as the forecasting analysts interact with the forecasting system control and parameter maintenance functions and the historical demand database. The forecasted demand information is fed to the MPS, and management reports are generated by the system. These forecasts are eventually modified based on the final consensus forecast developed at the S&OP meeting. It is obvious that, given the forecasting system complexity and high degree of manual intervention, there will be breakdowns within the system as well as conditions in which nonoptimum forecasts will be created by the forecasting analysts. These breakdowns could serve as a source of Lean Six Sigma projects or may be found to be the root causes of certain operational problems investigated by the Lean Six Sigma improvement team.

Forecast-Modeling Capabilities

Some key elements of effective forecasting systems are shown in Table 3.1. These system elements facilitate building the forecasting model based on demand history. The first important element is use of "actual item demand rather than shipment history." This concept is important because the shipment history is calculated as "actual customer demand minus what is not available to ship to the customer, that is, backordered items." Forecasts based on shipment history will underestimate demand for out-of-stock items, resulting in inadequate inventory levels to meet actual customer demand. This situation causes extra shipment and handling costs since the original customer order could not be completed using on-hand inventory. "Seasonality of demand" is another important

1. Actual demand, orders, shipments by item/location
2. Three or more seasons (years)
3. Forecast horizon at 12 months for all items
4. User-specified periodicity
5. Subjective adjustments allowed by management
6. Trend, seasonal, and other models
7. Tracking and automatic parameter estimation
8. Ability to change forecast at group level

Table 3.1 Elements of Forecasting Systems

consideration when building forecasting models. If items have seasonal demand patterns, at least three or more years of monthly demand data will be required to build the forecasting model. This is because seasonal indices must be estimated by month year-over-year to ensure that they accurately represent the increase or decrease in demand for the month of interest. For products without seasonal demand patterns, 12 to 24 months of demand history will be sufficient to build the forecasting model. Advanced systems allow the forecasting analyst to specify the periodicity of the time series. This option would be used if the product demand had clearly defined highs and lows during the year. Another way to manage seasonal demand would be to break the year into seasonal and nonseasonal demand components. The "time horizon of the forecast" will depend on the specific planning requirements, which the forecasting model will help estimate. As an example, forecasts required to estimate the purchase of new manufacturing lines or equipment must be made 6 to 24 months into the future. However, forecasts for a specific product can be made at its cumulative lead-time if manufacturing capacity constraints are satisfied. In addition to building forecasting models using historical data and system- or analyst-specified parameters, adjustments by management are possible if conditions that affect demand should suddenly change. As an example, if a major customer is expected to leave, the impact of this demand stream can be simulated using the modeling software. Correspondingly, if a major customer is expected to be brought on, then the additional demand can be entered into the model by time period.

The demand forecast must also be built using the correct modeling format. There are many types of forecasting models. These major modeling types will be discussed later in this chapter. When forecasts are made and matched to actual demand for the forecasted time period, forecasting accuracy (or error) statistics can be calculated by the system. Forecasting error statistics are shown later in this chapter in Figure 3.15. In addition to tracking forecasting accuracy, these systems also have the capability to change modeling parameter levels to improve forecasting accuracy over time. However, the accuracy limits are dependent on the historical database as well as actual demand. Finally, the forecasting analyst can build the aggregate forecast for an item and location from the bottom up or forecast at a product group level and linearly allocate the aggregate forecast down to the item by location level. The latter practice may be the most accurate since it has been shown that forecast accuracy increases as demand is aggregated upwards through the organization. Forecasting is done on a unit basis and converted using standard costing information into the financial forecast and eventually into the high-level revenues forecast of the organization.

The goal of the organization is to agree on a "one-number" forecast for every item and location. This "one number" represents the collective information the organization currently has regarding demand for the item out into the immediate future. Most organizations start with the financial forecast, that is, what sales level is required to achieve strategic goals and objectives. The sales numbers are broken down to the product group level where the projected quantities are discussed at the S&OP meeting. Adjustments are made to the numbers if they are unrealistic, but then other actions must be taken to ensure that the organization's original financial goals and objectives are achieved during the fiscal year. If this situation occurs, product's marketing plans are revised to close the projected sales gap or operational costs reduced to meet financial goals and objectives.

Forecasting at Which Level?

How should forecasts be made by the organization? Figure 3.5 shows that the forecast can be broken down by time, geography, and product group. The best approach is to initially develop a top-down forecast at the business unit level by product group and geographical location, followed by a reconciled bottom-up forecast using the S&OP process as shown in Table 3.2. Adjustments are made by product group to reconcile the top-down and bottom-up estimates. This methodology ensures that the organization is working to "one number" based on the best available information. There is another advantage to high-level demand aggregation. An aggregate forecast has higher accuracy than at lower levels since

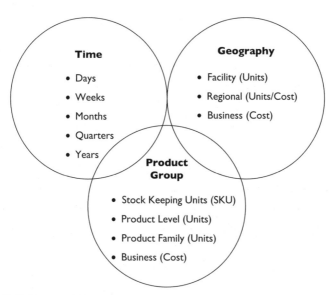

Figure 3.5 Demand Aggregation

the forecasted quantities both positive and negative at each time period will tend cancel each other. The reconciliation between top-down and bottom-up forecasts takes place within the hierarchy of the system forecast. This hierarchy is based on the required decision level and process. Decision level relates to where the forecast will be used in the organization. Is it an aggregate forecast across the entire organization? Or is it localized to just one facility? Each decision level in the organization requires specific forecasting information typified by the time horizon of the forecasted item as well as its aggregation level. The decision process is determined by the forecast breakdown.

1. Business unit one-number goal
2. Broken down by product group
3. Forecast with goal applied to group
4. Forecast linearly applied to item and location level
5. Forecast adjusted by product group at S&OP meeting
6. Forecast linearly applied to item/location level

Table 3.2 Forecast Reconciliation

The initial financial forecast is broken down linearly and allocated across each product group down to an item and location. This allocation is also converted to quantities by item and location. The bottom-up forecast makes modifications to the top-down forecast from an operational viewpoint. But this could cause problems when linearly applied to lower-level items, especially by their location. Since the demand forecast is used by operations as well as finance, the management reports usually contain both unit forecasts and using standard costing information, cost-of-goods sold (COGS). These modifications are coordinated through the S&OP team by consensus. A good practice would be to estimate capacity requirements and financial forecasts at the product group level, but separately forecast individual products by location to verify that there are no major discrepancies between the two methods. In a similar manner forecasts by geographical location are more accurate at higher levels, but from an operational perspective individual forecasts at each location are required in order to ensure that customer demand is satisfied at an item's service level.

Potential Problems

Process breakdowns eventually occur if the original sales and operational plans supporting the financial forecasts were arbitrary. These situations occur if an accurate consensus forecast was never reached by the organization. Process breakdowns drive inventory investment and operational costs higher. Customer service levels will also degrade. Examples of typical process breakdowns include building nonexistent orders, creating sales incentives that distort current and future demand patterns, and failure to execute manufacturing schedules. Any of these issues can serve as the basis for Lean Six Sigma improvement projects. Building nonexistent orders occurs when customers change demand quantities or delivery schedules, but the information is not effectively communicated to the organization. Distortion of sales incentives can have a major impact on inventory investment. One poor practice is to encourage customers through pricing discounts to buy excess quantities over and above their immediate requirements. This practice lowers gross operating margins for the supplier and demand for the product in subsequent time periods. Failure to execute forecasted manufacturing schedules wastes capacity and increases cost. These are simple examples in which the failure to develop accurate demand forecasts or properly execute forecasted schedules can adversely impact an organization.

Developing a Consensus Forecast

The S&OP team develops the consensus forecast using information from the strategic plan, the current system forecast, MRPII information, available manufacturing capacity, and current inventory levels. The revenue projections from the strategic plan are converted to item unit demand using standard cost information. These estimates of item unit demand are modified using historical demand information as well as the latest information that may impact sales, such as the gain or loss of major orders. Members of the S&OP team then agree on the consensus forecast using an approach shown in Tables 3.3 and 3.4. The S&OP team reconciles financial and operational forecasts to determine the final manufacturing

	Qtr 1	Qtr 2	Qtr 3	Qtr 4
Strategic Plan (+5% Over Baseline)	900	2,100	3,100	3,900
System Forecast (Last Year History)	855	1,995	2,945	3,705
Consensus Forecast	**1,000**	**2,000**	**3,000**	**4,000**
MPS (Production)	1,100	2,200	3,200	4,200

Table 3.3 Revenue Forecast for XYZ

Calculation Order		Qtr 1	Qtr 2	Qtr 3	Qtr 4
4	Beginning Inventory	100	200	400	600
5	MPS (Production)	1,100	2,200	3,200	4,200
3	Available to Ship	1,200	2,400	3,600	4,800
1	**Forecast**	**1,000**	**2,000**	**3,000**	**4,000**
2	Ending Inventory	200	400	600	800

Table 3.4 Operations Plan for XYZ

schedule and ending inventory quantities. The manufacturing schedule is adjusted based on variations in capacity and supply over the planning horizon as well as other system constraints. From an inventory investment viewpoint, a critical output from the S&OP process is setting the end of period inventory level. In addition to financial considerations, the ending inventory targets depend on manufacturing-system capacity constraints. System constraints impacting capacity may include seasonal sales, plant shutdowns, or anticipated work stoppages. These constraints impact the ability of the organization to build products to meet anticipated customer demand in the immediate future and may require inventory investment.

In Table 3.3, the S&OP team begins the demand/supply discussion using the strategic forecast as its basis. The example shown in Table 3.3 shows that the strategic forecast must be 5% over the previous year for a hypothetical product, "xyz." The calculations in Table 3.3 are shown in units for simplicity. The required 5% increase over the prior year is compared to actual sales by quarter (or month) for historically equivalent time periods. In the example, the annualized consensus forecast for product "xyz" is 10,000 units. This quantity is 5% over the previous year's annualized unit sales, but the actual quarterly quantities differ from both the strategic plan and last year's forecast. The S&OP team makes adjustments to the forecasted numbers to smooth demand relative to available system capacity. In the final analysis, some product groups may have a lower consensus forecast while others have a higher forecast. However, the overall forecasted annualized quantity across the product mix must meet the projected financial forecast set by senior management, that is, 5% over last year's sales. Referring to Table 3.3, the goal of the S&OP team is to develop the quarterly manufacturing schedule for product "xyz." This MPS demand is also shown in Table 3.4 on line 5 and is equivalent to the S&OP team's consensus forecast shown on line 1 in Table 3.4 and ending minus beginning inventory. In the second step of the calculations, the ending inventory is determined based on financial and operational considerations. Ending inventory targets vary by quarter and depend on demand and available system capacity. The third step adds, by quarter, the demand forecast to the ending inventory to determine the "available to ship quantity" for product "xyz." Netting out the beginning inventory calculated in the fourth step produces the final quarterly manufacturing schedule for product "xyz" as shown in the fifth step for the calculation. Step 5 shows that the original MPS, which was based on the consensus forecast, must be updated to reflect inventory estimates.

Master Production Schedule (MPS)

The master production schedule (MPS) creates the manufacturing schedule using the updated MPS demand forecast, approved production plan, current estimates of material availability, and estimates of system capacity, as well as current backlogs of customer orders. The MPS manufacturing schedule is downloaded to the MRPII system for dissemination throughout the supply chain. The MRPII converts the MPS manufacturing schedule using the bill-of-material (BOM) information to determine "what" and "how much" to build. The "when to build" each dependent demand item is determined by the lead-time offset and build sequence relative to the final product's target shipment date. Based on the cumulative lead-time offset, the time fence for the production order is established in the schedule. This time fence, as shown in Figure 3.6, is the cumulative lead-time along the critical path representing the sequence of operations necessary to build the independent demand item using the components specified by the BOM.

The forecasting schedule is projected into the future by the MPS. As time moves forward and approaches the product's cumulative lead-time, the MRPII system begins to "firm up" or freeze the product's manufacturing schedule. The process of firming up the schedule involves updates to the original forecasted quantities within the MPS. Within this frozen window or time fence, it is important not to change the schedule because dependent demand items have been scheduled throughout the supply chain. Organizations that break into their frozen schedule cause numerous operational problems. As an example, products that have already been promised may be delayed because their dependent items or machine time has been assigned to the product that just displaced them in the schedule.

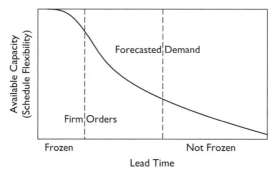

Figure 3.6 Time Fence

The result may be expediting orders that were originally scheduled but were not manufactured on time. Order expediting is very costly for the organization. This situation exists when there are not clear-cut management policies regarding schedule changes. In these situations, sales, in response to customer orders, inserts new orders into the schedule within their normal cumulative lead-time. The products that were just inserted into the schedule consume resources and capacity from those which have already been promised for shipment to customers. This is the worst of both worlds in that customers are dissatisfied and manufacturing costs are higher than necessary. In conclusion, products should be scheduled at or beyond their cumulative lead-time. If customers require more flexibility, then the organization must reduce lead-times through the proper means, that is, Lean methods, Six Sigma, or automation.

Estimating Customer Demand

The MPS is designed as a "push" scheduling system. Once the MPS forecast for the independent demand item is released, manufacturing schedules are determined forward in time for each dependent demand item based on its lead-time offset and position in the manufacturing sequence. The manufacturing sequence is determined by the BOM hierarchy and process design. A major problem with "push" systems is an inability to dynamically respond to fluctuations in external demand or internal operational issues that impact lead-time. If external demand patterns change, capacity may be wasted or too much inventory built for products whose manufacturing schedules have changed. Forecasting may also be problematic. Figure 3.7 shows the effects of system forecasting error in a push system. The variation (error %) is cumulative from one stage to another in the process.

The problem with most supply chains is that their reliance on passing information "over the wall" to each other results in long lead-times and distortion in the cumulative forecast error since forecasting variances are additive at every step of the supply chain. Figure 3.7 shows the cumulative result of simply adding error percentages at every step in the supply chain. At a more advanced level, this analysis would be made using forecasting variances that would be added linearly across each step of the supply chain. Depending on the number of steps in the supply chain as well as the magnitude of the error, there will eventually be a breakdown in the ability of the supply chain to accurately transmit demand information. In addition, cumulative forecasting error increases with the standard deviation of the unit demand as well as lead-time.

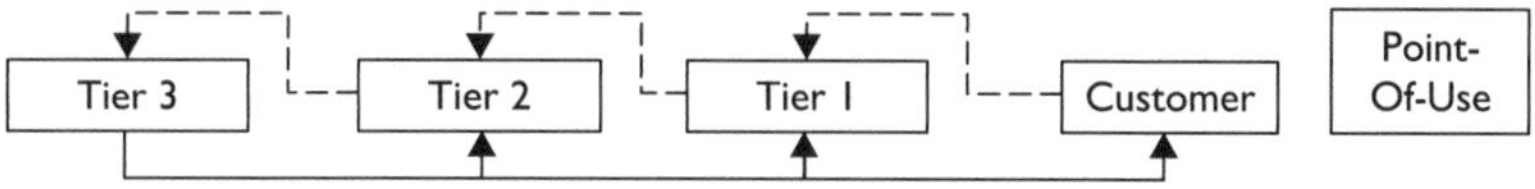

				Point- Of-Use
Tier 3	Tier 2	Tier 1	Customer	

	Tier 3	Tier 2	Tier 1	Customer	Point-Of-Use
+/– 10% Error	60/140	66/134	72/128	80/120	N/A
+/– 20% Error	Breakdown	Breakdown	44/156	60/140	N/A
+/– 30% Error	Breakdown	Breakdown	16/184	40/160	N/A

Assumptions: Actual Demand = 100 Units, Normal Distribution, Error = 1 Standard Deviation
i.e., 10% = 10 units

Figure 3.7 Current Paradigm (Push)

Operational issues impacting lead-time, including material and labor shortages, machine breakdowns, and poor quality, may also have a negative impact on schedule attainment. Operational issues are exacerbated in complex systems using "push" scheduling. In these situations, there will be increased levels of raw material, work-in-process (WIP), and finished goods inventories due to scheduling changes that disrupt the build schedule of work-in-process (WIP) inventory. In these situations, WIP inventory may not be available to build products just added to the schedule, or it may have already been built for products that have been displaced from the manufacturing schedule. In addition, when schedule changes are frequent, machine setup costs increase due to the increased frequency of machine changeovers. Lean Six Sigma projects have been successfully applied to eliminate these process breakdowns by ensuring that procedures and policies accurately reflect organizational requirements and are easy for everyone within the organization to adhere to over time.

In contrast, "pull" scheduling systems are driven by external customer demand patterns. These systems are called "pull" systems because the manufacturing schedule is directly linked to external customer demand (demand pull) and pulled backwards through the system. This is shown in Figure 3.8. Linkage to the external customer is accomplished by providing visibility across the supply chain to customer *point-of sales (POS)* information. POS visibility is provided using either visual controls alone or a combination of IT technology and visual controls to capture information in real time. The manufacturing schedule alignment associated with pull systems results in lower supply chain cost, reduced inventory investment,

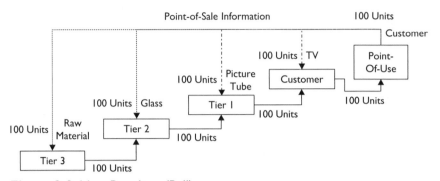

Figure 3.8 New Paradigm (Pull)

and higher customer service levels. Although many organizations must still use MPS and MRPII systems (job shops), hybridized systems have been created to obtain at least some of the operational advantages of pull systems. In these hybridized systems, portions of the MRP are disabled and visual shop floor controls are substituted to maintain operational flow within work cells on the shop floor in real time. This allows a dynamic response by the work cells to scheduling changes or disruptions in the work flow by the work cells to either manufacturing schedule changes or unforeseen operational issues.

Product Forecasting

There are many different types of forecasting methods. Some are very qualitative and rely on management opinion and consensus, while others are very quantitative and require the help of professional forecasting analysts. The more common qualitative methods include management opinion, jury of executive opinion, and sales force composites. Quantitative methods include moving average, exponential smoothing, regression, time-series decomposition, and more advanced *autoregressive moving average (ARIMA)* models. Quantitative methods require that data be collected to build the model. A minimum of 12 months of demand data is required for simple analyses and at least 36 months is required to build a seasonal model. Quantitative methods will be discussed in more detail since they are most commonly used to forecast demand. Prior to developing a forecasting model, it is important to understand the objectives of the forecast including the units in which the forecast will be made as well as the forecasting time horizon.

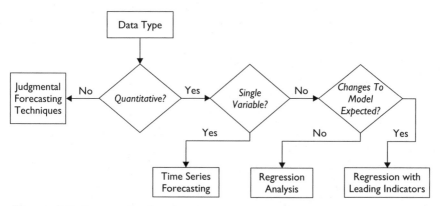

Figure 3.9 Forecasting Roadmap

Figure 3.9 shows several more advanced forecasting methods, which can be used to build forecasting models. These advanced models use several independent variables rather than just the historical pattern of the time series (that is, lagged dependent variables and forecasting error terms). These independent variables contain information useful to explain demand patterns of the "dependent variable." As an example, historical product sales may have been shown to depend on real disposable income and employment levels as well as previous sales levels (lagged dependent variable component). This information may help build a forecasting model that is more accurate than reliance on demand history alone. Since every item and location has a unique demand pattern, there will be as many models as there are products and locations. In order to more effectively manage these large number of models, it is good practice to stratify the entire product population based on coefficient of demand variation (standard deviation of demand divided by average demand) and a second metric such as unit volume sales. This method is shown in Figure 3.10. Stratification allows the forecasting analyst to focus and actively manage time series having high volumes (or another relative importance criterion) and low demand variation. Low demand variation means a good model can be built, whereas high demand-variation situations are difficult to model well. The MRPII system is used to automatically manage demand of those items having low volume and variation using a minimum/maximum inventory model (min/max models will be discussed in Chapter 6). On the other hand, if demand variation is large relative to the average demand for the time series, special models must

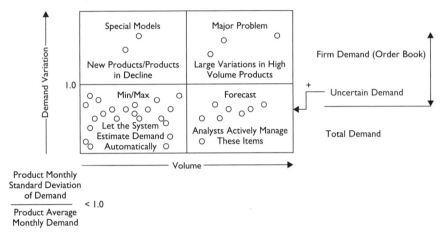

Figure 3.10 Stratify Item Forecasts

be developed to forecast demand. In summary, quantitative time-series models can be developed for products with stable demand patterns, but special modeling techniques are required for other situations.

Time-series methods are used to create forecasting models when a single variable is to be forecast using combinations of demand and forecasting error from previous time periods. As an example, a model could be built to forecast sales next month based on a model that differentially weights sales realized this month versus previous months. Time-series methods do not require complicated analysis since the system software optimizes the model's weighting parameters based on historical demand. This minimizes the resultant forecast error relative to the model fit versus its historical baseline. The exact methods of developing these forecasting models will not be discussed in this book, but at a high level, exponential smoothing is used by these models. Also, most organizations use exponential smoothing models to automatically forecast product sales. If the product has no trend or seasonal pattern, "simple exponential smoothing" models are used with a single weighting parameter to build the forecasting model. If the time series has an obvious trend, but not a seasonal pattern, then "double exponential smoothing" (also called Holt's method) is used to build the forecasting model using two weighting parameters. In the case of a product having both a trend as well as a seasonal pattern, "triple exponential smoothing" (also called Winter's method) is used to build a forecasting model using three weighting parameters.

"Moving average" models can be used to predict the average demand of a product over time, but seasonality cannot be modeled using this method. "Time-series decomposition" methods can also be used in a manner similar to triple exponential smoothing to decompose the time series into its trend and seasonal components. Autoregressive moving average (ARIMA) models are complex and are not routinely used to forecast product sales.

Regression-based methods build forecasting models using one or more independent variables and in some cases use autoregressive components of the dependent variable. As an example, future sales might demand on sales from previous time periods (autoregressive component) as well as one or more independent variables such as real disposable income and customer demographics. Some of these independent variables might be leading, or lagging, variables. Leading indicators are variables whose levels we know today and which have been shown to influence the dependent variable in the future. As an example, higher fuel prices might be a leading indicator of motor vehicle sales in the future. In this example, the user would purchase information on leading economic indicators to produce more accurate forecasts.

Experience and education are required to accurately construct and analyze forecasting models. Some organizations use very advanced forecasting methods, such as regression models built using econometric variables, that is, leading and lagging economic indicators as well as several independent variables. However, most manufacturing organizations use the simpler time-series models to forecast product sales rather than regression-based methods. Time-series forecasts of product sales can be routinely calculated each month, by the forecasting system, for all products or items by their location. In these time-series models, forecasts are built, in an iterative manner, using parameters to smooth historical demand. There are three basic concepts that define and are used to build a time-series model. The first is the forecast time horizon, which is the period of time into the future for which the forecast is designed to be operative. The second is the forecast time interval or time unit for which forecasts are prepared, that is, month or year. The third is the time-ordered sequence of historical demand, by time interval, starting from the oldest to most recent demand observation. Using this information, a model can be built to predict future product demand.

In recent years, information technology has been deployed across supply chains to make demand information available in real time across every part of the supply chain. As an example, in retail supply chains, demand information is accumulated in real time as each item is scanned at the checkout counter. The accumulated demand information is sent to the supply base every day. This allows them to schedule replacement items without having to create forecasts with their associated forecasting errors. In more advanced versions of these systems, suppliers can use this information to manage their inventories at the customer locations in real time. In these situations, the supplier can analyze external customer demand versus available inventory to maximize sales and operating margins by making sure product is available for sale. In Figure 3.8, the supply chain replaces only what was actually sold to an external customer, effectively eliminating item level forecasts.

Creating Sales Forecasts

After a suitable forecasting model has been chosen, the next step is to create the sales forecast on a timely basis, that is, monthly. Creation of the aggregate sales forecast is an integral part of the S&OP process and receives direction and information from all participants of the S&OP team. These participants include marketing, sales, operations, finance, distribution, purchasing, and materials planning. Marketing (if responsible for the forecast) creates the forecast plan for the S&OP meeting as shown in Table 3.5. This forecasting process requires extensive data on historical customer demand at an item and location level of detail. Using aggregated demand information provided by marketing and sales based on S&OP recommendations, the forecasting department updates the original system forecast and electronically feeds these updated forecasts to the MPS.

Complications will occur if the forecasting model must be readjusted because of changes in a product's demand pattern over its life cycle. A typical product life cycle is shown in Figure 3.11. Reviewing

1. Identify markets and sales drivers
2. Create sales models
3. Collect and analyze data
4. Estimate market share
5. Estimate sales
6. Update as information changes

Table 3.5 Forecasting Tasks

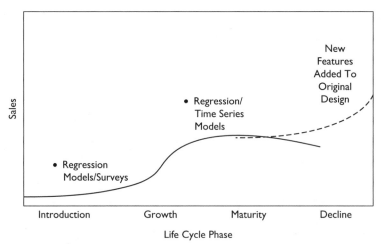

Figure 3.11 Typical Product Life Cycle

Figure 3.11 we see how a product moves through four phases of its life cycle. In the introduction phase, if the marketing research was accurate, demand increases rapidly for the new product. However, if the product is a completely new design, it may be difficult to estimate its demand prior to the introduction stage. On the other hand, if the new product's demand pattern is similar to previously released products, it might be possible to accurately predict its future demand based on its demand history. Forecasting systems match their forecast to the product's historical demand pattern by automatically modifying the model's smoothing parameters. In addition to automatic parameter modification, the forecasting analyst can truncate the historical basis of the time series. This would be done if significant differences in demand existed between more recent versus past time periods. In the more complicated regression-based models, independent variables and their coefficients (parameters) must be analyzed and modified by the system/analyst. If a product's position within its life cycle has not been fully considered by the forecasting analyst or if the historical basis of the time series changes, then the resultant forecasts will be poor.

The more the organization can remove products from the forecasting process altogether, the higher the cumulative forecasting accuracy will be. This can be done either by contract or by using electronic systems to acquire real-time demand information.

Forecasts must be made for products whose demand cannot be made firm. However, stratification of products prior to forecasting will allow the organization to concentrate scarce analytical resources on products requiring an analyst's attention. This is because they are either high volume, having a large impact on the organization, or have high demand variation, making it difficult to use standard forecasting models to predict future demand. High demand variation is defined as a ratio greater than "1" of monthly demand standard deviation divided by monthly average demand. Low-volume products with stable and low variation in demand can be forecast automatically by the system using a min/max system. Min/max systems use a product's reorder point to signal that additional quantities of the product must be built. New products or current models having declining sales volumes or unknown demand patterns must be handled using more complicated models.

Creating sales forecasts requires that the organization segment its markets in order to understand the major elements driving product sales within a given segment. Once the sales drivers are understood, sales models can be created to predict future sales. Often, the product's demand history is well known and the forecast can be standardized and routinely forecast. However, this may not always be true. In fact, in many organizations such as those selling industrial equipment and having a relatively few number of customers, it may be prudent to build sales estimates using a macroeconomic model and to work directly with their major customers to understand their future demand. These models would most likely be regression-based rather than time-series-based. In either case, collection of accurate data that reflects forecasting objectives is critical to building good predictive models.

The types of forecasting models used to predict product demand vary as the product moves through its life cycle. In the introduction phase of the life cycle, when demand information is sparse, relevant information from the direct customer is critical to estimating product demand. At this point, accurate market research will greatly minimize the risk of poor demand estimation. Regression-based models and customer survey information are normally the most useful methods employed to estimate product demand at this stage of the product's life cycle. As the product enters its growth stage and maturity stages, standard time-series models are used to estimate product demand. If product sales also depend on several independent variables, regression-based methods are required to estimate product demand; otherwise, simple time-series methods will be adequate. In the decline stage of the product life cycle, standard mathematical techniques can also be used unless the historical baseline is no longer relevant.

Quantitative Forecasts

Quantitative forecasting methods vary by organization. The most common models use exponential smoothing (time series) models. These models can be automatically updated by the forecasting system using historical data and smoothing parameter optimization algorithms. Because they can be automatically updated, they are especially useful in organizations forecasting thousands of products each month. Time-series models, depending on seasonal patterns, require several years of month-over-month demand data. Since the most common forecasting methods are time-series models, these will be discussed in more detail.

Time-Series Models (Exponential Smoothing)

A time-series model is based on demand information collected in sequence at equal intervals over time. These models can be decomposed into four major components. These are the level component or average historical demand, a trend component showing upward or downward movement of demand over time, a seasonal component showing seasonal (periodic) patterns, and a cyclical component showing longer-range periodic patterns around the trend component. Any variation unexplained by the time-series model is considered an irregular component (forecasting error).

Simple Plot/Monthly Sales

It is always useful to plot the time-series data prior to detailed analysis of the time-series pattern and selection of the final forecasting model. As an example, Figure 3.12 shows a time series of monthly product sales. A close examination of the pattern reveals an increasing upward trend in sales. There also appears to be seasonality in the sales pattern and the overall variation in sales appears to increase over time. The periodic and increasing demand pattern suggests using a time-series model, which will capture both trend and seasonal information. There are two forecasting methods that can model these patterns. The first is a decomposition algorithm and the second is a three-parameter exponential smoothing model, also called Winter's method. Without getting into technical details, the decomposition algorithm breaks the time series down into the level, seasonal, cyclical, and irregular components to build a simple regression model of historical demand. This regression model is extrapolated into the future and adjusted for seasonality to create the forecast. Winter's method uses three smoothing parameters to build the model

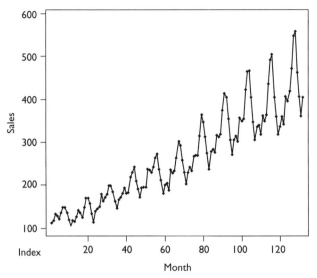

Figure 3.12 Original Time Series (Sales Versus Time)

from historical demand. Using either algorithm, a model is fit to the historical pattern and a demand extrapolated (forecast) into future time periods. The results that were derived using both the decomposition and Winter's algorithms are shown in Figures 3.13 and 3.14 respectively,

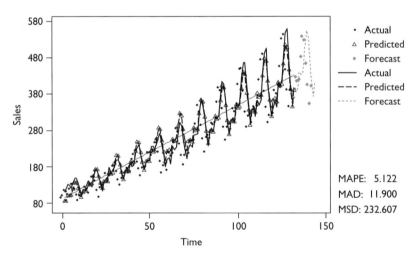

Figure 3.13 Decomposition Fit for Sales

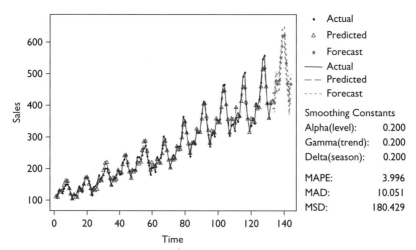

Figure 3.14 Winter's Multiplicative Model for Sales

using Minitab's software. Both models adequately fit the sales pattern (have low forecast error, that is, 4% to 5%).

Measuring Forecasting Error

Forecasting-error calculations are made after the time series has been forecast and demand for the corresponding forecasted time period has occurred. A systematic analysis of forecasting errors will show the forecasting analyst where improvements can be made to the forecasting model or system. Products having the largest forecast errors and importance to the organization should be the focus of Lean Six Sigma improvement projects. This error analysis is conducted at each forecast level by item and location, product group, and so on. In Figures 3.13 and 3.14, the error percentages are 5.1% and 3.9% respectively. These error percentages were calculated using the *mean absolute error percentage (MAPE)*. The MAPE averages the absolute value of the error percentage at each time period. The MAPE is useful in comparison of different time series since the MAPE is volume adjusted by the calculation, that is, percentage error for each time series. Two other error statistics are provided by Minitab. These are the *mean absolute deviation (MAD)* and the mean square deviation *(MSD)*. The MAD simply calculates the average difference (deviations) between actual versus forecasted demand across all historical time periods. The problem with the MAD is that positive and

	December[*]	January	February	March	April	May
Forecast		150	110	200	100	120
Actual		160	100	150	120	110
Error		(10)	10	50	(20)	10

[*]Forecast lead-time is 30 days (1 month forecast time horizon)

$$\text{Error}_t = \text{Forecast}_t - \text{Actual}_t \begin{cases} \text{Percent Error}_t = \dfrac{\text{Forecast}_t - \text{Actual}_t}{\text{Actual}_t} \times 100 \\\\ \dfrac{\text{Root Mean}}{\text{Squared Error}} = \sqrt{\dfrac{\sum\limits_{t=1}^{n}(F_t - A_t)^2}{n}} \end{cases}$$

Figure 3.15 Calculating Forecasting Error

negative forecasting errors will cancel each other, artificially reducing the forecasting error estimate. The MSD is calculated as the average squared differences (deviations) between actual versus forecasted demand across all historical time periods. The MSD is not volume adjusted, but the positive and negative forecasting errors do not cancel each other since these are "squared" in the formula. The square root of the MSD is the *root mean squared error (RMSE)*. This is a very useful statistic because it is used in safety-stock calculations in lieu of a standard deviation of demand. If the forecast is highly accurate relative to the historical standard deviation of demand, the safety stock can be lowered if the RMSE is used rather than the historical standard deviation. Figure 3.15 shows how to calculate the mean absolute percentage error (MAPE) and root mean squared error (RMSE) statistics. Table 3.6 shows, by organizational level, the magnitude of the forecasting error. Notice that forecast error decreases as forecasts are aggregated to higher organizational levels.

Corporate forecast	2%–5%
Product group forecast	5%–15%
Product line forecast	10%–20%
Product forecast	10%–25%

Table 3.6 Forecasting Benchmarks

Forecasting Software

The basic forecasting algorithms used by software packages include exponential smoothing models, Box-Jenkins; also called autoregressive integrated moving average models (ARIMA), as well as regression-based models. Exponential smoothing models use previous values of the time series to develop forecasts. Regression-based models (either multiple or Cohrane-Orutt) can use independent variables as well as previous values of the time series (lagged dependent variables) to build the forecast model. At a more advanced level, software packages can be modified to take into account marketing promotions and seasonality as well as holidays, which may impact future product demand. Using sophisticated algorithms, forecasts are made at an item and location level. These item-level forecasts are aggregated upwards through the organization to attain aggregate forecasts at product group, business unit, and organizational levels. After the forecasting model is created and evaluated against realized demand, various statistics are produced to allow the user to fully evaluate the model's effectiveness. A few of these statistics, such as mean absolute percentage error (MAPE) and root mean squared deviation (RMSD), were discussed earlier in this chapter. Other statistics are more advanced and are not discussed in this book.

Impact of New Product Forecasts

Even under the best circumstances, it is difficult to accurately and precisely estimate new product demand. One complicating factor, making demand difficult to estimate, is that demand patterns of new products may be erratic in the introduction phase of their life cycle due to extensive promotional activity by marketing and sales. Unfortunately, inaccurate forecasts of new products have major detrimental effects on inventory investment as well as operational efficiency. These negative impacts are most easily seen in the form of excess and obsolete inventory. Excess inventory represents multiples of lead-time quantities which, if large enough, may never sell due to product obsolescence or internal damage associated with handling and storage of the product. If the product does not sell at all, it would be categorized as obsolete. Unfortunately, few organizations do the necessary due diligence to fully investigate new product sales based on demographic factors and marketing research.

Estimating sales demand for a new product is difficult because it is impacted by numerous external factors. Marketing research is necessary to develop relationships between the variables driving demand for

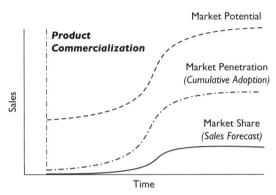

Figure 3.16 Developing New Product Forecasts

new products and projected demand for the product. For this reason, good sales forecasts depend on effective marketing research. Figure 3.16 shows the information required to estimate demand for a new product. The process begins with an analysis of market potential and progressively works towards the sales forecast. Estimating the market potential for a new product is a complicated series of tasks involving analyses of competitive and regulatory environments, pricing levels, promotional options, market locations, and product features. Environmental factors include population demographics, the local culture, technology level, and competitors as well as local laws and regulations.

Survey and statistical methods as well as current market share data are used in the marketing analysis to predict which customers will purchase a new product. In marketing analysis, customers are segmented and their purchase intentions are correlated to product features and price to determine purchase probabilities. Product awareness is determined using survey and statistical techniques. Marketing determines the proportion of customers, by market segment, which are aware of the new product. This awareness is caused by advertising, public relations, and related methods. The projected product-usage rate is estimated using the historical number of units purchased per buyer. This usage rate is estimated over the time period of interest. Advertising strategies and sales force incentives are used to exploit marketing opportunities and close identified gaps between actual and projected sales.

In addition to marketing's impact on the new product sales forecast, its activities often negatively impact inventory investment through excessive product proliferation. Excessive product proliferation increases both the number of bill-of-material (BOM) components as well as independent items (finished products). This situation results in operational inefficiencies due to excessive material handling, purchase of components, and increased production setup costs. These conditions cause higher inventory investment of raw materials, work-in-process (WIP), and finished goods. The root causes of many Lean Six Sigma projects associated with excess and obsolete inventory can be traced to problems with product proliferation.

Lean Six Sigma Forecasting Applications

Forecasts are necessary to provide information to the organization relative to expected demand on the system; this allows the organization to match its resources to the expected demand to ensure customer satisfaction and optimum operational costs. If properly executed, forecasting systems provide accurate demand estimates for local resources as well as expected operational efficiencies. Although forecasting methods are used in very diverse industries and applications they have several common characteristics. These characteristics include data acquisition to make the forecast, data analysis to build the forecasting model, selection of the specific forecasting model used to create the forecast, and use of the forecasting model's information to make decisions concerning future demand. Understanding the basis of forecasting allows the Lean Six Sigma improvement team to identify projects and analyze the root causes for poor forecasting.

Some real-world examples will be used to show how data is acquired by systems to be used in their forecasting models. Colleges must forecast not only total student enrollment, but also enrollment by specific student categories based on many diverse demographic factors. Some of the information used in these forecasts, relative to student acceptance rates by demographic factors, is historical. Other information is unique to each incoming student class. Colleges use multiple regression models to create the forecasting models that will predict acceptance probability. Hospitals forecast patient needs by required service type as well as length of hospital stay. These factors impact available bed capacity in the hospital. Historical databases are used to create regression models that help the hospital plan staffing levels and facility upgrades to meet patient demand over time. Using a different example, retail stores estimate product demand based on customer buying preferences as well as the local demographics of the population close to each store. Complicating these product forecasts is the fact that retail organizations also promote products through advertising. This distorts the demand patterns for each item and store. Capacity forecasts are also required for police, fire, and almost any other system where demand must be estimated in an environment of limited resources.

In manufacturing, forecasts are used in a number of ways to ensure that adequate resources are available to satisfy customer demand. Depending on the specific industry, forecasts may be projected out more than 20 years. These situations would include industries requiring heavy capital expenditure or infrastructure that takes a long time to build, such as the railroad or telecommunication industries. In manufacturing, strategic forecasts are important since the decision to build a new facility depends on projections of demand out 5 to 20 years into the future. As the time frame comes into an intermediate range (which differs by industry), decisions are made to actually build facilities or expand current ones and purchase the equipment necessary to satisfy expected customer demand. The intermediate timeframe varies by industry, but is normally 18 to 36 months out into the future. As the forecast time horizon approaches the cumulative lead-time necessary to build the product, demand is aggregated, at a business unit level, by the master production schedule (MPS) and passed to the master requirements systems (MRPII). Once the product has been scheduled at a particular facility and time, local production activity control (PAC) is exercised by the facility. Throughout this process there are many possible breakdowns given the number of operations as well as their complexity. Lean Six Sigma black belts routinely work to eliminate these breakdowns using Lean Six Sigma tools and methods. However, the ability to easily recognize the relationships between process breakdowns is an advantage that black belts and green belts who are cross-trained in forecasting, material requirements planning, purchasing, and logistics have over those not properly trained since understanding these systems requires extensive training and experience.

The types of forecasting models used in these diverse systems include time-series models based on historical demand of the dependent variable, regression-based models in which the dependent variables are forecast using one or more independent variables as well as lagged values of the independent variable. These methods were discussed earlier in this chapter. The data for the models varies from single-unit estimates to those using aggregated demand. However, unlike manufacturing applications in which demand is aggregated by product line and facility, it may not be possible to aggregate demand in some systems. As an example, it is difficult to aggregate demand for a hospital since demand is specific to the demographic of the local population. However, various degrees of aggregation may still be possible in these systems. The important point is that common forecasting methods are used across very diverse industries and this fact can be exploited by Lean Six Sigma black belt and green belts to identify and eliminate breakdowns caused by their forecasting systems.

At a high level, the goals of each of the diverse systems are the same. Each system attempts to ensure that adequate capacity is available to meet demand on its system. However, it is interesting to note that capacity differs by system. In manufacturing, excess capacity is evident in the form of inventory or idle machines and workers. In hospitals capacity is driven by available staffing levels as well as by the skill mix of the staff. In each system, the forecast is probabilistic, so there will always be a probability that capacity may not be available if demand increases beyond the original forecast.

There are many examples of Lean Six Sigma project applications in demand management and forecasting. Demand management itself is heavily dependent on the VOC information since to fully understand the basis of customer demand, it must be segmented by market segment and customer needs and value expectations determined by marketing research. The more an organization can leverage VOC information to firm up their customer's demand, the lower the overall forecasting variation will be since those items associated with firm demand can be taken out of the monthly forecasts. Using the forecasting error statistic, or even better actual inventory investment estimates, Lean Six Sigma black belts can work back to the root causes of high inventory investment. These causes will be associated with lead-time or demand issues at a high level. Those root causes which are the result of demand issues can be systematically investigated by the improvement team. Those root causes associated with operational breakdowns of the current process can be attacked by Lean Six Sigma black belts. Those which require changes to the system design can be attacked using the design-for-Six-Sigma (DFFS) project approach.

Issues resulting from process breakdowns of the current system design could be impacted by any portion of the forecasting process. As an example, the historical customer demand patterns could be incorrectly specified due to the inclusion of shipment rather than actual demand history. The demand information may not have been correctly posted at its actual location of demand, that is, a particular distribution center. Or it may have been incorrectly aggregated by the forecasting analysts. Any of these issues could be the root causes discovered through Lean Six Sigma projects. Also, the forecasting model may have been incorrectly specified by the analyst. An excellent application of Lean Six Sigma tools in these situations would be the application of experimental design methods to optimize forecasting parameter levels to minimize forecasting error. This is a kind of sensitivity analysis in which parameter levels are successively modified, but instead of changing several levels of the three smoothing parameters (Winter's model), the analyst changes just two or three levels, in an optimized manner, since the algorithm is not probabilistic.

Lean Six Sigma projects can also be initiated at the functional interface between the forecasting department and other organization functions if there are breakdowns in these areas. However, the most straightforward approach for forecasting analysts who become Lean Six Sigma black belts or green belts is to systematically work down the forecasting error percentages over time in conjunction with moving demand off to order book, that is, making it firm. In the case of new product development, where forecasting models are more regression-based given the uncertain nature of forecasts for these products, Lean Six Sigma tools and methods are directly applicable. The process of customer segmentation, extracting VOC information, and projecting customer demand out over time lends itself very well to advanced lean Six Sigma statistical tools and methods. Also, these tools and methods can be used to eliminate the obvious reasons for excess and obsolete inventory. The chronic reasons for excess and obsolete inventory can be removed by a system redesign of the new product introduction process using DFSS.

Summary

Poor demand management increases inventory investment, resulting in excess and obsolete inventory. How to effectively manage demand is a complicated process. Some organizations are make-to-order environments in which demand is made firm by a contract prior to component sourcing or manufacturing. Other organizations are make-to-stock systems. In make-to-stock systems, finished goods inventories are built in anticipation of sales. Finished goods inventories are required due to the large number of products being manufactured in an environment having limited manufacturing capacity. Mathematical forecasting methods are heavily used in these make-to-stock environments since customer demand is unknown. In these situations, the concept of "demand management" becomes important. Demand management implies forecasting prioritization and continuous improvement over time, that is, systematically reduced forecasting error over time. System complexity has forced organizations to share real-time demand information with customers and suppliers. Real-time information is obtained through a shared information technology (IT) infrastructure as well as significant reductions in the order cycle using Lean methods. All other things being equal, lead-time reductions are a good way to decrease variations associated with estimates of demand. Improvements to demand management systems will continue to increase as organizations move towards "pull" versus "push" scheduling systems.

Lean Six Sigma tools and methods are routinely used to improve demand management system performance either up-front by identifying problems with the forecasting system or bottom-up through Lean Six Sigma black belts and green belts working their improvement projects. Lean Six Sigma tools are also useful to simplify these complex systems and aid forecasting analysts in building forecasting models. The tool sets are also very useful in developing accurate estimates of new product sales. In Chapter 10, examples show how Lean Six Sigma has attacked problems associated with demand management and forecasting.

Chapter 4

Lead-Time Impact on Lean Six Sigma Projects

Key Objectives

After reading this chapter, you should understand how lead-time analysis and reduction is useful to simplify supply chain systems, reduce inventory investment, and help identify Lean Six Sigma projects, as well as understanding the following concepts:

1. Why creating a value stream map (quantitative process map) and breaking operations down into smaller time elements is a powerful way to identify areas within the process where lead-time can be reduced to lower cost and improve quality.
2. Why identifying the system's critical path and its time variation is necessary to effective lead-time reduction.
3. Why using the system's takt time allows the team to see the wasted resources in the current system.
4. Why lead-time reduction is one of the major ways to improve operational performance as well as reduce inventory investment.
5. How to use tools such as value stream maps to identify and eliminate unnecessary process operations, bottleneck management to increase system throughput, application of mixed-model scheduling, application of transfer batch methods, and deployment of Lean, JIT, and QRM techniques to reduce lead-time.

Why Reducing Lead-Time Is Important

Lead-time is the time required to perform all operations in the process up to the point under analysis. As an example, the lead-time for a finished product includes the time to receive raw materials and components and the time to manufacture intermediate components and assemblies as well as the time to assemble the final assembly. At a higher level, relative to the external customer, it is the time between receiving a customer order and receipt by the customer of the same order. Figure 4.1 shows how the components of lead-time can be broken down at an operational level to aid subsequent analysis. These individual components of lead-time include order preparation time (setup), queue time (waiting), processing time (doing the work), move/transportation time (taking the work to the next operation), receiving (accepting the work at the next operation) and inspection time (ensuring that the work is correct prior to setup at the next operation). An indirect measurement of lead-time is inventory levels at each operation throughout the process. Inventory levels are necessary because of uncertainties of internal or external supply. It is also important to note that lead-time has a probability distribution including a central location (or average lead-time) as well as a variation in lead-time (or standard deviation). As process complexity increases, cumulative lead-time increases on average, but the lead-time variation also increases as the square root of the cumulative variation of all operations along the critical path. Figure 4.2 shows how lead-time is magnified as the number of process steps increase. This concept will be discussed later in this chapter

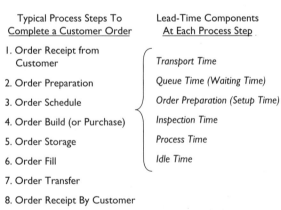

Figure 4.1 Lead-Time Components

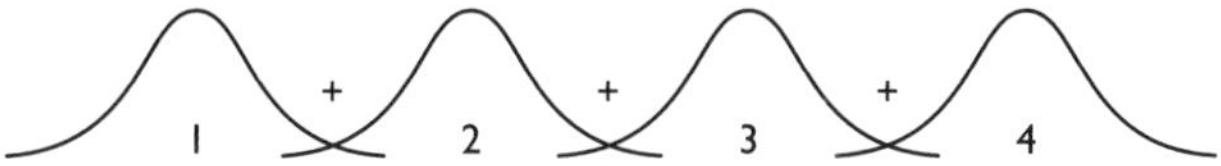

Operations on the critical path

$$\text{Total Lead-Time} = \sum_{i=1}^{4} \text{Lead-Time} + \sum_{i=1}^{4} \text{Lead-Time Variances}$$

Step 1 $\mu = 10$ days $\sigma = 5$ days	Average Lead-Time = $10 + 2 + 20 + 5 = 37$ Days
Step 2 $\mu = 2$ days $\sigma = 1$ days	Standard = Square $\left\{ 5^2 + 1^2 + 10^2 + 2^2 \right\} = 11.4$ Days
Step 3 $\mu = 20$ days $\sigma = 10$ days	Deviation of Root
Step 4 $\mu = 5$ days $\sigma = 2$ days	Lead-Times

Total Time Through the Network Will Vary Between ~ 14 to 59 Days

Figure 4.2 Lead-Time Estimation

to show the importance of reducing average lead-time and its variation along the system's critical path.

Long lead-times increase operational inefficiencies across a supply chain and are also often associated with root causes of many Lean Six Sigma projects deployed within supply chains. Lead-time issues will usually be seen as high inventory levels or low customer service levels. Table 4.1 lists many examples of process breakdowns that cause lead-times to be long or highly variable. Lean Six Sigma projects are often created around these issues. As an example, when customers must wait for material deliveries or work centers must wait for components and a part, system throughput is reduced, increasing lead-time as well as per-unit transaction costs. When work is batched, as will be shown later

■ Waiting	■ Ambiguous goals
■ Transport	■ Poorly designed procedures
■ Non-value-added activities	■ Outdated technology
■ Serial versus parallel processes	■ Lack of information
■ Batch work	■ Poor communication
■ Excessive controls	■ Limited coordination
■ Unnecessary transfer of materials	■ Ineffective training
■ Scrap/rework	■ Complexity

Table 4.1 Some Reasons for Lead-Time Variation

in this chapter, lead-times increase and quality levels may be lower. Excessive controls typified by long approval and signoff cycles slow the flow of work materials and information through a process. Scrap and rework require that work be made up to cover operational losses resulting in longer lead-time. Finally, poor communication, ambiguous goals, and outdated procedures and training will ensure that the work is incorrectly completed or the wrong work is done, resulting in poor quality, high per-unit costs, and longer lead-times. Each of the reasons listed in Table 4.1 has been shown to be associated with one or more Lean Six Sigma projects. Their ultimate solutions ranged from simple value stream mapping (VSM) of the process to the use of advanced statistical analysis tools. VSM will be discussed later in this chapter. Lead-time analysis and reduction are important because, when demand estimates are reasonably accurate, that is, in the 25% error range at an item and location level, the impact of lead-time alone on inventory levels may represent more than 70% to 90% of inventory investment. For this reason, it makes sense that application of Lean Six Sigma methods is critical for driving down inventory investment levels by eliminating process breakdowns that drive lead-time issues.

To accurately set inventory levels within the supply chain, proper lead-time definitions are very important since it is not unusual for different organizations within the same supply chain to have different lead-time definitions. As an example, if a supplier and customer simply specify lead-time as 35 days, what does this number mean in practice? If the 35 days are defined as calendar days, then the total lead-time is 35 days or five calendar weeks because a week has seven calendar days. However, if the 35 days are defined as working days, then the total lead-time is approximately 40% longer, or seven weeks, since there are five working days per week. This example, while simple, is quite common in a supply chain. This situation results in conflicting viewpoints of on-time delivery since no one can agree there is a problem in the first place. In this example, the impact on inventory investment is significant, that is, 40% impact.

The longer the lead-time, the higher will be the required inventory investment necessary to satisfy anticipated customer demand to prevent customer service levels going lower than target. As an example, assume the average finished goods inventory level has a linear consumption rate. This implies that the average inventory level at the midpoint of the order cycle is one-half of the order quantity Q (Q is calculated as the *economic order quantity*, that is, *EOQ*). The length of the order cycle is the time period between receipts of two sequential orders. As an example, I could receive an order on the first of every month (making a 30 days order cycle). The lead-time in this example refers to the time period between placing an order and its receipt. Lead-time in this discussion is less than the order cycle and is usually called the Reorder Point (ROP). Because Q is based on one month's usage (30 calendar days), the average inventory quantity will be equal to 15 calendar days supply or ½ Q at the midpoint of the order cycle. On the other hand, if the lead-time is 20 calendar days, then the average inventory level will be just 10 days supply. Thus, to properly manage inventory investment, it is important to understand the impact of the order cycle lead-time and lead-time on required inventory investment. This discussion does not include safety-stock inventory levels, which are set based on the reorder point lead-time as well as variations from expected demand.

Lead-time is impacted by many contributing factors other than its definition given by the supplier and customer. Many of these factors are listed in Table 4.1. But there are other contributing factors (or root causes) existing at a level higher than those listed in Table 4.1. These include large lot sizes, low on-time delivery, and schedule changes, to name a few. To show how lead-time (and inventory investment) can be significantly increased despite having firm lead-time agreements, consider how misinterpretations relative to lot size can occur in practice and impact lead-time and inventory investment. If a supplier guarantees a lead-time of one month (30 calendar days), but requires a minimum purchase lot size of two months (60 calendar days), the effective lead-time from the customer's perspective is two months supply and the average inventory investment is actually 30 days versus 15 days supply. It is important to incorporate these lead-time considerations into the inventory analysis and final inventory model since there are other complicating issues impacting inventory investment. The purpose of this discussion was to reiterate the importance of proper lead-time definition to estimate the required inventory investment necessary to ensure the target customer service levels. Also, in Chapter 10, lead-time will be a major input into the analysis of the inventory model as well as a focus for identification of Lean Six Sigma improvement projects. From an operational viewpoint, our goal is to eliminate unnecessary lead-time components in order to reduce cycle time, and improve the overall quality level of the process.

Building an inventory model requires understanding the manufacturing process lead-times and other critical operational inputs and metrics. Many of these are shown in Table 4.3. It is also important to understand spatial relationships between operations including sequential and parallel work tasks. This also allows calculation of cumulative lead-time by summing the completion time of each operation along the critical path of the network as well as the variation (standard deviation) of lead-time for all tasks. A simplified example of these concepts is shown in Figure 4.2. In Figure 4.2, the completion times for each operation are assumed to be normally distributed. However, there are many other distribution assumptions that may be applicable. In these situations simulation software is required to complete the lead-time analysis. It can be shown that the minimum lead-time through the network lies on its "critical path." Thus, to properly analyze cumulative lead-time, attention must be focused on every task's expected and standard deviation of completion time along the system's critical path. This results in a distribution of lead-times at every operation rather than a single lead-time value and a distribution of total lead-times across all operations. As an example, in Figure 4.2, the average lead-time, over many simulated runs through the network, is 37 days, but any given simulated run through the network varies between 14 and 59 days 95% of the time.

Reductions in lead-time will result from elimination of one or more of the process break downs listed in Table 4.1 or application of one or more of the methods listed in Table 4.2. As an example, lead-time can be reduced if we can identify and eliminate areas in the process where unnecessary waiting, movement, and batching of work occur. Also, to the extent we can eliminate unnecessary process controls, clarify procedures and work instructions, and communicate more effectively, lead-time will be reduced in the system. Finally, lead-time will be lowered if we can reduce system complexity and unnecessary tasks as well as situations that create scrap and rework of materials. System complexity can be reduced using process analysis tools and methods such as value stream maps (VSM). In fact, mapping of a process using a value stream map (VSM) is one of the best ways to understand the complex operational relationships impacting lead-time. As the VSM is being analyzed by the Lean Six Sigma improvement team, every operation in the process is broken into time

1. Value stream map (eliminate steps and operations)
2. Manage bottlenecks (maximize flow)
3. Mixed model operations (minimize setups)
4. Transfer batches (direct lead-time reduction)
5. Apply Lean, just-in-time (JIT), and Quick Response methods (QRM)

Table 4.2 Lead-Time Reduction Methods

elements by work task to identify process waste. As the VSM analysis proceeds, each work task is correlated to the customer value it creates. Understanding what the external customer considers important (of value) has been shown to enhance the ability of the Lean Six Sigma improvement team to differentiate nonessential operations from those which are important to the customer. VSM will be discussed in more detail later in this chapter. In summary, it is important to understand the components of lead-time including contributing factors which lengthen lead-time since lead-time is a major driver of supply chain cycle time which negatively impacts cost, quality and inventory investment. As a general rule the Lean Six Sigma improvement team should focus lead-time reduction efforts on the critical path as its first priority.

Lead-Time Reduction

The list of causes for lead-time variation, shown in Table 4.1, represents the root causes of many Lean Six Sigma projects we have found over the past 12 years. Often, the reasons for the process breakdowns could be grouped into a small number of categories. Interestingly, many of these categories are also called "red flag" conditions in mistake-proofing literature. Red flag means that when these conditions are present the process will exhibit higher error rates due to error conditions and their associated mistakes. Some of the more important categories will be discussed to show how they impact lead-time.

Waiting for any reason directly increases lead-time. If the waiting time can be studied and broken into smaller time elements, opportunities to reduce lead-time will become apparent. Also, since it takes time to move materials or information from one place to another, reducing or eliminating transportation time will positively impact lead-time. Elimination of non-value-adding work tasks and activities will directly save time since they do not have to be performed in the modified system. Activities scheduled in parallel rather than sequentially (or serially) can be done in less time or the maximum time of either activity. It has been shown "transfer batch" systems are much more effective than "process batch" systems relative to lead-time reduction (this concept will be discussed later in this chapter). The application of excessive controls increases lead-time since work cannot be sent to the next operation until it is approved to move. An example would be required sign-offs by supervisors or inspectors. Unnecessarily moving materials or information also increases lead-time. Scrap and rework increase lead-time directly in proportion to the number of units scrapped or reworked. Ambiguous goals, poorly designed procedures, outdated technology, lack of information, poor communication, limited coordination, ineffective training, and unnecessary process or design complexity increase lead-time due to mistakes which require doing the work a second time. Process or product complexity will also increase lead-time. Each of these categories could serve as a good source to identify Lean Six Sigma improvement projects since they are close to the root causes of long lead-time.

Based, in part, on information contained in Figure 4.2 and Table 4.1 as well as Lean principles, there are five major methods commonly used to reduce lead-time. These methods have been applied over the past 30 years across many industries with excellent results. But some methods are easier to implement than others. As an example, process simplification and implementation of a transfer batch system are easier to implement in an organization than either a mixed-model production system or a full-scale Lean system, which requires significant changes in product and process design. The characteristics of the five methods are listed in Table 4.2. We will now discuss each method in more detail.

Value Stream Mapping (VSM)

The operations within a process can be broken into three categories based on the concept of "customer value." These are value-adding (VA), business-value-adding (BVA), and non-value-adding (NVA) operations. Value-adding (VA) operations create the product, service attributes, and features the customer desires. All other operations increase process complexity resulting in higher cost and cycle time. These other operations can be broken into two subcategories. The first category consists of business-value-added (BVA) operations. BVA operations are often hidden from the customer, but they are necessary to ensure that the product or service offerings meet external customer requirements. Examples are back-office invoice processing, updating IT systems, inspection activities, and other tasks necessary to run the process. BVA operations exist because of technological barriers that prohibit their elimination. Although BVA operations may be necessary in the short term, they should eventually be eliminated from the process when technically feasible. The second value category consists of operations that are completely unnecessary from either the internal or external customer viewpoint, that is, non-value-adding (NVA). NVA operations should be eliminated immediately from the process.

An operational assessment is used to distinguish the three categories from each other. An operational assessment analyzes a process starting from customer requirements and working backward through the process

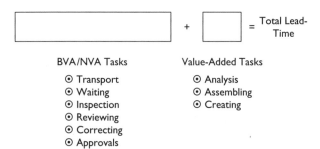

BVA/NVA Tasks   Value-Added Tasks
⊙ Transport   ⊙ Analysis
⊙ Waiting   ⊙ Assembling
⊙ Inspection   ⊙ Creating
⊙ Reviewing
⊙ Correcting
⊙ Approvals

Figure 4.3 Simplifying the Process

toward supplier inputs. This operational analysis takes the form of a highly quantified process map showing all process operations as well as their inter-relationships. Customer value is mapped back through the process in these maps. For this reason it is called a "value stream map (VSM)." The value stream map shows operational inter-relationships including rework loops, inspection points, and interruptions of material or information flow within the process. Figure 4.3 qualitatively shows the logical result of process simplification through elimination of NVA tasks to reduce overall lead-time.

How is a value stream map created by the value-stream-mapping team? The first step is to identify facilitators who will guide the team through data collection and creation of the map. These facilitators lead the team though a simplified training session showing them how to collect data from the process and build the value stream map. The team constructs the value stream map in a visible location on the wall (usually the wall is covered using inexpensive Kraft paper, that is, brown paper) so that people can easily make necessary modifications to the map. A "brown paper" map is shown in Figure 4.4. The map is created using

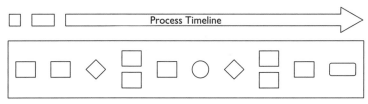

Attach data collection forms, management reports, inspection forms, procedures and so on below value stream map

Figure 4.4 "Brown Paper" Mapping

"sticky notes," which quantify each operation. These sticky notes are spatially arranged on the wall so that the team can identify relationships between each operation. In addition to operational metrics quantifying material flow, the team pastes copies of all management reports just below the sticky notes describing each operation. These management reports include product forecasts, material control forms, inspection reports, testing reports, and schedules as well as other relevant information. As people build the value stream map, facilitators walk the actual process verifying that it operates as people think it does.

The VSM shows the time duration, inventory (or other capacity metrics), setup time, batch sizes, quality levels, and machine uptimes for each operation within the process. Some of these operational metrics were listed in Table 2.3 and the full listing is shown in Table 4.3. The VSM also shows spatial relationships and rework loops between operations. The quantified map identifies the system bottleneck operation as well as capacity-constrained resources within the process. The VSM team uses the current state map as a guide to identify what should be changed in the process to simplify it (future state). The future state map shows the process with all its non-value-adding operations removed. The value-adding operations remaining in the process are eventually optimized relative to cycle time, inventory level, setup time, batch size, quality level, and machine set-up time as well as other relevant operational metrics.

To create the future state map, the team conducts a "what-if" analysis of the current process. This brainstorming session asked the question, "What would be reasonable improvements in the key operational metrics if we applied Lean tools to the current process?" Part of the VSM exercise

---

1. VA/NVA/BVA
2. Production rate (units/minute)
3. Scrap percentage
4. Rework percentage
5. Downtime percentage
6. Capacity (units/minute)
7. Setup time (minutes)
8. Inventory (units)
9. Floor area

Table 4.3 Key Process Metrics

Figure 4.5 Posting Process Information

is identification of areas for operational improvement. As the VSM is being built, the team places colored tags on the areas of opportunity or concern as shown in Figure 4.5. "Green" colored tags indicate process strength. "Red" colored tags indicate process weakness. "Yellow" colored tags indicate that additional information is required or clarification of the operation's metrics is necessary. Relevant Lean tools include calculation of the manufacturing unit cycle time based on external customer demand (takt time will be discussed in later in this chapter), application of standardized methods to work, leveling of work loads, application of mistake-proofing strategies, reducing setup time, application of preventive maintenance methods, and improvements to quality using Six Sigma methods. In the future map, the system should have more scheduling flexibility since batch sizes and setup times become smaller and quality and machine uptime become higher. Also, operations will be balanced relative to each other to maintain system takt time. Several of these concepts were discussed in Chapter 2. In addition to the Lean concepts discussed in Chapter 2, there are other key methods that help simplify a process, reduce lead-time, and reduce inventory investment.

Managing Bottlenecks

Bottleneck resources must be managed effectively to minimize lead-time along the network's critical path to maximize material or information flow through the system. Figure 4.6 shows several variations on this concept. In bottleneck management, material or information flow is balanced for each operation relative to its bottleneck resource. This is because

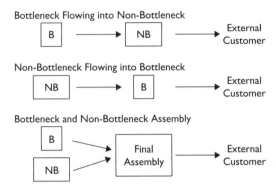

Figure 4.6 Bottleneck Management

the system cannot convert material and labor faster than the bottleneck throughput rate. If upstream operations are run at a production rate higher than the bottleneck's rate, excess inventory will build up in the system. On the other hand, if upstream operations are operated at a rate less than the bottleneck resource, the flow through the process will be suboptimized because the bottleneck will be "starved for work," resulting in reduced system throughput. Figure 4.6 shows four common examples. Reviewing the first example, if two machines are in series, with the first machine (upstream from the second machine) producing 100 units of work per minute, but the subsequent machine (downstream) producing just 50 units of work per minute, the overall production rate is fixed at 50 units per minute by the second machine. The output from the process is 50 units per minute (neglecting addition of second machines running in parallel or running the second machine more than one shift). In this example, the first machine is utilized only 50% of the time, but the second machine is utilized 100%. To manage flow relative to the bottleneck resource, that is, the second machine, the effective output per unit time is set at 50 units per minute (or 1.2 seconds per unit, that is, the required takt time). This operational practice effectively balances material flow through the two machines. The result is a net reduction in work-in-process (WIP) inventory in the system. These examples are also applicable to service industries relative to material or information flows.

Mixed-Model Operations

Operations scheduled using a mixed production model disaggregates the manufacturing schedule into smaller time increments to increase the manufacturing schedule frequency of a product. Disaggregating the manufacturing schedule is done by sequentially building a portion of each product's total schedule by alternating the manufacturing sequence of products having similar product and process designs. The similarity is accomplished through a methodical series of focused changes to product and process design. The resultant design commonality effectively results in reductions of machine or process setup cost (economic lot size) effectively increasing the frequency at which the product can be built. As an example, the modified build sequence of four products "w, x, y, and z," might be "wxyz wxyz wxyz wxyz" rather than the original "wwww xxxx yyyy zzzz" sequence. The mixed-model scheduling methodology results in significant lead-time reductions as well as enhanced scheduling flexibility. As an example, in the original scenario, we may have built each of the four products "w," "x," "y," and "z" by the week. In this scenario, the lead-time for any product would be four weeks. During this four-week lead-time, there may have been an increase or decrease in external customer demand. The operational result would have been either a product shortage (lost sales) or excess product (higher inventory). However, in the mixed-model production scenario, the lead-time would be effectively reduced to just one week since the manufacturing schedule dynamically responds to changes in external customer demand. The mixed-model scheduling strategy results in a better match of the manufacturing schedule to external customer demand patterns. The result will be higher product sales and lower inventory investment.

However, the ability to effectively implement a mixed-model scheduling system is highly dependent on reducing machine setup times (and costs of setup) through commonality of product and process designs to facilitate rapid machine changeovers. To summarize, the important characteristics of a mixed-model scheduling system are commonality of lower-level components in the product design, commonality of the process design due in part to product design commonality, which reduces setup times for all products allowing disaggregating of the manufacturing build schedule. Deployment of mixed-model scheduling systems reduce work-in-process inventories (WIP) and lead-time since the manufacturing system dynamically responds to changes in external customer demand.

Transfer Batches

"Transfer batches" are another effective lead-time reduction method. Using the "transfer batch" concept, a unit of production is transferred to the next operation as soon as it is built rather than batched and transferred later to the downstream operation. Lead-time reductions of approximately

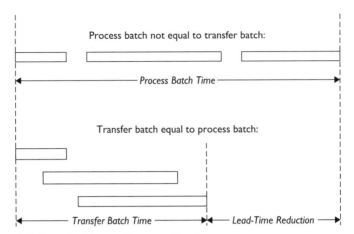

Figure 4.7 Transfer Versus Process Batches

66% are possible using this method. As an example, Figure 4.7 shows three operations in series doing different work on a component. The production rate for each operation is 100 components per minute. The total time to process the 100 components through the three operations (total lead-time through the process) will be different for the "process batch" versus "transfer batch" strategies. Using the process batch system, the total lead-time through all three operations would be 300 minutes, that is, 100 minutes + 100 minutes + 100 minutes = 300 minutes. However, using a transfer batch system, in which a unit of work is immediately transferred to the next operation after the work is performed, will require a total time of 102 minutes. This is a lead-time reduction of ~66%. Of course not every system lends itself neatly to a transfer batch process design, but on the other hand many processes do lend themselves to this method; however, opportunities are lost through inaction.

Integration of Lean Tools and Methods

A Lean system is an integrated set of activities designed to produce continuous flow through a process. In a Lean system, lead-time reductions contribute to continuous flow through the system. However, as mentioned in Chapter 2, Lean systems require day-to-day operational stability as well as many other system improvements to ensure sustainable lead-time reductions. However, external customer demand patterns will impact the improved system to varying degrees. As an example, some industries are characterized by large variations in external customer demand. In these industries, it may be difficult to match manufacturing schedules to variable external demand patterns. The Toyota Production System (TPS) is an often quoted classic application of a Lean system. However, variations in external demand are low in the Toyota Production System. Monthly variation of units (vehicle) demand has been estimated to be in the range of +/− 10%. Because of the stable external demand pattern, the TPS has evolved many state-of-the-art operational improvements including operational stability. These improvements link manufacturing operations, balance material flow across the supply chain, emphasize preventive rather than emergency maintenance, and emphasize reduced incoming and in-process material lot sizes, reduced setup times, multifunctional workers, and significantly improved product quality through modifications of product and process design.

Lean initiatives have been ongoing for several decades and there are many documented case studies of resultant successes and failures. In Chapter 2, Table 2.1 shows typical operational benefits commonly derived from Lean projects. In particular, inventory reductions of 25% to 75% are common. Lean activities, typically represented as Kaizen events, are usually successful in significantly improving operational metrics at a local work-team level. However, a problem with many Lean activities is that they are not strategically deployed throughout the organization. Instead, they are deployed as a series of isolated Kaizen events, which, although useful in generating operational improvements at a local level, are not easily seen at the senior management level as improvements to higher-level financial metrics, including reductions in inventory investment. Strategic alignment of a lean initiative is necessary to fully realize the potential productivity benefits for your organization.

The three major operational elements characterizing Lean systems are just-in-time (JIT) operations, robust quality systems, and creation of standardized work and visual controls. These elements were discussed in detail in Chapter 2 and are shown in Figure 2.3. As previously mentioned, robust quality and stable operations are achieved through a combination of work standardization, mistake proofing, and modifications of product and process designs. Work standardization is particularly important because it documents work flows and procedures. This ensures that work is done the same way all the time to reduce process variation which causes processing errors and defects. Work standardization also lower manufacturing cost. Mistake proofing is a system of simple and effective techniques that detect error conditions and malfunctions within the manufacturing process. These simple controls are automatic and do not require human intervention. Mistake proofing is also usually inexpensive. These three operational elements must be fully integrated with each other to achieve the full benefits of a Lean system. These operational elements also contribute to level–loading the manufacturing schedule facilitating the continuous flow of material and information through the system.

The seven tools, shown in Table 4.4, are used to support the three major operational elements and integrate operational improvements over time to expand operational capability across the supply chain. These seven tools are capacity planning, process mapping and analysis, calculation of takt time, work standardization, mistake proofing, visual controls, and creation of performance measurements. The first tool, capacity planning, is important to meet manufacturing schedules. If capacity is not sufficient to meet the manufacturing schedule, process breakdowns occur resulting in order expediting and higher cost. On the other hand, maintaining excess capacity in the system will increase cost. The second key tool is process mapping which is usually applied in the form of a value stream map (VSM). Process mapping is a useful input to the takt time calculation. It also identifies process waste. Takt time calculations show the production rate (or drum beat) the system must maintain to meet external demand. All system resources should be balanced to the takt time. Most systems require resources in excess of optimum prior to the application of Lean methods to maintain their manufacturing schedules. In other words, the process may initially consume resources that are greater than optimum due to rework loops, maintenance issues, and demand variation and so on until a Lean system has been fully implemented by the organization. The long-term goal should be to balance manufacturing flow to meet the required takt time using an optimum input resource level. We will discuss takt time calculation later in this chapter.

Visual controls, which were discussed in Chapter 2, are very effective at maintaining operational control since real-time system status can easily be seen by the local work team. As an example, many manufacturing

---

1. Capacity planning (resources to meet demand)
2. Process mapping (operational spatial relationship)
3. Takt time calculation (production per time)
4. Standardization of work (used to balance flow)
5. Mistake proofing (stabilize takt time)
6. Visual controls (maintain flow)
7. Performance measurements (constant improvement)

Table 4.4 Building a Lean Program

organizations mark off places where inventory must be placed for easy access by manufacturing personnel. In these systems, it is easy to see if the inventory is available simply by looking at its assigned location on the manufacturing floor. Another example is the use of display boards showing system status. In these systems, manufacturing metric boards display hourly production and quality status. As a final example of visual controls, in service operations such as call centers visual controls are used to display incoming calls, calls not answered on time, and average time to answer a call, in order to shift agents from one call category to another to maintain efficiency as well as customer service levels.

Reducing Lead-Time (Example)

We have discussed many ways to reduce a system's lead-time once the basic operational elements are clearly defined and spatially organized to create a system network or value stream map of the process. This network is equivalent to a *project evaluation and review technique (PERT)* chart in which every operation is characterized by average duration and standard deviation of each work task. Using these statistics, the overall lead-time can be calculated, in a probabilistic sense, across the network. In particular, the lead-time along the system's critical path must be estimated carefully. In addition to building and quantifying the network, every operation should be characterized relative to customer value. This methodology provides the improvement team with a significant amount of information that is useful to reduce lead-time along the critical path.

In Figure 4.1, operations associated with unnecessary movement, waiting, work setup, inspection, and idle time are clear candidates for immediate elimination.

Figure 4.8 shows a simple example of this concept. Initially, the product is assembled in 7.83 minutes (470 seconds) using the current process design. This total assembly time represents all the operations within the process. These include value-adding as well as non-value-adding operations. The goal of the Lean Six Sigma improvement team is to maintain manufacturing rates (later to be called the system takt time) and systematically reduce process complexity and resource requirements into the system over time. This methodology will decrease system lead-time, decrease cost, and improve quality. As an example, using the quantified process map shown in Figure 4.8 as a guide, it is apparent that operations "L" and "M" should be eliminated because they are NVA activities related to inspection and the unnecessary movement

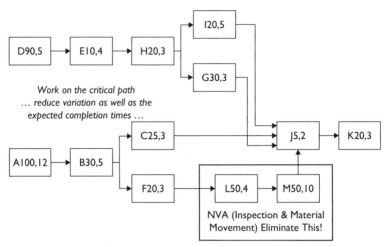

Figure 4.8 Current Process

of materials. These operations represent 100 seconds of time. However, the assumption for doing this is that immediate operational improvements can be put in place to allow elimination of these non-value-adding tasks.

Takt Time/Resource Calculation

A second assumption is that the required daily production requirement is 600 units. These 600 units must be assembled over an 8-hour shift (480 minutes). The 480 minutes is reduced by 60 minutes to account for lunch and rest breaks to provide an operational availability time of 420 minutes or 25,200 seconds available to manufacture 600 units. Takt time (time to complete one unit) is calculated using Figures 4.8 and 4.9; the takt time "T" is calculated as the time available to manufacture one unit. This is calculated as 25,000 seconds divided by 600 units to yield 42 seconds per unit. This means every 42 seconds a unit must be manufactured by the process. The takt time is also called the "drum beat" of the system. Using the takt time of 42 seconds per unit, the theoretical (minimum) number of operations (also people or workstations) is calculated as 9 workstations. It is usually very difficult to achieve the calculated minimum number of workstations because of the spatial relationships between operations and their work tasks. Figure 4.10 and Table 4.5 show that 12 work stations are currently necessary to maintain

1. Using lean methods NVA (100 seconds) was eliminated reducing assembly time per unit to 370 seconds.

2. The required cycle time (takt time) per unit to achieve the required output is:

$$T = \frac{\text{Production Time Per Day}}{\text{Output Per Day}} = \frac{60\ \text{Sec/min} \times 420\ \text{min/day}}{600\ \text{Units Per Day}} = \frac{25{,}200}{600} = 42\ \text{Sec/unit}$$

3. The theoretical minimum required process steps/stations (operators) is:

$$S = \frac{\text{Time}}{T} = \frac{370\ \text{Seconds}}{42\ \text{Sec/unit}} = 8.8 \sim 9\ \text{Steps/Stations (Operators)}$$

Figure 4.9 Solution

the takt time of yz sec/unit (in the absence of major changes as to how the work is performed or the spatial relationships between operations). The required 12 workstations versus theoretical minimum of 9 workstations results in an initial operational efficiency of just 73.4%.

Implementing Process Improvements

Calculating the baseline process efficiency of 73.4% is the first step in the analysis. Over time, the resources required to maintain the system takt time are progressively reduced through systematic application of Lean Six Sigma improvements projects. Some common improvement

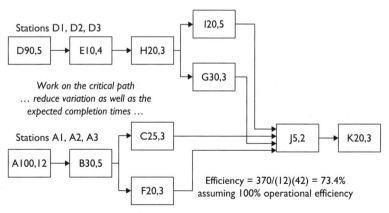

Figure 4.10 Operation Balancing

Station	Task Time	Idle Time
A1	33.33	8.67
A2	33.33	8.67
A3	33.33	8.67
C/F	40.00	2.00
G	30.00	12.00
D1	30.00	12.00
D2	30.00	12.00
D3	30.00	12.00
E	30.00	12.00
H	25.00	17.00
I	20.00	22.00
J/K	25.00	17.00
Total	360.00 seconds	177.00 seconds

Table 4.5 Data Summary

tools and methods that will help reduce resources and improve quality are creation of Lean Six Sigma problem-solving groups, deployment of Lean Six Sigma green belts within the process, cross-functional training of employees, deployment of total preventive maintenance (TPM) systems, deployment of Lean methods including 5-S, elimination of unnecessary tasks, equipment upgrades, and improved process layouts. To the greatest extent possible, Lean methods should be used to simplify and stabilize the process prior to attempting to improve it using Six Sigma or other quality improvement methods.

Table 4.6 lists several additional ways to improve operational efficiency. Excess capacity ensures that the system is flexible and can respond to changes in customer demand or unexpected problems. However, maintenance of excess capacity might not be possible in industries in which capital is very expensive. High-quality maintenance and effective training will ensure that work tasks are done right the first time, eliminating rework and scrap problems, which increase lead-time and increase cost. Depending on the manufacturing volume, system complexity, and external demand patterns, it may be possible to implement quick response systems (QRS) in which bills of material are partially collapsed to provide local work cells with the ability to dynamically respond to minor changes

- Maintain excess capacity in the system
- Improve quality, maintenance, and training
- Create product family processes (collapse MRP)
- Use multifunction equipment
- Measure lead-time, quality, and uptime
- Simplify processes
- Use multiskilled workers
- Only accept orders you can complete
- Make to order with no excess
- Partner with a few suppliers
- Share demand data in real time

Table 4.6 Ways to Improve Operational Efficiency

in either work flow due to manufacturing problems or external customer demand. The more multifunctional equipment can be used in portions of the process, the more robust the system will be relative to problems since if one machine fails another machine can take its place. The ability to improve operational effectiveness also depends directly on an organization's ability to measure key metrics related to time, quality, and cost (as well as other factors). Process simplification always reduces lead-time and cost and improves quality. Multiskilled workers allow reductions in total direct labor and increase the ability of the organization to respond dynamically to changes in either work flow or external customer demand. Many problems related to high operational costs are due to abrupt changes to the manufacturing schedule, which result in excess direct labor cost and problems related to work setup including increases in rework, scrap, and expediting expenses. Making orders according to schedule is important since not making them on schedule results in expediting expenses and many other hidden costs. Partnering with a small number of suppliers reduces system complexity and cost. Finally, understanding external customer demand and creating realistic manufacturing schedules based on this analysis is crucial to ensuring customer satisfaction and maintaining continuous flow through the system.

Maintaining Excess Capacity

It is important to discuss maintenance of excess capacity in a system in more detail. The fact that system capacity is an important contributor to

Machines	% Capacity	Arrival Rate	Service Rate	Units In Queue	% Idle Time	Waiting Time
1	57%	20 units/hr	35 units/hr	0.76	43.0%	3 minutes
"	85%	30 units/hr	"	5.14	14.0%	10 minutes
"	94%	33 units/hr	"	15.56	6.0%	28 minutes
"	97%	34 units/hr	"	33.62	3.0%	58 minutes
2	85%	30 units/hr	"	.19	40.0%	.38 minutes
"	94%	33 units/hr	"	.27	35.9%	.38 minutes
"	97%	34 units/hr	"	.30	34.6%	.48 minutes

Table 4.7 Queuing Analysis Study of Capacity Utilization

lead-time reduction is not really obvious to most people. In fact, few people realize that the impact of capacity on lead-time is a nonlinear relationship. That is, small changes in capacity have a disproportionate impact on lead-time. This situation is exacerbated as the system becomes more complex. The queuing analysis summary shown in Table 4.7 demonstrates the impact of capacity utilization on lead-time. Table 4.7 shows that increased capacity utilization at a bottleneck, typical of the end-of-the month push, results in an exponential increase in cycle time (waiting time queue). Improvements in quality, maintenance, and training ensure that the work is done right the first time, minimizing wasted material and labor will benefit capacity utilization.

Summary

Since lead-time is a major factor in setting inventory investment targets and directly impacts operational efficiency, attacking the root causes for long lead-time will always be a good source of Lean Six Sigma improvement projects. Therefore, understanding the time components and major factors contributing to lead-time is important. The best way to reduce lead-time is through an integrated and strategically aligned Lean Six Sigma deployment rather than a series of isolated Kaizen events or other improvement efforts.

A sequential deployment of successive Lean tools and methods is required to build organizational competence to ensure the Lean initiative will be effective over time. As an example, standardized operations and high quality are necessary to ensure a stable process and maintain its takt time. A stable process is also required to implement Six Sigma quality improvements since the process baselines must be estimated for Six Sigma projects. Takt time adherence is also the philosophical basis of continuous flow through a system. Adherence to takt time also drives continuous improvement in the system as excess resources are identified and eliminated from the process.

Chapter 5

Lean Six Sigma Applications to Materials Requirements Planning (MRPII)

Key Objectives

After reading this chapter, you should understand why master requirements planning (MRPII) and master production schedule (MPS) systems are important elements of a supply chain, and understand the following concepts:

1. Why understanding your organization's MRPII and MPS systems will enable your Lean Six Sigma "belts" to identify improvement projects in receiving, production activity control, purchasing, and inventory investment are related to MRPII system areas.

2. Why understanding the comparative advantages and disadvantages between make-to-stock, assemble-to-order, and make-to-order systems will allow your improvement team to identify opportunities to both improve internal processes and reduce inventory investment.

3. Why analyzing system capacity, especially at bottlenecks and system-constrained resources, will enable your team to increase material throughput rates to increase revenue, and the advantages of moving from push to pull scheduling systems to reduce operational costs and inventory and improve quality.

Material Requirements Planning (MRPII)

Materials requirements planning (MRPII) systems use demand and inventory information and other relevant system information to develop manufacturing schedules for all products including their subcomponents. The purpose of the MRPII system is to balance available system capacity against expected demand across the supply chain. System imbalances due to inaccurate demand projections or inefficient use of capacity will contribute to operational breakdowns and excessive inventory investment as well as low customer service levels. Figure 5.1 shows the interrelationships between MRPII elements and related system functions. The planning process starts with the organizational strategic plan. The strategic plan is disaggregated by product group/facility based on expected product demand over time. The expected demand is fed to the master production schedule (MPS). The MPS integrates demand information with the rough-cut capacity-planning module to feed the material requirements planning module (MRPII). Inventory status and bill-of-material (BOM) information are also fed to the MRPII module. The BOM is linked to the current revision level of the product's design.

The MPS impacts the MRPII system in the areas of material planning, receipt of materials and subcomponents, and purchasing, and through the production activity control (PAC) activities. This is shown in Figure 5.2. As the MPS changes over time due to fluctuations in external demand,

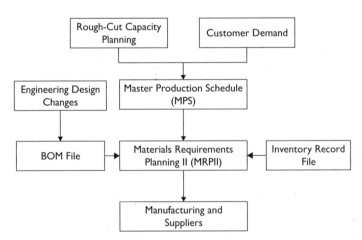

Figure 5.1 MRPII System Overview

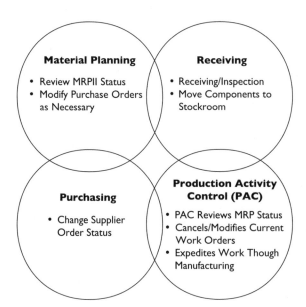

Figure 5.2 MPS Changes Impact MRPII

material planners review the MRPII status. This review also includes open supplier purchase orders. Manufacturing schedule changes may necessitate changes to the supplier purchase order subcomponent quantities. Receiving will then be impacted, which necessitates modifications to the receiving plans. Finally, the PAC may have to modify or cancel current work orders or expedite them through the system.

There will be process breakdowns because of the complexity of these systems including numerous manual interventions by material-planning analysts. These breakdowns include missing or inaccurate information within the system. This missing or inaccurate information impacts inventory, capacity, purchasing, and related organizational functions. However, Lean Six Sigma projects can be identified and deployed around these process breakdowns. Alternatively, these process breakdowns may eventually be found to be the root causes of other Lean Six Sigma projects. There are several ways to analyze these MRPII or MPS breakdowns starting from observed gaps in operational performance or more proactive analysis using inventory models and targeting areas having excessive levels of inventory investment (low inventory turns) for analysis by the Lean Six Sigma improvement team.

MRPII High-Level Functions

MRPII has several high-level functions to coordinate the manufacture of products. These tasks are accomplished by effective planning and control of all system resources. Closed-loop coordination by the system ensures integration of material requirements for the entire business enterprise across diverse supply chain functions such as manufacturing, marketing, finance, engineering, and purchasing. In addition, the MRPII functionality facilitates coordination of the supply chain through use of three key reports. These reports are "planned orders," "order release notices," and "open order changes." The planned orders report provides information on the release of an order at a future date based on subcomponent cumulative lead-time offset. This report helps ensure that all materials and subcomponents, at least in theory, are available at the right time and place to execute the master production plan (MPS). The order release notices report provides information on execution of the planned order function. Open order changes are required to reschedule the manufacture of products due to demand and lead-time variation. However, excessive demand variation can wreck havoc on the MRPII system causing poor coordination of materials and resources throughout the supply chain.

Another critical function of the MRPII system is to coordinate the manufacturing plan for products over their forecasting time horizon. In this context MRPII is both a planning and control system. Control and planning can be evaluated at "long-range," "medium-range," and "short-range" time horizons. These operational planning horizons are shown in Figure 5.3. Long-range planning goes out as far as 2 to 10 years or more depending on the specific industry. It includes strategic, aggregate, and facility/process planning. Medium-range planning goes out between 3 to 24 months. It impacts the MPS and related order-scheduling processes. Short-range planning is coincident with weekly and daily planning and roughly corresponds to subcomponent lead-time. It impacts both weekly and daily workforce scheduling. Each planning level is concerned with managing labor and resource capacity to meet projected demand at its relevant planning time horizon. Planning and control, in the context of a scarce resource environment, force prioritization of the projected work schedule. Inefficient scheduling wastes capacity and increases lead-time and cost. Also, effective real-time coordination may be poor because the MRPII build schedule must be simultaneously executed for several hundred or thousands of subcomponents. This is a major reason for the

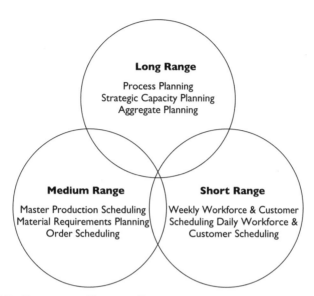

Figure 5.3 Operations Planning Overview

occurrence of excess or insufficient inventory. Generally, this situation has a negative impact on inventory investment. It is difficult for these complex and integrated systems to respond to real-time changes in the manufacturing environment.

Planning and control of work depends on the specific industry. Figure 5.4 breaks planning and control along two dimensions. These

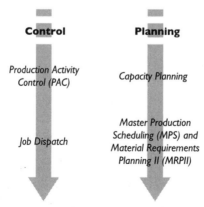

Figure 5.4 Planning Versus Control

Figure 5.5 Production Planning

dimensions are related to planning/control versus scheduling/job dispatch. Planning and control will follow one of three types of workflow patterns depending on the specific industry. These planning systems are shown in Figure 5.5 and cross-referenced by workflow pattern. The three workflow patterns are *make-to-stock, assemble-to-order,* and *make-to-order.*

Make-to-stock workflows build finished goods inventory. This is usually caused by insufficient manufacturing capacity to build all products on demand. In these environments, the product mix is often large and diverse. In make-to-stock systems, the MPS pushes the demand forecast of the product through the MRPII system. Problems occur when changes in external demand patterns are inefficiently matched by the modified manufacturing schedule. These schedule mismatches may result in late orders and excessive inventory. To mitigate this problem, Lean systems attempt to substitute real-time control of schedules in place of the MRPII system. The critical characteristic of these Lean systems is visual control of manufacturing by matching resource capacity to demand in real time. These visual controls are called Kanban systems. Kanban systems may be either manual (physical cards) or electronic (hybrid MRPII and card systems).

Assemble-to-order workflows are designed to minimize finished goods inventories by differentiating the final product as late as possible in the manufacturing build sequence. The ability to accomplish this goal is directly dependent on the degree of subcomponent design/process commonality. The product mix, in these make-to-order systems, is often less diverse than in the finished goods system. However, raw material, subcomponent, and work-in-process (WIP) inventories still exist due to variations in demand and lead-time.

The third pattern, make-to-order workflow, results in minimum levels of finished goods inventory. This is because demand estimates are based on firm external customer orders or contracts rather than a product forecast. However, make-to-order systems require raw material subcomponent and WIP inventory since lead-times are not zero. Deployment of Lean systems and bottleneck management will minimize inventory investment in all these systems.

Capacity Requirements

Capacity can be thought of as the specific labor and machine resources available at a particular time and location to build a product. Inventory can be thought of as reserve capacity since labor, materials, and machine time has been invested to manufacture the inventoried subcomponent or product. There are three common types of capacity. These are "design capacity," "available capacity," and "actual capacity." Design capacity is the maximum system capacity (quantity per unit time). It is calculated without counting the losses incurred by inefficient scheduling, material shortages, scrap, rework, or other process issues. One way to think about design capacity is to imagine a day in which everything in the system works well. Available capacity is the average design capacity impacted by everything except losses due to scrap or rework. The third capacity designation is actual capacity. Actual capacity takes into account losses due to scrap and rework as well as commonly recurring manufacturing problems. Actual capacity is estimated at the facility, work center, and machine level. The MRPII system uses these actual capacity estimates

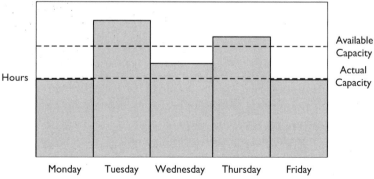

Figure 5.6 Capacity

Figure 5.7 Bottleneck Identification

to develop the manufacturing schedule for each product line. Material planners may modify the manufacturing schedules including labor and equipment requirements based on any changes.

Older MRPII systems use backward scheduling to *infinite load* the system resources relative to the manufacturing schedule. Infinite loading the system resources results in unbalanced workflow because some resources manufacture at a faster rate than others. Practically, the system cannot manufacture finished products at a rate higher than the slowest resource, but newer MRPII systems can *finite load* the system. In finite loading, system capacity is allocated so that work loads do not exceed available work center capacity. This ensures a higher probability of schedule completion. More advanced systems using optimization algorithms schedule work relative to the system bottleneck resource. This ensures continuous material flow since the slowest operation in the system is the bottleneck resource. In Figure 5.7, stage 2 is the bottleneck resource because its hourly production rate is less than either the upstream or downstream operations. The rest of the system resources are balanced relative to the capacity of the bottleneck resource. In Figure 5.7, the production rates of stages 1 and 3 would be set at 2,500 units per hour. Although finite loading attempts to balance work flow and capacity through the system, external demand and lead-time must be fairly stable over the planning horizon. This is particularly true relative to lead-time. Optimization of the system requires flexible workers, plants, and processes.

Increasing Capacity

Capacity utilization varies by industry. In process industries such as paper, chemical refining, and so on, characterized by heavy capital expenditure, capacity exceeds 95% of available design capacity. In discrete-part-manufacturing industries, optimum capacity should vary between 80% and 95% of design capacity. There are many operational strategies that will increase capacity. Many of these strategies

Manufacturing	Service
■ Match Resources to Demand	■ Match Resources to Demand
■ Encourage Self-Service	■ Encourage Self-Service
■ Replicate Lessons Learned	■ Replicate Lessons Learned
■ Remove Operational Inefficiencies	■ Remove Operational Inefficiencies
■ Modify Product Design	■ Modify Service Package
■ Cross-Train Workers	■ Cross-Train Workers
■ Facility Modification	■ Facility Modification

Table 5.1 Increasing Capacity

are short-term and relatively easy to implement while others are more capital-intensive and longer-term. Table 5.1 lists some of these strategies. Although the current discussion is focused on manufacturing, it has analogous applications to service industries. Table 4.7, in Chapter 4, showed lead-time increase nonlinearly as available capacity utilization increased in the system.

Many of the strategies listed in Table 5.1 are also found in Lean Six Sigma deployments. To elaborate more on several of these strategies, simple modifications to equipment will often result in higher capacity. Elimination of manufacturing rework and scrap will directly increase capacity because fewer products must be manufactured to meet the original manufacturing schedule. Design changes to the product or process, especially simplifications in design, will reduce the number of components that must be scheduled and built. Although there are many other ways to increase capacity, these examples will suffice to show how capacity can be increased in most systems. Inexpensively increasing capacity does not force an organization into building inventory to satisfy the product schedule. One last comment is important: these improvements must be made to the bottleneck and capacity-constrained resources. In particular, the system bottleneck controls the workflow through the system. By definition any other resource in the system has excess capacity relative to the constraining resources. However, it is also important to analyze the impact of adding capacity at the bottleneck resource on other capacity-constrained portions of the system since one of them will become the next bottleneck.

In addition to process improvements, increases in actual system capacity result from both higher levels of organizational experience, that is, experience-curve and volume increases, such as economies of scale. These situations shift a facility's optimum operating level as shown in Figure 5.8. It has been shown that, as an organization gains experience building a product, the time to build the product and the manufacturing cost decrease. This is due to a myriad of activities that unfold over time to improve product and process design. All other things being equal,

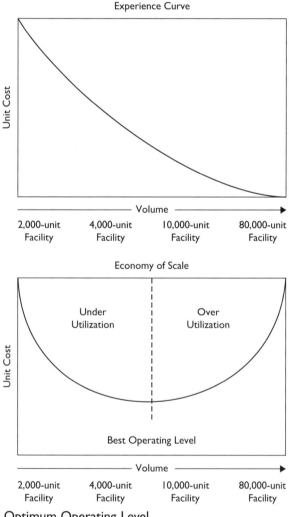

Figure 5.8 Optimum Operating Level

increases in demand and its corresponding volume increases allow an organization to allocate overhead costs over larger production quantities, effectively lowering operational cost. Higher production volumes also allow more effective utilization of capacity.

MRPII-Specific Functions

Using the MPS, the MRPII system creates production schedules identifying specific subcomponents and materials required to manufacture the product (end item). The MRPII module calculates exact quantities of each subcomponent using the BOM file, inventory file, and subcomponent build schedule. Critical information includes the subcomponent release dates and the materials required for their assembly. Subcomponent release dates are calculated using cumulative lead-time offsets working backward from the required shipment date of each product. This is called backward scheduling and is associated with "demand push" systems. The BOM file, through the MRPII system, serves as the link between the finished product (end items or independent demand items) and subcomponents (dependent demand items). The material planners reviewing the manufacturing schedule, created by the MRPII system, make adjustments to the schedule if demand patterns or resource availability change.

Bill-of-Materials File (BOM)

The BOM file specifies the product structure tree. The BOM includes information on materials, parts, subcomponents, production sequences, and subassemblies as well as their hierarchy to build the finished product. Figure 5.9 shows an example of assembling an automobile.

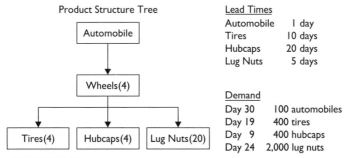

Figure 5.9 BOM Structure

To construct the BOM, engineering develops the automobile's build sequence based on its product design. In this simple example, each automobile requires four wheels. Each wheel requires four tires, four hubcaps, and 20 lug nuts to bolt the wheels to the automobile. Using the BOM file as well as the demand quantity from the MPS, that is, 100 automobiles, the MRPII module explodes requirements across the supply chain. These requirements are now a multiple of 100.

Production Activity Control (PAC)

PAC executes the MPS at the shop-floor level using either a push or pull system. Push systems drive workflow using MRPII lead-time offsets and the MPS (product forecast). Pushing work through the system results in an inability to dynamically respond to variations in either external demand or resource availability. This situation results in inefficient changes to the production schedule causing order expediting as well as breaks in the current manufacturing schedule. Manufacturing schedule changes increase the number of machine setups and waste system capacity because of an inability to easily modify subcomponent schedules when the product forecast changes or resources become unavailable. Figure 5.10 shows how PAC has been adapted to perform in push and pull systems, which are very different environments. Notably, PAC has been modified to work in low-volume environments using visual control methods. These systems use Kanban or modified versions, that is, Kanban integrated with MRPII to respond dynamically to changes in external demand and resource availability.

Push Systems

- Work is driven by MRP lead-time offsets driven by the MPS (*Forecast*).
- Unnecessary work is performed as demand fluctuates or system components fail.
- System capacity is wasted.

Pull (Flow) Systems

- Work is performed in response to actual demand.
- Only required work is performed based on need.

Figure 5.10 PAC Environments

Scheduling Techniques

There are three basic types of scheduling systems. "Backward scheduling" and "forward scheduling" systems are driven by MRPII. "Flow control scheduling" is driven by visual, that is, Kanban or other simple real-time control systems. Backward scheduling uses MRPII to establish order completion dates based on MPS demand quantities that have been offset by calculating the cumulative lead-time of all subcomponents necessary to build the finished product. Forward scheduling starts with the first operation and then, using cumulative lead-time offsets, forward-builds an estimate of the order completion time. Flow control scheduling uses visual signals (Kanban) as well as other Lean methods to balance and smooth material flow through the system according to the calculated takt time based on external customer demand in real time.

Changes in the Master Production Schedule (MPS) require the MRPII system to cancel or open purchase orders. In some cases, orders must be expedited if there is resource unavailability. Making these changes to the schedule is a costly and time-consuming process often resulting in overall schedule erosion due to lost capacity. This makes the MRPII a poor system to control flow through the process. On the other hand, Kanban is an excellent flow control system but requires manual intervention for planning and execution. Also, it becomes unwieldy as the number of products increases or if demand varies significantly. MRPII is a good planning tool but a poor execution tool, and Kanban is a good execution tool but a poor planning tool; therefore, these systems are sometimes integrated to ensure smooth scheduling in complex MRPII and MPS environments. These hybrid systems integrate MRPII with Kanban flow control and will be discussed later in this chapter.

Scheduling of work is dependent on the types of products being manufactured as well as their process design. Job shops make to order in relatively low volumes. Demand patterns in these environments are discontinuous and irregular. Product routings tend to be variable from one product to another in these environments. Other process characteristics include push scheduling using MPS and MRPII systems, resulting in long lead-times to process orders. In contrast, intermediate- or high-volume manufacturing environments are characterized by make-to-stock or assemble-to-order systems. Demand in these systems is more continuous than in job shop environments, but these systems are also characterized by push scheduling systems. Pull scheduling systems can be implemented over time if demand variation is low. Hybrid systems have also been created by collapsing the bill-of-material (BOM) to provide more scheduling autonomy at a local work-cell level. Depending on the effectiveness of the scheduling system in intermediate- and high-volume environments, the work-in-process lead-time through the system could be longer or shorter.

Relative to scheduling in either distribution center or manufacturing environments, several inputs go into the scheduling decision including specific customer requirements, that is, prioritization of orders as well as the availability of people, materials, and equipment to manufacture or assemble the order. Effective scheduling is important since many orders or jobs compete for these limited resources. Additional factors that enter into the scheduling decision, in addition to resource status, are the complex inter-relationships between resources and demand on the system including resource efficiencies, probabilistic arrival and completion times at work centers, and operational factors such as setup times, worker training, and the rework and scrap associated with each type of job as well as available capacity of each work center.

Forward (push) scheduling and backward scheduling (reverse of forward scheduling but still a push system) systems were discussed earlier in this chapter. These systems are called infinite-loading systems since they assume that the capacity status of the system relative to work centers has not changed since the schedule was originally calculated by the MRPII system. A better system to schedule work at a work center or cell is a pull system since the material flow through each work cell remains balanced relative to external customer demand, that is, the system's takt time, as well as the real-time capacity of associated work cells. However, the ability to fully implement a pull system in MPS and MRPII environments is limited where there are hundreds or thousands of products each with different product designs.

However, there are other scheduling systems that attempt to capture the real-time capacity available in the system but with varying degrees of effectiveness. Scheduling according to the real-time capacity of the system is important if a job cannot be completed due to unexpected interruptions in available system capacity including labor, materials, and equipment; why start the job and leave it half finished? Scheduling systems that take into account the limited capacity of the system include finite-loading systems (but still in push environments), management of bottlenecks (constraint management) and just-in-time systems (JIT). JIT systems are normally part of larger Lean deployments. In Chapter 2, Figure 2.6 shows

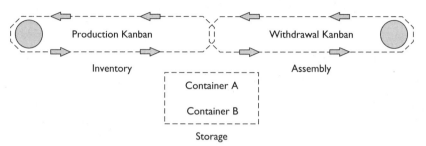

Figure 5.11 Kanban Systems

that Lean deployments require high-quality JIT systems as well as standardized operations to maintain scheduling control of the system.

What Is Kanban?

Kanban is a Japanese word meaning "card" or "signal." Accurate estimates of lead-time and demand are important to determine how many card sets are required to minimize WIP in discrete operations. There are three basic types of Kanban cards. These are the "production card," the "move card," and the "supplier card." A generic Kanban system using these cards is shown in Figure 5.11. The production card authorizes the replenishment work cell to manufacture a subcomponent to the specified container quantity. The container quantity is proportional to the required demand over lead-time divided by the Kanban quantity. The move card signals the upstream replenishment work cell to move a specified quantity of an item downstream to the consuming work cell. The supplier card signals the subcomponent cell to obtain a specified quantity from an external supplier. Kanban quantities are multiples of the container quantities. Each container represents the minimum operational production lot size.

Reorder Points

The reorder point is the time when a new order is placed for Q_i units. It is the sum of the expected number of subcomponents required during the order lead-time (time to receive an order once it has been requested) and safety-stock quantity. A graphical depiction of the reorder point is shown in Figure 5.12. Figure 5.12 is also a graphical depiction of a perpetual inventory model (PIM). The PIM model will be discussed in detail in Chapters 6 and 10. In Chapter 10 an example of the PIM model will be discussed using simulated data. Relative to the reorder point calculation, assume that the lead-time to receive a new order for subcomponents

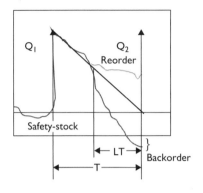

Q is in units and is determined to minimize total costs associated with ordering and holding the model

Reorder is the reorder point in units

Reorder = dLT + $z\sigma_t$

d is the average daily demand

LT is the lead-time in days (place order to receive items) ...adjust to monthly level in I_{opt} calculation

z is the number of standard deviations for a specified service level

σ_t = standard deviation of demand and lead-time in months

$$\text{Inventory}_{Optimum} = 1/2 \ \text{Inventory}_{During\ Order\ Cycle} + \text{Safety Stock} = 1/2Q + [\ z*\ \sigma_t]$$

Figure 5.12 Perpetual Inventory Model

once requested by the MRPII system is 10 days, and an average of 10 subcomponents per day will be required during the 10-day lead-time. Also assume the required safety-stock quantity is 30 subcomponents. The calculated reorder point = (10*10) + 30 =130 subcomponents. At this reorder point, the calculated order quantity is calculated as Q_i minus the available inventory quantity. Depending on how the order quantity has been specified within the MRPII system, Q_i might equal the EOQ or be determined dynamically as the inventory quantity required to bring the inventory up to an amount Q_1 as shown in Figure 5.12. Given information relative to annual demand, ordering costs, inventory holding costs, and related variables, the EOQ system calculates the maximum inventory levels, average inventory levels, annual inventory holding costs, and annual ordering costs as well as the EOQ quantity. Calculation of the EOQ is shown in Figure 5.13.

Calculating reorder points and order quantities from a manufacturing viewpoint requires a modification of the EOQ formula. Unlike EOQ calculations, which assume orders are placed with external suppliers and arrive as one lot or batch, manufacturing systems supply inventory over the year at a predetermined rate, that is, once every several weeks. In these economic production-lot-size (EPQ) systems, inventory is added to the system at a constant rate due to manufacturing and is removed from the system at a constant rate due to external customer orders. Given information relative to annual demand, setup costs, inventory holding costs,

$$Q_i = \sqrt{\frac{2 \text{ (Annual Unit Demand) (Cost per Order)}}{\text{Annual Holding Cost per Unit}}}$$

Figure 5.13 Economic Order Quantity (EOQ)

and related variables, the EPQ system calculates the maximum inventory levels, average inventory levels, annual inventory holding costs, and annual setup costs as well as the EPQ quantity. Calculation of the EPQ is shown in Figure 5.14.

Reorder points are also the basis of Kanban systems. The major difference between MRPII reorder points versus Kanban quantities is visual control of the reorder point quantity using Kanban cards in the latter systems. Figure 5.15 shows that the Kanban quantity is based on a PIM model since it uses lead-time, expected demand, and safety-stock estimates to set the Kanban quantity. A minor difference is that in the Kanban system, the safety stock is estimated as percentage of total inventory level and systematically lowered as system performance improves; however it could also have been set using the formula in Chapter 6, Table 6.1, which is equivalent to the one shown in Figure 5.12 as well as the actual safety-stock formula shown in Chapter 6, Figure 6.4. Classical Kanban systems require manual intervention to set the number of cards, supplier quantities, and subcomponent levels if external demand patterns change.

Hybrid MRPII Systems

Hybrid systems have been created to implement Kanban in MPS/MRPII environments using subcomponents from each system. These hybrid systems are sometimes called "MRPII-Driven Kanban." They are built using integrated EDI, MRPII, bar coding, and warehouse management systems *(WMS)* as well as other related systems. Hybrid MRPII systems are designed to allow the MRPII system to verify preliminary Kanban lot

$$Q_i = \sqrt{\frac{2 \text{ (Annual Unit Demand) (Cost per Order)}}{(1 - \text{Annual Demand/Annual Production Rate) (Annual Holding Cost per Unit)}}}$$

Figure 5.14 Economic Production Quantity (EPQ)

$$Kanbans = \frac{Expected\ Unit\ Demand\ During\ Lead\text{-}Time + Safety\ Stock}{Container\ Size}$$

Figure 5.15 Kanban Calculation

sizes against accumulated demand and then electronically transfer these requirements to the manufacturing work cells in real time. This enables real-time capacity and resource requirements planning. Important items of information used by the modified system to estimate Kanban lot size and manufacturing schedules are expected average daily demand for the subcomponent; its lead-time to manufacture; the number of sub-components the container will hold, that is, container size; the required safety-stock level (as shown in Figures 5.12 and 6.4 as well as Table 6.1); calculated direct labor requirements; operational efficiencies; and MRPII gross requirements net on available inventory.

Figure 5.16 shows a generic version of an MRPII/Kanban system and how these required inputs are used by the system. The MRPII gross requirements are used to estimate work-cell labor requirements as well as the Kanban lot size. The estimated Kanban lot size should meet expected MRPII demand for the time period under review. If the Kanban lot size cannot be increased sufficiently to meet projected MRPII demand for the time period under review, then the algorithm should create an exception or error message for the analyst. A key advantage of these hybrid systems is electronic access to order quantity information across the entire supply chain. In these systems, order quantity information is electronically translated into Kanban quantities and multiple container equivalents as the system generates the MRPII schedule. To ensure proper system functioning, it is important to have relatively smooth demand patterns. For this reason, the best candidates for these systems are normally high-volume products having excellent subcomponent quality history and relatively stable demand patterns.

Lean Six Sigma MRPII Applications

The best way to ensure strategic alignment of your projects is to work backward from the high-level problem identified by your project charter into the system under analysis to understand the root causes driving the problem. Using this problem-solving methodology in the context of the 10-Step Solution Process will allow Lean Six Sigma black belts and green belts to find the specific reasons for process breakdowns in order

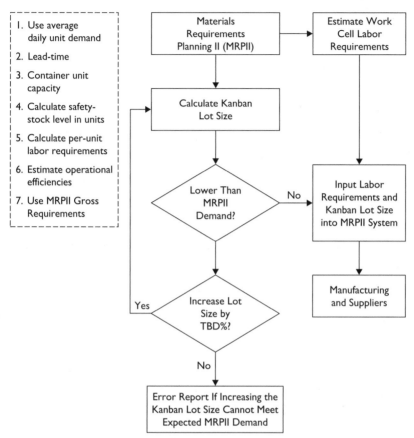

Figure 5.16 Hybrid MRPII Systems

to eliminate these higher-level problems. Lean Six Sigma improvements have been undertaken throughout the MRPII function and its related systems. Table 5.2 lists several areas where Lean Six Sigma projects can be identified by your improvement team. These areas of opportunity might serve as the basis for your project's problem statement and objective. As an example, in the case of "incorrect purchase orders," where does the real problem lie? Is it with the external supplier who failed to manufacture on time or schedule correctly? Or is it with the customer's organization and its systems? "Cancelled or modified work orders" is another example in which there was a communication breakdown within the supply chain. Why was the work order cancelled? How chronic is this problem? What is the business impact of this problem? "Expedited orders"

1. Reduce incorrect purchase orders	7. Improve labor and material inefficiencies
2. Reduce canceled or modified work orders	8. Reduce BOM errors
3. Reduce expedited work orders	9. Increase capacity
4. Reduce receiving inspection issues	10. Increase scheduling efficiency
5. Reduce unnecessary movement of materials	11. Improve order release and planned order report accuracy
6. Improve inventory accuracy	

Table 5.2 Lean Six Sigma MRPII Applications

are another example where available system capacity is reduced due to the increased number of setups. System costs are also greatly increased by expediting orders. The balance of the process breakdowns listed in Table 5.2 decrease system productivity by increasing lead-time and cost and decreasing quality levels. These are all areas where your Lean Six Sigma improvement team can create improvement projects.

Identifying the extent of process breakdowns listed in Table 5.2 as well as reducing their levels will yield significant improvement of operational efficiencies as well as reductions in inventory investment. The process breakdowns listed in Table 5.2 might also be identified through a root cause analysis of higher-level operational problems. Also, there may be several lower-level root causes responsible for the problems listed in Table 5.2. These lower-level root causes might be associated with incorrect order status, bill-of-material (BOM) issues, and problems with capacity utilization, availability of labor or materials, poor order scheduling, quality problems, or other breakdowns within the process.

Summary

Most manufacturing organizations must use MPS and MRPII systems because many products must be built under limited capacity across geographically dispersed supply chains. In fact, sometimes several thousands of different products must be built across these supply chains. Unfortunately, the planning and control systems used to push external demand through these systems use a cumulative lead-time off-set which makes them inflexible to changes in external demand and resource availability. However, recently, the concept of MRPII-driven Kanban has been developed to enable local control of work-cells in push environments using MPS and MRPII systems. In these hybrid systems, the BOM is collapsed to a local work-cell level to facilitate local control. Local control is gained as work cells schedule their work based on real-time resource availability and actual demand status using visual systems such as Kanban cards at a local level to move work between work cells.

Over the past several years, Lean Six Sigma teams have made significant improvements throughout the material-planning function including its MRPII, MPS, and related systems. In addition to significant reductions in inventory investment, productivity has been enhanced by the deployment of Lean Six Sigma projects in these areas. Using Lean Six Sigma problem-solving methodology in the context of the 10-Step Solution Process will allow Lean Six Sigma black belts and green belts to find the specific reasons for process breakdowns the MPS, MRPII and their related systems.

Chapter 6

Identifying Lean Six Sigma Projects Using Inventory Models

Key Objectives

After reading this chapter, you should understand inventory basics and how inventory models are useful to identify Lean Six Sigma improvement projects, as well as understanding the following concepts:

1. Why Lean Six Sigma black belts and green belts should calculate optimum inventory quantities to identify where Lean Six Sigma projects should be strategically deployed by the organization to reduce inventory investment and improve key supply chain metrics.
2. Why identification and elimination of excess and obsolete inventory investment by Lean Six Sigma "belts" is an important first step to improve supply chain performance.
3. Why lead-time, demand variation, and service levels are the major factors that determine inventory investment levels and how you use this information to identify Lean Six Sigma improvement projects.

Inventory Basics

The purpose of this chapter is to review inventory basics related to inventory valuation, common inventory models, and how the perpetual inventory model (PIM) can be used to create a simple inventory model which can be used to analyze inventory investment and Lean Six Sigma improvement projects. Miscellaneous inventory issues are discussed at the end of the chapter and will be useful in Lean Six Sigma project identification to minimize inventory investment. As we build models to analyze inventory investment, it is important to understand the valuation basis of each inventory type. This is important because, in the inventory model, the average on-hand quantity is calculated as the difference between the optimum minus actual average on-hand inventory quantity. This implies the average on-hand inventory quantity must be accurate and its actual value determined prior to taking any improvement action.

Types of Inventories

Inventory is stored capacity. From this perspective every organization has inventory in the form of materials and labor. However, our discussion will be from a manufacturing perspective. In this context, inventory will be discussed from the perspective of raw materials, work-in-process (WIP), finished goods, and supplies inventory, also called *materials, repairs, and operating supplies (MRO)*. Raw materials are purchased from outside the organization and eventually converted by the manufacturing process using machines and labor into WIP and finished goods inventories. WIP inventories include partially or fully converted raw materials including direct labor as well as a portion of assigned overhead. Finished goods inventories are completed products that are made up of materials, direct labor representing the standard costs the product accumulated as it was built, and assigned overhead. Finished goods may be immediately sold or placed in inventory. If they are placed in inventory, finished goods incur many incidental costs. These incidental costs include storage and handling costs, direct labor costs to move the materials, incidental charges due to interest costs related to investing in the materials, and shrinkage costs due to theft, damage, or spoilage depending on the specific type of inventory. MRO inventory includes materials that are necessary to maintain the facility but are not directly used in the manufacturing process.

Inventories are extremely important to organizations since most organizations do not have enough manufacturing capacity to provide goods (or services) on demand. This is usually the case where an organization has a large number of products. Inventories exist because, from an investment perspective, it is less costly to invest in them than to expand system capacity for the time period under consideration. Other reasons to create inventories include unanticipated disruptions in the supply chain. These disruptions can occur for a variety of reasons including weather, labor issues, natural disasters, and seasonal demand patterns. However, as mentioned earlier, the necessity of creating and maintaining inventories exposes an organization to a variety of cost issues, which may degrade inventory value. Degradation of inventory value may result in major financial loss to an organization.

If degradation of inventory value is not adequately accounted for by the organization, the balance sheet will be overstated relative to actual inventory value. What this means is that inventories are overvalued in the sense they cannot be used for one of several reasons including being out-of-date, damaged, stolen or some other reason. In the majority of cases, overstatement of inventory value is caused by a high percentage of excess or obsolete products. Excess inventories are a problem because they cannot be sold at the normal selling price. Obsolete inventory cannot be sold at all. In these situations, cash reserves are kept by the organization to write off the value of the excess and obsolete inventory. This asset value write-off is shown as a loss on the profit and loss (P/L) statement and has the effect of decreasing an organization's gross profit margin. However, the existence of excess and obsolete inventories is usually ignored because many organizations feel external pressure to show the very best financial performance every fiscal quarter and do not want to report inventory valuation issues. However, this is a fraudulent practice in which investors, through the organization's annual report, are led to believe the organization's profitability is higher than actual. One purpose of this book is to encourage organizations to implement best-in-class systems to prevent excess and obsolete inventories from occurring in the first place by deploying Lean Six Sigma improvement projects to systematically eliminate the problems causing inventory write-offs.

Measuring Inventory Quantities and Errors

Inventory must be measured (counted and its value determined) because it is an asset. There are many ways to measure inventory, with some being easier than others depending on the operational design of the inventory system. At a high level, inventories are commonly measured and controlled using either perpetual or periodic inventory systems. Perpetual inventory control systems record inventory transactions, that is, receipt and shipments, as they occur (the inventory calculation depends on periodic refreshing of the system's database, that is, frequency of shipments and receipts). In the periodic inventory system, the "purchases" and "beginning inventory" accounts are adjusted at the end of each period to calculate cost-of-goods sold (COGS). In these systems, COGS is calculated as "beginning inventory and purchases" minus "ending inventory." In both inventory systems the ending inventory quantities should reflect the cumulative effect of receipts and shipments versus the beginning inventory quantities.

Because supply chains are complex systems, of which inventory systems are a subcomponent, errors can occur throughout the process. The supply chain process at a high level includes customer ordering, receipt of materials to build products, storage of materials, manufacturing the products, warehousing the products and picking the customer order in the warehouse (order picking), shipment of the order, and receipt of the order by the customer. Throughout this process, materials can be lost, damaged, mixed up with similar items, or miscounted at any step. These types of mistakes will directly impact inventory valuation estimates. To identify and eliminate the root causes associated with inventory valuation, periodic physical inventory counts must be made throughout the fiscal year. These are called "cycle counts." Cycle counts are useful in identification of incorrect inventory quantities. Inventory problems identified through the cycle counting system can serve as the basis for Lean Six Sigma improvement projects. The goal of these projects is to prevent discrepancies in inventory quantities due to the many operational breakdowns within the supply chain. Progressive organizations have various strategies to optimally conduct cycle counts. These range from full Lean implementation so that inventories are minimized and easily visible to other more complicated systems using statistical sampling methods. Common cycle-counting systems will be discussed in Chapter 7.

Physical Control of Assets

Physical control of inventory, from an operational perspective, passes to the buyer upon physical receipt of the inventory. However, from a financial (and legal) perspective, control depends on contractual terms defining the transfer of the title for the goods to the buyer. However, in most situations physical control upon receipt of the material is a useful practical definition of inventory control unless the seller and purchaser have clearly defined another scenario. In many situations, legal receipt of inventory is specified as goods shipped free-on-board (FOB). FOB means the point in the process when the buyer takes possession of the goods: at the moment of transfer by the seller to the carrier (FOB shipment point) or upon arrival at the buyer's distribution center (FOB destination). There are also other systems that describe terms and conditions relative to inventory ownership. Special financing agreements depend on common practices of the particular supply chain. In these agreements, there are many variations of physical versus legal control of the inventory, resulting in timing differences relative to inventory valuation. One common example is the use of consignment inventory based on special financing agreements. Consignment inventories will be discussed in Chapter 7.

Cost Flow Assumptions

Valuation of an inventory depends on which goods are to be valued, how they will be valued, that is, which costs to include in their valuation, and the cost flow assumption to be adopted by the organization. Cost flow assumptions include *last-in-first-out (LIFO), first in-last-out (FIFO)*, and average-cost models. In FIFO systems, the assumption is that materials are used in the order of purchase. The inventory that is left is always the most recently purchased by the organization. FIFO follows the physical flow of material through the process. The FIFO COGS calculation is equivalent in both perpetual and periodic review inventory systems. In a LIFO valuation system, COGS is calculated using the cost of the most recently purchased materials rather than the oldest. This may better match costs to their revenue, providing a more accurate COGS calculation. However, the valuation using the LIFO system may differ between a perpetual and periodic inventory system if pricing changes occur during the time period under consideration. The advantage to using a LIFO system is that COGS will be higher if material costs are increasing over time. This improves cash flow by lowering income taxes on net sales. Disadvantages of the LIFO system include lower reported earnings and understatement of total inventory value since older inventory is left on the balance sheet. Average-cost valuation (also called weighted-average-cost method) estimates COGS using an inventory value that reflects the total current inventory investment divided by the inventory quantity. This is the valuation method least subject to distortions in inventory value.

Fraudulent Valuation of Inventory Assets

Inventory valuation fraud exists due to process breakdowns within an organization's inventory and financial reporting systems. Process breakdowns include missing or nonexistent inventory quantities as well as maintenance of valuations at levels that do not accurately reflect the likelihood of excess and obsolete inventory being sold. One of the major reasons for missing or nonexistent inventory quantities involves incorrectly estimating shipment or receipt quantities or not entering transactions into the system on a timely basis. As an example, if the receipt of materials is delayed relative to their shipment, revenue will appear higher than it actually is for the time period in question. In this situation, payment by the buyer to the supplier may be intentionally delayed for a period of time or the payment terms may be misstated to external auditors. Fraud can also be created when inventory quantities are double-counted or excess and obsolete inventory is valued at full purchase cost rather than discounted due to reasonable risk to their valuation. The worst situations occur when missing or nonexistent inventory is recorded as an asset.

Unfortunately it is difficult to determine whether the process has broken down or people have intentionally manipulated the system for unethical reasons. This is why organizations must be vigilant in their continuous improvement efforts. An analytical way to detect fraud (or just a poorly performing inventory system) is to look for changes in inventory valuation by product category from period to period or to benchmark comparable systems. The best way to identify breakdowns in the inventory system is to "walk the process" and create a value stream map. This value stream map should show every inventory transaction relating to the movement of material and information throughout the system. In particular, it should show the operational and information system issues that impact inventory valuation and are not visible on management reports.

Building System Models

There are many ways to model and analyze a supply chain system. Process maps can be drawn of the system and if quantified will serve as a very good basis upon which to improve the process. Alternatively, an analysis of the system, using graphical or statistical tools, might reveal important relationships between the system's inputs ("X's") and their associated outputs ("Y's"). However, if an improvement team can go one step further by creating system models allowing evaluation of how the system dynamically responds to changing levels of the input variables, they will truly understand the DMAIC relationship $Y = f(X)$. Understanding the underlying functional relationship $Y = f(X)$ will allow the team to set the levels of the input variables ("X's") to optimize the output variable level ("Y's"). Simulation of the $Y = f(X)$ model, by compressing time between events, that is, speeding up the process so that system events occur very rapidly, allows a better understanding of how a system changes dynamically through time as the input variable levels change. Event statistics can be analyzed offline after the simulation is complete. On the downside, simulation models are not 100% accurate, may be costly to create, and may be difficult for other people in the organization to comprehend if the model has not been well documented by the improvement team. However, since inventory systems are relatively simple to model and off-the-shelf software or Microsoft Excel is readily available to create and interpret these models, simulation is a good tool to analyze supply chain subsystems including those associated with inventory.

System Design

What is a system? There are many definitions, but the one we will use is the following: "A system is a set of activities, resources, and control systems that interact to transform inputs such as material or information from suppliers into outputs used by a customer." A simple example is the SIPOC (Supplier-Input Boundary-Process-Output Boundary-Customer). This high-level map was first described in Chapter 1. Another example is the value stream map described in Chapters 2 and 4. In either case the system map should be quantified relative to its inputs and outputs. If the transformation relationship is also known, that is, $Y = f(X)$, then a simulation model can be built to analyze the dynamic characteristics of the system under study.

In addition to identification of system inputs and outputs, the allowable ranges over which the variables are allowed to vary must be specified so that the final results will not contradict system reality. In some cases this means knowing the probability density function of the input variable (how the variable is distributed). However, in the inventory models we will study, simplifications can be made, since the simulation results can be easily checked against current inventory and service levels by item and location. In other words, if the service level is already below target we are not going to reduce inventory without subsequent analysis. The advantage of using an inventory model is the ability of the Lean Six Sigma improvement team to evaluate the impact of changes in lead-time and demand on required (or optimum) inventory investment.

Data Collection and Analysis

Data collection and analysis must be strictly linked to the project objectives as specified by the original team project charter. It is important for the Lean Six Sigma team to stay focused on the original project goals and objectives because supply chain systems can be very complex and contain numerous functional components and subcomponents, The specific tools and methods related to data analysis will be discussed in detail in

Chapter 8. Relative to the practical aspects of data collection, this process could be very easy or difficult depending on the availability of data relevant to providing information necessary to answer the questions of the project charter.

The data collection process could be very fast and simple if the project is strictly confined to inventory improvement within the current system and does not touch external suppliers or require very accurate estimates of lead-time or operational data that may be difficult to obtain. However, if the project is focused on issues such as on-time supplier delivery, then data collection and analysis could be very complex and it could take a relatively long time to understand the underlying issues. As an example, a supplier on-time delivery report might show that a supplier has poor on-time performance, but the supplier may disagree. This situation would require a *measurement system analysis (MSA)* prior to starting the actual improvement project. In the MSA, the first step would be to verify that the measurement system is accurate and precise to the required degree prior to collecting data for the inventory model. In fact, if the information obtained from the IT systems, relative to on-time delivery, is in doubt, the improvement team must "walk the process" from "order release" by the MRPII system to "order receipt" by the customer to identify the cumulative lead-time components on which the supplier is being measured by the on-time delivery report. The difficulties associated with data collection in supply chain systems will vary by project and should not be underestimated by the Lean Six Sigma improvement team.

Building a Model

The best system model is the one which allows the improvement team to answer its original project charter questions using the minimum number of variables. Another consideration might include eventual integration of the model into other supply chain subsystems. In this situation, the organization would want to specify standardized software from which to create the system model. In this book we use simple Microsoft Excel models.

Analyzing Model Outputs

Typical output metrics include inventory investment, inventory turns, cycle time, utilization of machine and labor resources, cost, and system capacity. There could be many others depending on the type of simulation model being used by the team. Many of the newer models allow the user to change the model by varying input levels or removing parts of the system that are clearly not adding value. As previously mentioned, analysis of the system model and its simulation results will vary depending on the team's original project charter and the questions the team was required to answer by their senior management. Relative to supply chain and inventory models in particular, the 12 metrics shown in Chapter 1 in Figure 1.2 are a good list from which to build the system model.

As an example, in Figure 1.2, we see that the 12 key supply chain metrics include cash investment, profit and loss, inventory turns, on-time delivery performance, forecasting accuracy, lead-time, unplanned orders, schedule changes, data accuracy issues, material availability, and excess and obsolete inventory. Some of these metrics such as lead-time and forecasting accuracy clearly drive some of the higher-level metrics such as inventory investment. On the other hand, second-level metrics such as unplanned orders, schedule changes, data accuracy, and material availability issues clearly drive lead-time. In the background are issues related to measurement error, which impacts all metrics and related process inputs and outputs. In Chapter 8 we will discuss the components of measurement error using supply chain examples. The Lean Six Sigma improvement team must understand these complex metric inter-relationships to improve supply chain performance. The best way to develop this understanding is by building a simple model to analyze inter-relationships between the metrics.

Inventory Models

As you will recall, the basis of this book is to provide information on basic supply chain concepts as well as the analytical tools necessary to build and analyze simple inventory or other system models to identify operational breakdowns within the supply chain. Analysis of operational breakdowns will identify areas where Lean Six Sigma projects can be deployed to improve operational performance. Chapter 8 will discuss the analytical tools necessary to build and interpret inventory and supply chain models. An applied example will be presented in Chapter 10. The focus of these discussions will be on make-to-stock applications with emphasis on the "perpetual inventory model" since this is one of the most common inventory models. However, the make-to-order model will be discussed briefly. Although these simple models are not 100% accurate, they provide sufficient understanding of an inventory system to facilitate analyses useful in determining the best ways to reduce inventory investment and improve supply chain performance. Also, overly complex models are not always better. As an example, if the model is too complex, it may be expensive to maintain over time and the analytical skill level of people creating and interpreting the model results must be very high. Alternatively, an organization could retain major consulting firms to do the inventory and supply chain analysis, but at a very high cost.

Building a model helps everyone to understand underlying relationships between key supply chain metrics and inventory investment. Figure 6.1 is a qualitative representation of an inventory model. Once the model is created, it is used to answer questions relative to the 12 metrics shown in Figure 1.1 as well as the original questions the team was asked to answer. The model also provides management with visibility to the entire inventory population at an item and location level. The improvement team will be able to balance the inventory level by item and location using this information. Based on the balance analysis, inventory reductions may be possible if the organization is currently achieving its customer service targets, but has excess inventory quantities. But, very high levels of excess and obsolete inventory may limit potential reductions in inventory investment. Additional reductions in inventory investment over and above those obtained using a balance analysis might also be achieved by reducing variations in lead-time and demand.

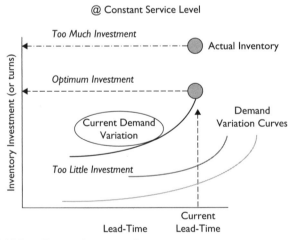

Figure 6.1 What Drives Inventory?

As an example, lead-time reduction projects would eventually focus on contributing factors such as lot size, order quantity, and missed deliveries, to name a few. Demand management projects would reduce forecasting error as well as related issues.

Using inventory investment as a key metric to understand supply chain performance, the optimum inventory quantity for an item and location is determined based on its target service level, lead-time, and demand variation. A graphical view of this concept is shown in Figure 6.1. Referring to Figure 6.1, actual inventory investment is usually higher or lower than the calculated optimum level. If it is higher than the optimum inventory investment and service targets that are being met in practice, inventory investment can usually be decreased (on average). On the other hand, if inventory investment is lower than the calculated optimum, it is usually necessary to increase it (on average) to achieve the required customer service-level target. Figure 6.1 also shows how specific reductions in lead-time and demand variation can lower inventory investment as we move down any of the three demand curves by reducing lead-time or jump from one demand curve to another at a constant lead-time.

Arbitrary reductions in inventory will have a negative impact on manufacturing schedule performance. This occurs because inventory is used to support manufacturing operations and maintain operational independence relative to variation in external demand, resource availability, and on-time delivery. Inventory can be found in the form of raw materials, work-in-process subcomponents, and finished goods. Inventory is also used in manufacturing support activities such as machine maintenance, machine repair, and necessary operating supplies. To specifically determine where to add or reduce inventory investment, optimum unit-inventory levels are calculated by item (subcomponents as well as finished products). These optimum quantities are compared to actual investment. This analysis often shows an imbalance across the inventory population. This is shown in Figure 6.2. If an imbalance does exist, it may be possible to reduce investment for some items that exceed the optimum quantity; but, have very high per-unit service levels. However, items below the calculated optimum inventory quantity require additional investment to meet their target per-unit service level.

The optimum inventory quantity is estimated item by item for every stocking location using a Microsoft Excel model. This Excel model uses the target service level of the appropriate inventory class as well as average and standard deviation of lead-time and demand to calculate the optimum inventory quantity for every item and location.

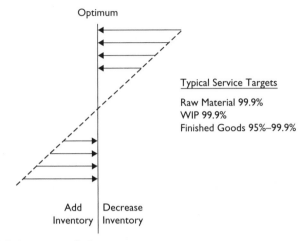

Figure 6.2 Inventory Balance

The optimum investment level for the entire inventory is calculated by summing each item's optimum quantity and its corresponding investment level. This analysis provides the organization with the information necessary to reduce inventory investment by item and location without adversely impacting the customer service levels. The model also shows where improvements in lead-time and demand variation are necessary to reduce optimum inventory investment levels. It should be noted that service levels may be different for the various inventory classes, that is, raw materials, WIP, and finished goods. Also, service-level targets can be expressed in several forms and measurement units as shown in Figure 6.3.

Perpetual Inventory Model (Fixed Order Quantity Model)

The perpetual inventory model (PIM), shown in Chapter 5 in Figure 5.12, is a very common inventory model. Information describing the PIM can be found in most operations management textbooks. It is characterized

- Customer ⟶ Orders, Lines, Units, Dollars (95%–99.9%)
- Work-In-Process (WIP) ⟶ Units (99.9%)
- Raw Materials (RM) ⟶ Units (99.9%)

Figure 6.3 Typical Service-Level Targets

by use of an order quantity Q_1. Q_1 is calculated analytically using a minimum cost function based on inventory holding cost, setup cost, lead-time, customer unit demand, and target service level. It should be noted that there are more advanced models in which Q is optimized relative to many other factors depending on modeling assumptions. In the PIM model, inventory is perpetually monitored by the MRPII system over the order cycle T. A new order for Q_2 units is released when the inventory level reaches the reorder point. The danger of a stock-out normally occurs during lead-time LT, but this risk is covered by the safety-stock inventory. The perpetual inventory model will be used to estimate optimum inventory quantities for raw materials, WIP, and finished goods items in subsequent chapters of this book since the emphasis is on make-to-stock systems.

How Do We Create the PIM Model?

The optimum inventory quantity is calculated for every inventory item and location based on actual lead-time, demand history, and the unit service-level target as shown in Table 6.1. This calculated optimum quantity is compared to the current on-hand inventory quantity, that is, "actual." In Step 1, the calculated difference between the actual versus optimum quantity for each item and location will be either "higher" than the calculated optimum quantity (excess) or "less" than the optimum quantity. In the latter case, additional inventory investment is required to achieve the unit service-level target. Once this "actual versus optimum" calculation is made, the item and location information is aggregated over the entire inventory population. These basic steps are shown in Table 6.1. In calculating excess inventory in Step 1, it is important to obtain a precise estimate of the actual on-hand inventory quantity. As an example, assuming a linear consumption rate, inventory will be higher

1. Excess Inventory = $\text{Inventory}_{Actual} - \text{Inventory}_{Optimum}$
2. $\text{Inventory}_{Optimum} = \text{Inventory}_{Average\ Demand\ During\ Order\ Cycle} + \text{Safety Stock}$
3. Safety Stock = Service $\text{Constant}_k * \sigma t$
4. If (1) is positive decrease inventory…if negative add inventory.
5. Excess inventory is calculated for every item and then totaled to the product and business levels using Excel pivot tables.

Table 6.1 Inventory Analysis Algorithm

at the beginning than at the end of the order cycle. In this scenario, the best estimate of average on-hand inventory is at the mid-point of the order cycle or ½ Q_1 (assuming a linear depletion rate). But every situation must be analyzed separately. The calculation of ½ Q_1 is an important estimated quantity because excess inventory is calculated using it as a baseline. The second important parameter in Table 6.1 is the service-level target. It is in units. Although there is often a strong correlation between line and order fill, these correlations must be verified through subsequent analysis if management requires order fill information relative to inventory investment.

Safety-Stock Calculation

In Steps 2 and 3 of Table 6.1, the optimum inventory quantity is composed of two components based on the perpetual inventory model. The first component is "the average demand during the order cycle." This quantity is calculated as ½ Q_1. Q_1 is the inventory quantity satisfying expected demand over the order cycle assuming a linear consumption rate. The second model component sets the safety-stock level of inventory based on uncertainty of lead-time and unit demand using the target customer unit service-level. In the safety-stock calculation, lead-time is the time period between the reorder point and receipt of the reorder quantity. In make-to-order or advanced Lean systems some of the terms of the safety-stock formula become zero. As an example, in make-to-order systems, demand variation is assumed to be zero because quantities are specified by contract, but lead-time will vary. So in any inventory system, because either lead-time, demand, or both may vary, inventory safety stock will be required for raw materials and WIP inventories. The formula shown in Figure 6.4 is used to build the safety-stock portion of the inventory model according to Step 3 in Table 6.1. The terms of the safety-stock formula are defined in Figure 6.5. As mentioned above, depending on the inventory system, some of these terms may be assumed as zero.

$$\sigma_t = \sqrt{\left[\overline{LT} * \sigma^2_d\right] + \left[\overline{D}^2 * \sigma^2_L\right]}$$

$\sigma^2_d = 0$ ——————▶ *Make to Order*

$\sigma^2_{LT} = 0$ ——————▶ *Typical Simplification*

Figure 6.4 Overall Standard Deviation

$\overline{LT}$ = The average lead-time for item in days.

σ^2_L = Standard deviation for item lead-time in days.

$\overline{D}^2$ = Average demand for item over a month in units.

σ^2_d = Standard deviation for item demand over month in units.
If the forecast error is small, the RMSD is substituted
into the formula that is, good forecast accuracy lowers demand
variation and safety stock.

Figure 6.5 Definitions

Bottleneck Integration

Make-to-stock models are characterized by unknown demand and variable lead-time. Relative to make-to-stock systems, capacity constraints require that finished goods inventories be built in anticipation of customer demand since actual demand patterns consist of fixed, promotional, seasonal, trend, and random demand components that periodically exceed available capacity. Calculating required safety-stock levels for independent demand items associated with finished goods inventories is a straightforward process since they are estimated independently of system constraints when they are warehoused. This is because they can be built over time. However, calculation of safety-stock levels for dependent demand items typified by raw material and WIP inventories is more complicated since they must be estimated relative to a specific system constraint called the "bottleneck resource." The bottleneck resource determines the system's throughput rate and work center raw material and WIP inventory levels. The manufacturing schedule of the bottleneck is directly linked to external customer demand, that is, the demand for finished goods. This linkage forms the basis of the takt time calculation discussed in Chapter 4. Safety-stock quantities at operations upstream from the bottleneck resource are calculated to ensure that the capacity of the downstream bottleneck resource is protected from interruptions in material flow due to manufacturing issues.

Make-to-order models are characterized by known demand, but variable lead-time. The most efficient way to control inventory in these systems is to focus directly on the bottleneck resource. In these systems, lead-time variability requires that safety stock be held at upstream operations impacting the bottleneck. Figure 6.6 shows how inventory investment increases from raw materials through WIP due to cumulative lead-time of the process. To estimate inventory investment in make-to-order

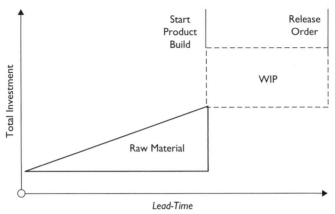

Figure 6.6 Make-to-Order Model

systems, resource spatial relationships, including the critical path through the manufacturing network, must be accurately modeled to include system constraints as well as its bottleneck resource.

A simple inventory model for a make-to-stock system can be created using Excel spreadsheets containing the minimum information shown in Table 6.2. This information is sufficient to satisfy the equations described in Table 6.1 as well as in Figure 6.4. After the inventory model is constructed by the Lean Six Sigma improvement team, sensitivity analyses can be conducted on the model's inputs and outputs to facilitate project identification and improvement efforts. It is important the model should be constructed in a manner that will answer your team's original questions and identify improvement opportunities relative to the key financial and operational metrics including the 12 metrics listed in Figure 1.1. Construction of a simple inventory model will be discussed in Chapter 9.

1. Lead-time information

2. Demand information

3. Service-level information

4. Descriptive information to allow segmenting the database, that is, product class, customer, supplier, and so on

5. Information related to forecast accuracy

6. Information related to backorders, obsolete inventory, and so on

Table 6.2 Minimum Required Information

Other Types of Inventory Models

There are many different inventory systems and models which reflect their characteristics. The PIM system is one of the most common inventory models since organizations have automated systems to track inventory receipt and shipment transactions. The following inventory models are presented to provide additional modeling concepts and ideas for your Lean Six Sigma team. In particular, the minimum/ maximum (min/max) system is particularly useful in situations where an item's demand variation is low allowing the item to be automatically tracked by the MRPII system. In Chapter 3 and in Figure 3.10 it was shown a min/max system frees up forecasting analysts by stratifying item forecasts.

Periodic Review System (Fixed Time Period)

In a periodic review inventory system, inventory is counted at predetermined frequencies and orders are placed for each item and location based on the quantity that has been depleted by demand during the order cycle. Unlike perpetual inventory systems, periodic review systems do not track daily demand in real time across the supply chain. This situation results in inventory stock-outs and lower customer service levels when demand increases above expected or forecasted levels. The length of the ordering cycle is calculated using economic order quantities (EOQs) including ordering costs. Another important consideration in these systems is the aggregation of items from the same supplier to reduce ordering costs. Developing the system involves balancing the economic order quantities and delivery frequencies to satisfy expected demand over the order cycle. To build this type of model, several variables must be analyzed for each item and location including annual requirements, purchase cost of each item, ordering costs, and ordering intervals.

Min/Max Inventory Systems

Minimum/maximum systems rely on a "two-bin" approach to inventory control. In the two-bin system, each bin contains an inventory quantity equal to the *reorder point quantity (RPQ)*, that is, when the first bin becomes empty it is time to place a new order equal to the RPQ. These min/max systems can be used by the MRPII system to automatically monitor inventory levels. This type of inventory system is particularly effective for managing low-volume items with low demand variation. Placing items on an automatic min/max system will reduce the workload of forecasting analysts. This will allow them to focus their attention on building forecasting models for product groups requiring their expertise.

This stratification concept, that is, demand variation versus sales volume, was shown in Chapter 3 in Figure 3.10. Min/max systems are also used in Lean inventory control systems, that is, Kanban quantities.

MRPII Inventory System

MRPII inventory systems are useful to control raw material and WIP inventories associated with dependent demand items. The advantage of MRPII inventory control systems is that orders for each item can be time-phased by the master production schedule (MPS), based on their required use. It is important in these systems to maintain accurate system parameters. These system parameters are related to end-item demand, lead-time for the item, and other relevant information used by the system to calculate time-phased inventory levels.

Optional Replenishment Inventory System

Optional replenishment systems are a special case of the periodic review system in which order quantities are automatically placed, but a decision is made whether the order should be placed at this time or delayed to a future date. In these systems the ordering process is expensive, hence the required controls relative to placing an order. The system variables defining this inventory control system are the review period, maximum inventory level, and the reorder point of the item.

Distribution Requirements Inventory System

Distribution Requirements Inventory (DRP) inventory systems accumulate lower-level item and location demand that has been placed on distribution centers within the system, and this lower-level demand is aggregated up to higher levels in the system. Using the aggregated information, orders are placed with suppliers and deliveries made to each distribution center based on demand requirements. DRP systems are used in systems that manage system inventory throughout the supply chain at several levels. An example would be warehousing lower-level demand items at centralized locations to minimize overall demand variation for the item to reduce inventory investment across the distribution system. This concept is shown in Chapter 7 in Figures 7.4 and 7.5.

Just-In-Time Inventory Systems

Just-in-time (JIT) inventory systems rely on the basic elements of a Lean system including operational stability, high quality, and JIT implementation.

JIT consists of level loading of customer demand on the system to stabilize the manufacturing schedule. These concepts were discussed in Chapter 2 and are shown in Table 2.2 as well as Figure 2.3. Using the expected demand on the system (which must be relatively stable over long periods of time), the Kanban container quantities are calculated using service level, lead-time, and demand (which has been broken down linearly to an hourly basis). Kanban quantities are based on the PIM formula, but lead-time and demand are assumed to be constant (except for a small safety-stock quantity as shown in Figure 5.15). Raw material and WIP inventory are controlled using Kanban quantities, which are moved within manufacturing operations and external suppliers. These concepts were discussed in Chapter 5 and a generic version of a Kanban system is shown in Figure 5.11.

Miscellaneous Inventory Topics

As the Lean Six Sigma improvement team builds and analyzes the inventory model, it may find that some components of high inventory investment are due to excess and obsolete inventory. A complicating factor related to excess and obsolete inventory is proliferation of the number of products. The greater the number of products, the higher the required inventory investment all other things equal. These three concepts may be important in the development of your inventory model, Lean Six Sigma project identification, the root cause analysis, or the eventual reductions in investment. Several other important topics such as distribution inventory management, third-party logistics, consignment inventory, and cycle-counting systems will be discussed in Chapter 7.

Excess/Obsolete (E&O) Inventory

Excess inventory consists of quantities such as raw materials, WIP, or finished goods inventory that exceed expected sales over a specified time period. However, every organization has its own specific definition. A common definition is, "inventory quantities exceeding 12 months supply or some multiple of lead-time." An important consideration in classification of excess inventory is how quickly the excess quantities can be sold. If demand for the item is strong, it may be possible to reduce incoming orders and quickly reduce excess inventory investment. However, if the item does not sell well, it may take years to reduce the item's inventory level. Our definition is that an excess quantity is when the inventory quantity difference, by item and location; between actual versus optimum

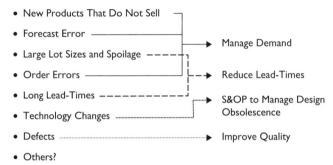

Figure 6.7 E&O Prevention

quantities is greater than zero and exceeds expected demand at the item's cumulative lead-time. This definition assumes inventory is not being built for future demand due to capacity constraints. The root causes for excess inventory vary. However, Figure 6.7 shows that common reasons include incorrect MRPII parameters including lot size, incorrect forecasts, ordering errors, long lead-times, and design and technology changes. Figure 6.7 also shows some ways to prevent the occurrence of obsolete inventory, but these are high-level categories. The Lean Six Sigma improvement team must investigate the actual root causes of excess and obsolete inventory to eliminate these chronic problems.

In contrast to excess inventory, obsolete inventory is material that has no current or future demand. For this reason it cannot be sold at its original selling price. Figure 6.8 shows that it may be possible to sell the obsolete inventory at lower prices or below standard cost to customers who may require limited quantities of the item. More recently, online auctions are used to dispose of obsolete and excess inventory. Also, it may be possible to sell the material through different distribution channels including overseas markets to different customers. In the case of raw materials

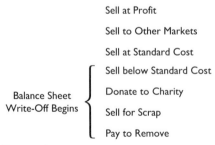

Figure 6.8 E&O Disposal

and components, it may be possible to return the material to the original supplier for credit. It may also be possible to scrap the obsolete inventory and write it off the financial balance sheet if it cannot be sold. The causes for obsolete inventory are similar to the causes of excess inventory. These common causes include design obsolescence, failed marketing programs, poor forecasts, and purchasing too much material, that is, large lot sizes, or taking advantage of quantity discounts. Figure 6.8 shows ways to prevent the occurrence of obsolete inventory.

Current inventory can be sold and effective actions taken to reduce its investment level over time. On the other hand, excess and obsolete inventory built up over time (legacy inventory) cannot be easily sold since its demand may be minimal or nonexistent. Considering the different improvement actions necessary to reduce current versus legacy inventory investment, it makes sense to break the inventory population into these components and separately calculate inventory turns and investment metrics to properly focus improvement actions on issues which are truly actionable by the team.

How do we reduce our exposure to excess and obsolete inventory? Reviewing the causes of excess and obsolete inventory listed in Figure 6.7, it becomes apparent that effective preventive actions must correspond to specific root causes. This requires investigation by the Lean Six Sigma improvement team as well as analysis of the situation using the 10-Step Solution Process described, in Chapter 1, Tables 1.1 and 1.3. As an example, E&O inventory associated with products that do not sell may indicate a problem with the new product forecasting process. Large lot sizes may indicate operational problems with suppliers. Order errors may indicate internal purchasing problems. These are just a few of the possible causes of E&O inventory. E&O preventive measures are described in Figure 6.7 and options for effective disposal are described in Figure 6.8. The most effective way for an organization to minimize creation of E&O inventory is continuous review of E&O status using cross-functional teams and immediate root cause analysis and elimination when E&O does occur. Visibility to excess and obsolete inventory levels should be a standard agenda item for S&OP meeting.

Product Complexity/Proliferation

Product proliferation increases over time if an organization fails to control the complexity of its product mix. This is due to the failure to obsolete or outsource older products whose demand has significantly decreased from their initial product introduction. To minimize product proliferation, the product mix should be reviewed by the S&OP team on a frequent basis. Products having very low annual demand should be eliminated or outsourced to reduce cost and maintain gross margin. High inventory investment will result from failure to control product proliferation due to subcomponent proliferation based on the product's bill-of-material (BOM) structure. This product proliferation also increases material handling and results in higher operational and inventory costs. Even if product proliferation is not a major problem for the organization, simplification of both product design and the overall product mix is important to maintain optimum inventory investment levels. Published case studies have consistently shown the benefits of design simplification. Reductions in the number of subcomponents may exceed 50% using a proven design-analysis methodology such as *design for manufacturability (DFM)*. Fewer subcomponents will reduce inventory investment. As a result, every organization should review its product design complexity and product mix on a regular basis.

Lean Six Sigma Inventory Applications

Table 6.3 lists several areas where Lean Six Sigma projects have been deployed over the past several years. Each of these areas of opportunity may have a direct impact on inventory investment for your organization. Accurate system parameters are important since required inventory is based on lead-time and demand variation as well as the

1. Ensuring system parameters are accurate.	5. Elimination of excess and obsolete inventory.
2. Using a quantified inventory model, showing relationships between inventory investment versus lead-time and demand variation.	6. Improving cycle-counting accuracy and efficiency.
3. Verification of inventory valuation by each asset classification	7. Reducing product complexity to simplify BOMs.
4. Identifying and eliminating the root causes of inventory discrepancies.	

Table 6.3 Lean Six Sigma Inventory Applications

target service levels. Inaccuracies in the parameter files can occur due to manual data entries or the failure to update the parameters as they change over time. Also, the parameter information may have been incorrectly specified in the first place. Lean Six Sigma improvement teams, working backward from the inventory investment analysis, may ultimately find problems with parameters that need to be investigated through root cause analysis and eliminated using one or more countermeasures. Critical relationships between inventory investment (and inventory turns) can be revealed using an inventory model for each inventory and each asset classification. Lean Six Sigma methods have also been very effective in elimination of excess and obsolete inventory by attacking the root causes for the high investment rather than just reducing the extent of the problem or writing off the excess for the balance sheet. Using Lean Six Sigma methods, the cycle counting and other material control systems can be process-mapped to identify material and information flows to simplify these systems over time. Lean Six Sigma methods as well as DFM can be used to reduce overall product complexity. This complexity drives inventory issues through the BOM hierarchy.

Summary

The unique characteristics of an organization's inventory system must be taken into account when developing an inventory model. Using these unique system characteristics as a baseline, the improvement team can integrate them into the inventory or supply chain model. There are many useful sources of information describing the mathematical basis of these system models. We presented the perpetual inventory model (PIM) to demonstrate the modeling concepts and how the Lean Six Sigma improvement team can identify projects to reduce chronic process failures. This discussion will be expanded in Chapter 10. In the PIM model, lead-time, demand variation, and service level are used to set the inventory investment targets. In earlier chapters, we identified some of the components of lead-time and demand variation, but the Lean Six Sigma improvement team must analyze their system to determine the specific reasons for inventory investment within their organization. We also discussed miscellaneous inventory issues. The purpose of this discussion was to alert the improvement team of their existence and provide a list of possible reasons for excess and obsolete inventory as well as some ideas on how to reduce their levels.

Chapter 7

Lean Supply Chains and Third-Party Logistics

Key Objectives

After reading this chapter, you should understand how Lean Six Sigma projects might be identified and deployed across your supply chain, as well as understanding the following concepts:

1. Why Lean Six Sigma deployment is more complex in a Lean supply chain than within a single organizational function.
2. The types of Lean Six Sigma projects that have been deployed in Lean supply chains over the past several years.
3. Why it is important to integrate third-party suppliers into your Lean Six Sigma initiative.

Lean Supply Chain Overview

A Lean supply chain is one in which all participants perform according to Lean principles, including level schedule loading using pull-based demand, deployment of continuous improvement activities, maintenance of sufficient (even excess) capacity to satisfy external demand, strict schedule adherence to optimize profit for all participants across the supply chain, and establishment of long-term reciprocal relationships among all participants. Establishing a Lean supply chain is important for an organization. This is because the supply chain will eventually become noncompetitive if it fails to systematically improve its supply chain performance. This is an important goal because supply chain costs often range between 50%–80% of all money spent by an organization. Another reason to improve operational performance with a high sense of urgency is the fact that customers demand cost reductions, on-time delivery, and higher quality based on the competitive environment. Organizations should also develop internal competencies that competitors cannot easily emulate to remain competitive over time. The 10-Step Solution Process can be applied very effectively to improve supply chain performance using Lean Six Sigma and operations research tools and methods.

At an operational level, Lean supply chains are characterized by rate-based demand and production smoothing through strict takt time adherence. Takt time is defined as the ratio of time available per day to produce the required schedule, that is, time per unit. As an example, if there are 480 minutes per day and the required manufacturing schedule is 48 units, the takt time is 10 minutes per unit. This means the system is configured to produce exactly one unit every 10 minutes. A second characteristic of a Lean supply chain is use of a mixed-model production schedule. A mixed-model production schedule minimizes process waste by reducing variations in lead-time and demand through changes to manufacturing schedules. These scheduling changes are enabled through modifications in product and process design. In mixed-model scheduling the level of design commonality and modularity is significantly increased, ensuring that products are scheduled in groups having similar process and design features. A third characteristic of a Lean supply chain is use of "demand pull" rather than "demand push" scheduling systems. Demand pull scheduling directly aligns the productive capacity of the supply chain to external customer demand patterns. In these systems, products are built based only on customer orders in real time, that is, up-front forecasting does not drive the schedule except at a high capacity allocation level. Unfortunately, not every organization has been able to fully implement Lean supply chain concepts due to infrastructure issues related to IT systems and constraints in product and process designs. In these systems, organizations must rely on the master production scheduling (MPS) and material requirement planning (MRPII) systems to build components and products. However, these situations make control of the system in real time difficult. To improve the local shop-floor control of orders, hybrid Kanban systems have been developed to integrate MPS/MRPII and Kanban systems. These hybrid systems as well as basic Lean methods were discussed in Chapters 2 and 4.

Lean supply chains are characterized by the ability to respond dynamically to changing customer demand as well as efficient asset utilization. These characteristics translate into high profitability and customer service for the organizations making up the Lean supply chain. The effectiveness and maturity of a Lean supply chain can be measured using the 20 important supply chain metrics listed in Chapter 1, Figure 1.2, as well as in Appendix I. The first 12 key metrics are operational in nature and useful in identification and deployment of Lean Six Sigma projects at a tactical level. The 12 key supply chain metrics should be systematically improved at an operational level by the Lean Six Sigma improvement team to drive overall supply chain financial performance and achieve customer service targets. The last 8 metrics are financial in nature and measure the overall financial effectiveness of a Lean supply chain. In particular, they measure the efficiency of asset utilization, including inventory and its impact on profitability. Using these 20 metrics, the degree of "Leanness" of a supply chain can be measured and benchmarked to ensure that operational improvements are aligned with business goals and objectives. As an example, in best-in-class Lean supply chains, inventory turns (efficiency of inventory utilization) exceed 10 and in some systems they approach 100, for example, Dell Computers. Return-on-assets (ROA) is another critical metric for best-in-class originations. The ROA ratio is in the double-digit range for such best-in-class Lean supply chains as 3M, Intel, The Home Depot, Johnson and Johnson, Wal-Mart, and Dell. Along this line of thought, many studies have shown that the key differentiators for superior supply chain performance include proper metric definition, metric linkage, and sustained operational performance to ensure strategic execution. Lean Six Sigma deployments have enabled many supply chains to improve operational performance as measured by one or more of the 20 important supply chain metrics.

Superior supply chain performance is based on specific core competencies inherent within an organization's supply chain. Core competencies may include improved product development, implementation of Lean Six Sigma practices, elimination of promotional activity to smooth demand, and simplification of product and process designs as well as other systems. Some supply chains achieve superior performance by excelling in one or more of these core competencies. In addition to the competencies already listed, Lean supply chains execute strategy well at an operational level. This ensures that sufficient capacity is available for the system to absorb demand variation in real time. These systems are also systematically improved to increase system flexibility. However, Lean supply chains do not arbitrarily increase asset levels to achieve system flexibility. This is especially true for inventory investment levels. On the contrary, products and processes are designed, using Lean methods, to ensure quick changeovers from one product to another using a variety of methods including, for example, mixed-model production, transfer batches, and system-constraint identification and management.

In these planning and control systems, customer demand is forecast at a strategic level, but operations are controlled at a local work-cell level using visual controls in real time. Supply chain planning and control are enhanced through linkage to external demand using advanced technology. The resultant operational efficiencies are seen in the form of superior return-on-asset (ROA) utilization efficiencies. In addition to ROA, gross margin return-on-assets (GMROI), calculated by dividing gross margin by average inventory investment, is a good indicator of the efficiency of inventory utilization. The higher the GMROI, the Leaner the supply chain and the more efficient it has become, assuming it is meeting its customer service and profitability targets. In addition to inventory, supply chain assets include fixed assets, vehicles and accounts receivables. These financial metrics are shown in Chapter 1, Figure 1.2, and defined in Appendix I.

From an operational perspective, Lean supply chains achieve lowest total system cost by shortening the length of the supply chain and reducing the number of suppliers in the system (reducing complexity). In parallel, actual external customer demand is available in real time across the entire supply chain to smooth daily production at each organizational interface within the supply chain. In contrast, dysfunctional supply chains do not synchronize demand and manufacturing schedules that cause numerous operational breakdowns across the supply chain. At a tactical level, key characteristics of Lean supply chains include working with the best-performing suppliers in a partnership relationship, ensuring that all participants have operational and technological strenghts that complement each other's capabilities, and implementing pull demand systems (by leveraging advanced technology) to smooth demand throughout the supply chain, as well as deploying Lean Six Sigma systems to stabilize and standardize all internal operations.

The organizational inter-relationships in supply chains are often complex. However, to improve processes linking the organization within the supply chain we must understand the key relationships between their operational metrics. It is also important to control the flow of material, labor, and information across the supply chain as well as critical activities of each organization within the supply chain.

Figure 7.1 shows a simplified flow chart of common organizational functions that are replicated across a supply chain. Functional boundaries lead to competing priorities and conflicting goals and objectives, which cause operational breakdowns and metric deterioration. Metric deterioration is also exacerbated by overall system complexity. Complexity is caused when businesses operate independently of each other and do not actively share information. System complexity increases lead-time and demand variation. The goals of Lean supply chains are to reduce system complexity and to deliver products and services at specified cost, defect-free, and on time across numerous business entities and organizational functions within the supply chain to the final customer.

To meet its basic functions, a Lean supply chain must translate customer requirements called critical-to-quality (CTQ) characteristics into internal process outputs, that is, key process output variables (KPOVs). KPOVs include operational metrics related to time, cost, and quality. The Lean Six Sigma improvement team works backward through the process to identify the major drivers of the KPOVs by understanding the functional relationships, that is, $Y = f(x)$, between key process input and output variables. The process for CTQ identification depends on the specific characteristics of the system under study. As an example, a critical CTQ is demand estimation. Demand estimates are a major building block of the organization's annual sales and operation planning (S&OP). This S&OP information is used to create and modify the organization's master production schedule (MPS). The MPS aggregates demand information and offsets the aggregated demand by time and location across the business. The aggregated MPS information is used by the master requirements planning module (MRPII) to time-phase the purchase of materials and manufacture of components based on lead-time offsets,

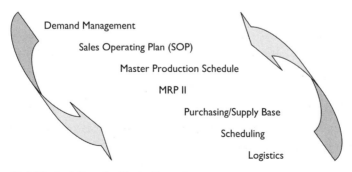

Figure 7.1 Typical Supply Chain Functions

bill-of-material (BOM), and inventory levels as well as other information. Once the manufacturing schedule has been executed, the products are either inventoried or shipped to the customer. Lean Six Sigma methods have proven very effective in investigating the root causes for failures in these systems as well as their elimination from the process. The specifics of demand estimation, the MPS, MRPII, and related topics were discussed in Chapters 3, 4, 5, and 6.

Another way to visualize a supply chain is as a series of functional "silos" as shown in Figure 7.2. Functional silos reduce communication between organizational functions, resulting in misalignment of goals and objectives. Misalignment creates many non-value-adding activities throughout the supply chain as well as "hidden factories," that is, rework loops. Operational metrics are also usually poorly aligned across each functional silo and between organizations. This situation makes process improvement difficult. As an example, when the logistics function is asked to lower inventory investment, but the sales function needs to increase sales, there may be an organizational conflict. It may be difficult to identify and work toward an optimum inventory investment level, necessary to support sales objectives, unless all functional silos work together. The reason for poor coordination is that information moves at a slower rate across organizational silos. Process improvements and inventory management can be difficult in these systems since lead-times will be longer and demand higher due to non-alignment between organizational functions. These system characteristics result in excess inventory and reduced customer service levels. In contrast to "silo" type organizations, Lean supply chains leverage advanced technology and Lean Six Sigma improvement projects to eliminate organizational barriers to improve process performance.

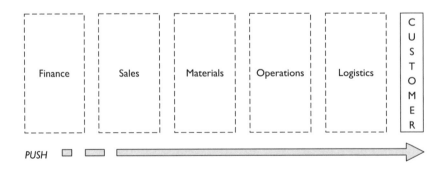

Stovepipe/Chimney/Silo … Lack of Integration … Conflicting Goals … Conflicting Priorities

Figure 7.2 Organizational Characteristics

The result will be common goals, objectives, metrics, and alignment of customer value across the supply chain. As an example, in Lean supply chains, the concept of account management by market segment replaces internally focused estimates of customer demand such as forecasting models. Central to this capability is deployment of electronic commerce technology and voice-of-the customer (VOC) methods. Electronic data interchange (EDI) allows real-time coordination between all participants within the Lean supply chain. The result is accurate, on-time orders and lower costs for the consumer. Intel, 3M, The Home Depot, Johnson and Johnson, Wal-Mart, and Dell's supply chains are classic examples of systems having high asset-utilization efficiencies by using real-time EDI technology. Taking this concept further, best-in-class organizations are also able to supply differentiated products, services, and information on demand due to effective demand management as well as reductions in lead-time.

An important differentiator of a Lean supply chain is high inventory utilization efficiency (turns). The first step in improving inventory utilization efficiency is backward linkage of key operational metrics from the external customer. This process of metric linkage continues through the entire supply chain organization by organization throughout all major functions. Figure 7.3 shows a high-level view of this concept in which the supply chain's functions are integrated using information technology (IT) and other enabling technologies. Enabling technologies use electronic commerce to capture customer demand and order status in real time, use marketing research methods to estimate product demand by market segment, and deploy operational initiatives such as

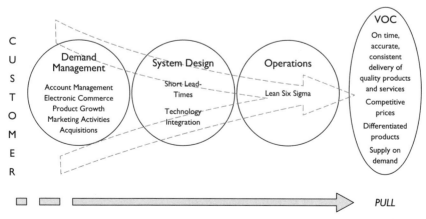

Figure 7.3 Integrated Supply Chains

Lean Six Sigma methods to simplify and standardize their systems. In these systems, complexity is also decreased, effectively reducing system variation. An integrated supply chain makes it easier to identify Lean Six Sigma improvement projects and eventually make the necessary process changes. In Lean supply chains the 20 important supply chain metrics are measured frequently against business goals and objectives to systematically improve organizational performance over time.

Distribution Systems

Distribution systems have become increasingly efficient through the use of advanced information technology (IT) and related systems that manage demand and supply across the supply chain. The deployment of integrated enterprise-wide systems such as SAP and Oracle has resulted in significant waste elimination and increased supply chain flexibility, improved quality, and lowered transaction costs. In parallel to the evolution of supply chain design has been the globalization of demand and supply across the world's diverse cultures and countries. Global supply chains have also evolved at a rapid pace within a multitude of operating system environments. In the global supply chain, customers have become more demanding and place a heavy reliance on supplier on-time delivery promises. Customer expectations and demand for on-time delivery also drive process improvement efforts. Improvement efforts not only rely on IT infrastructure for their effectiveness, but also on Lean Six Sigma tools and methods for success.

Although supply chain productivity has dramatically increased over the past several years, customers are demanding more operational efficiency and productivity in the form of lower per-unit costs and higher quality. This implies there will be constant pressure to do more to increase customer satisfaction. In the future, supply chains that have not implemented enterprise-wide IT upgrades to manage demand and supply or deployed Lean Six Sigma improvement initiatives will be replaced by those which are more efficient relative to cost, time, and quality. Customer demand for more differentiated product and services is also driving these changes in supply chain design. Global pressure for low-cost products and service differentiation has resulted in innovations of products and services creating entire new types of products and services. In fact, new approaches to match supply and demand have resulted in entirely new types of global supply chains.

In this global supply chain environment, outsourcing of work and partnering with other organizations are very commonplace. The consensus is that the work within a supply chain should be done by those who can do it best from an overall system optimization viewpoint. Outsourcing (also called third-party outsourcing) and insourcing (adding value adding and differentiated product and services to customers) have also been accelerated by the Internet and the emergence of new distribution channels. Examples of these new distribution channels include online shopping networks typified by Amazon.com, eBay, and similar business entities. An organization's productivity is directly linked to its supply chain efficiency since the purchase and distribution of materials and services in most organizations is more than 70% of their cost-of-goods sold (COGS). In parallel, customer expectations continue to increase. In this dynamic environment, the opportunity for deployment of Lean Six Sigma projects to further enhance productivity and quality is enormous.

To increase supply chain productivity and quality, Lean Six Sigma improvement initiatives have been deployed across numerous industries ranging from industrial products to retail chains as well as food distribution. In each of these supply chains Lean Six Sigma project opportunities have surfaced in every function within the supply chain. Typical Lean Six Sigma project applications are shown in Table 7.1. Project applications are associated with the major functional areas listed in Table 7.2. These project examples include reductions in direct and indirect labor, improvements in order accuracy as well as on-time delivery,

1. Managing labor expense/ overtime	11. Returned product
2. Ensuring on-time order shipment	12. On-time delivery
3. Order accuracy	13. Number of shipments
4. Inventory/data accuracy	14. Freight charges (inbound and outbound)
5. Eliminating damaged product	15. Inventory turns
6. Eliminating emergency orders	16. Percentage of obsolete and dead inventory
7. Eliminating order redeployment	17. Inventory shortages and overages
8. Eliminating backorders	18. Premium freight costs
9. Optimize inventory transfer	19. Product transfer between facilities
10. Reducing order lead-time	

Table 7.1 Key Areas to Apply Lean Six Sigma Projects

▪ Import/export	▪ MRPII/DRP
▪ EDI	▪ Internal inventory management
▪ Transportation analysis	▪ Materials handling
▪ Carrier management	▪ Packaging
▪ Load optimization	▪ Inventory management
▪ Fleet management/maintenance	▪ Light assembly
▪ Traffic routing	▪ Workload management
▪ Claims management	

Table 7.2 Typical Distribution Activities

improvements in inventory turns, reduction of premium freight charges, reductions in excess and obsolete inventory, and reduction of inventory shortages as well as returned goods. There are literally hundreds of other project opportunities within the various supply chain functions. Many of these will be discussed in either this chapter or later in Chapter 10 in the section entitled "Lean Six Sigma Applications."

How does Lean Six Sigma improve supply chain performance? Using "on-time delivery" as an example, as well as the 10-Step Solution Process, the first step of the solution process is to establish clear definitions and standards of "on-time delivery," that is, to define the problem well from the customer's viewpoint. The second step is to accurately and precisely measure on-time delivery performance to determine its current performance level. The gap in current versus target performance serves as the basis for the "on-time delivery" project. The team then works backward into the delivery process using various data analysis tools to determine the root causes for poor on-time delivery. As the root causes of poor on-time delivery are identified, countermeasures are applied to systematically improve process performance. The countermeasures are used to control the process to ensure that improvements are sustained over time. Over many Lean Six Sigma projects, a supply chain becomes more and more efficient and major metrics begin to improve both from an internal and external customer viewpoint. The supply chain evolves into a "Lean Six Sigma supply chain."

However, some supply chain problems are more difficult to solve. As an example, excess and obsolete inventory exists for many reasons. The first phase of the Lean Six Sigma project might be an immediate reduction in excess and obsolete inventory through online auctions, rework of materials, and so on. These activities provide a one-time business benefit.

However, eliminating the root causes for the problem can prove to be very difficult. As an example, if the excess inventory was caused by a failure of purchasing to negotiate smaller lot sizes, this situation may be easily corrected if the supplier agrees to a process change relative to lot size. But the supplier may not agree if the item has a relatively small usage. In fact, to economically manufacture the item, the supplier may have to run large lot sizes (in the absence of a Lean manufacturing system). Another example would be new product introductions. In new product introductions, demand forecasts tend to be overly optimistic. If this situation becomes a chronic problem, its solution may require a redesign of portions of the new product introduction process The improvement methodology employed in this process would be design for Six Sigma (DFSS). In the DFSS method, an organization redesigns its demand management process to reflect best-in-class tools and methods. In other words, supply chain projects touch several functions within an organization, and these functions may be dispersed over large geographical regions. The solution of the problem is innately more difficult than, say, optimizing a manufacturing machine in just one facility, which has been the "classic" Lean Six Sigma application.

Receipt of Customer Orders

Customer orders arrive into a distribution center in a variety of ways. Sometimes demand is entered into the system by an order entry person working from a centralized call center or electronically by electronic data exchange (EDI) between the customer and supplier's scheduling systems. If call centers are used to enter demand into the system, process breakdowns may occur because of the large number of manual transactions associated in these systems. As an example, a common characteristic of call centers is high employee turnover. High employee turnover contributes to process breakdowns because new employees are always on a learning curve and make mistakes at a higher frequency than more experienced employees. As a result there are always significant training issues within call centers that impact operational performance. EDI systems will always have a significantly higher quality level than manual systems. However, some call centers exist because the customer interaction is complex, which minimizes the opportunity to fully automate incoming transactions. In these high customer-interface systems, complexity arises because call center agents must ask for complicated information from the customer, perform work on the information, or pass it on to an automated system.

Call Center Operations

Call center productivity has increased dramatically over the past several years in large part due to automation of manual operations through IT investment. To the extent that there are manual operations in a call center, Lean Six Sigma improvements have significantly reduced operational costs, cycle time, and improved quality levels. However, significant productivity opportunities still remain given the complexity of these systems. As an example, in the order management process, Lean Six Sigma project identification begins at the customer/agent interface with its high degree of customer contact. At the customer/agent interface, the voice-of-the-customer (VOC) is translated into internally quantified operational metrics, that is, the big "Y's" of key process output variables (KPOVs). As shortfalls are found between required performance of KPOVs versus actual operational capability, Lean Six Sigma projects are created to close performance gaps.

There are many process breakdowns at this interface due to failure to fully quantify customer needs and requirements. Applying Lean Six Sigma concepts to the customer interface as well as other major functions of the call center will help to identify improvement opportunities in areas such as agent hiring, agent training, call escalation in cases where the agent cannot answer customer ordering questions. Other typical projects include reduction of call duration, measured as *average customer handling time (AHT)*, and improved efficiencies in agent staffing and scheduling.

Agent Hiring and Retention

Agent hiring and retention is a complicated process. This complexity contributes to excessive hiring costs due to many factors including poor retention rates and low hiring yields. In other Lean Six Sigma projects, poor retention rates were found to be caused by many factors including not correctly training agents (resulting in voluntary or involuntary terminations) and failure to adequately address the individual needs of agents on a timely basis. On the other hand, low hiring yields have often been found to be caused by unnecessarily complex hiring processes. These situations result in a failure to attract the most suitable agents, unrealistic screening exams that filter out potentially good agents, and a failure to adequately communicate job requirements and benefits to potential agents. There are often other contributing factors, but these are often the major reasons for chronic problems related to employee hiring and retention.

Over the past several years, Lean Six Sigma projects have also dramatically improved the human resource area of call centers. In one call center, agent turnover was more than 100% per year. It was costing the center millions of dollars to hire and train its workers. The ultimate solution was found to be an outdated corporate policy forcing terminations if people were absent or late more than a predetermined number of times. The final solution was to implement a more realistic and flexible work schedule. Within three months the call center's turnover rate was reduced to less than 25%.

Call Escalation

As customer questions concerning their orders are routed through an automated voice response system (VRS), they should be directed to an agent who has the necessary skills and knowledge to adequately handle customer questions. Skills-based routing systems are based on this simple concept. However, because of system breakdowns, the original intent (entitlement) of the system is not always achieved in practice. As customers call into the center, their call may not be efficiently handled since the agents who have been trained to handle this type of call are not available. This may force the attending agent to transfer or escalate the customer's call to a supervisor or another department within the call center. There may be many reasons for an escalated call. The customer may have been given incorrect contact information, the routing algorithm may be incorrect, or the agent may not have the necessary training or tools required to correctly answer the customer's questions. However, regardless of the issue, Lean Six Sigma projects have been successfully deployed to attack the root causes for these process problems.

Escalated calls are a major problem for call centers. Calls not resolved on the first attempt deteriorate productivity and customer satisfaction. In some call centers, the percentage of escalated calls exceeds 25%. This is a serious problem. If a call center receives a million incoming calls per year and 25% of the incoming calls must be rerouted to other agents or departments, a major productivity loss occurs. This is because 25% of the direct labor hours are wasted. In other words, if all calls were resolved the first time, we would need 25% less direct labor hours (or agents) for the current incoming call transaction volume. Customer satisfaction is also lowered as customers must wait to be routed to another agent with the required skills and knowledge. Many Lean Six Sigma projects have been deployed to reduce the incident rate of escalated calls. Some of the solutions to these projects included more effective call routing, identification of specific agent-training deficiencies, and providing the necessary tools so that agents could perform their duties.

Average Handling Time (AHT)/Service Levels

Once a customer has been properly routed to an agent to answer questions related to their order, the agent provides the customer with the required information or performs tasks necessary to answer customer questions according to a standardized script and procedures. In fact, this is true for most functions within the call center including incoming orders, outbound calls, billing questions (if applicable), and complaints as well as questions concerning technical support. Process breakdowns occur due to inadequate training, incorrect or outdated procedures, and irrelevant or hard-to-access information. Any process breakdown increases the length of the call (AHT), thereby reducing productivity, and lowers customer satisfaction.

Even moderate increases in call duration have a major negative impact on call center productivity. If the AHT has been designed to be 60 seconds, but the agents routinely exceed this target duration time by 30 seconds, the center productivity will be lowered by approximately 50%. Lean Six Sigma projects have been deployed in these areas to identify and eliminate the reasons for increases in AHT. A common solution to reduce AHT is to Lean out the process using 5-S techniques. In these projects, the Lean Six Sigma improvement team analyzes standardized agent scripts for each customer segment. These scripts consist of the sequence of customer deliverables, that is, key process output variables (KPOVs). These deliverables include the exact information that will be provided to the customer, and how long it should take to provide the information and alternate interaction pathways if the customer requires information the agent cannot provide. To execute the standardized script, the agent must be supplied with information, tools, and templates. This is where 5-S is very useful. In one Six Sigma project, a 20% reduction in call duration (AHT) for incoming calls resulted in a significant annualized direct labor cost savings. An additional benefit was higher agent satisfaction because agents were provided with adequate training and the tools to do their jobs successfully.

Agent Staffing and Scheduling (Utilization)

Staffing and scheduling of agents is critical to maintaining call center efficiency and customer service levels. Having the right agents in the right place at the right time ensures that the average customer waiting time and service targets are met in practice. Achieving this performance level requires an understanding of incoming call volume by time of day and week and required agent skill levels by customer market segment.

Queuing analysis is useful to model the dynamic relationships in systems in which work arrives in a probabilistic manner (distributed) and the service times are also probabilistic. Queuing models help set schedules in these systems since the required labor resources are determined based on customer demand and target service levels as well as agent utilization. Chapter 8 provides an example of using queuing analysis as well as Lean Six Sigma tools to improve agent scheduling.

Although many automated work scheduling tools are available to aid in this task, there always exist situations where the impact of variation caused by manual interfaces is a problem. As an example, accurate information regarding incoming call transaction volumes (demand) and labor (resource availability and capacity) are required to set work schedules. Also, to estimate resource availability and capacity, these automated scheduling systems use average handling time per call (AHT) and agent training (skill levels). Estimates of these metrics as well as other system parameters must be accurate to prevent operational breakdowns. Lean Six Sigma projects have proven very useful in reducing the variation found at the interface between people and the automated scheduling systems.

Reducing Customer Complaint

Lean Six Sigma methods are routinely used to identify, analyze, and reduce customer complaints to improve supply chain performance. Accurately capturing customer information when they call into the distribution center to complain about a problem is critical to long-term process improvement. There are many possible types of customer complaints due to supply chain complexity. Complaints having a high business or customer impact serve as good Lean Six Sigma projects. These complaints typically include order shortages or overages, damaged product, invoicing errors, or late shipments as well as many other problems which may be specific to an industry. A short list of common areas where complaints occur is shown in Table 7.1. Also complaints can occur in any of the operations shown in Table 7.2.

Electronic Data Interchange (EDI)

The advantage of EDI apart from its lower per-unit transaction cost, relative to systems having manual operations, is the fact that once optimally designed, quality levels in automated systems are very high compared with systems relying on manual interfaces between the customer and distribution center. EDI links the customer and distribution center functions using technology that enables information sharing. Information that is commonly transferred between customer and distribution centers includes order receipt and its ongoing status. In advanced EDI applications, customers can check supplier inventory levels, access special information related to sales promotions, and obtain other important information such as product performance specifications in real time. A simple example would be using a FedEx tracking number to determine where your order is located at any point in FedEx's system. EDI and other automated systems are major enablers of Lean Six Sigma supply chain productivity and quality improvement.

Transportation

The "transportation" function is a major component of distribution centers. Table 7.2 shows several major distribution activities. Several of these activities fall into the area of transportation, including import/export of products, handling of orders between facilities using electronic data interchange (EDI) and advanced shipping notification (ASN) systems, planning of optimum transportation routing, maintenance of equipment including vehicles as well as internal warehousing equipment such as forklifts, vehicle load management, direct labor management, and customer claims. Import and export operations involve building orders for export to other countries. How the export order is built may have an important impact on distribution efficiency and inventory investment. If the products composing the export order have long lead-times, the cumulative lead-time for the entire export order may be adversely impacted since some items must wait for others to be received into the distribution center before the order can be exported to the customer. Also, products assigned to the export order may be in short supply and be required by other orders. In these situations, products making up the export order may be "borrowed" to complete more immediate orders. This has the effect of further delaying the export order. Also, export orders are usually batched to fill out transport containers to reduce freight costs. This situation also adversely impacts order lead-time and an organization's inventory investment. On the import side of the distribution center, the receipt of imported items from the supplier may be delayed for similar reasons. In the implementation of an ASN system, the bulk of the integration work involves creation of IT interfaces between the supply chain's various software platforms. This requires extensive software engineering time and expense. However, the development of ASN system also requires the quality level of the items being received into the distribution center be very high. The reason is that these automated systems attempt to eliminate the direct labor of inspection and sorting of materials. This means the suppliers making up the system must be certified to ship the items defect-free. Also, internal warehousing operations must operate at very high quality levels. As an example, items must be correctly located within their assigned inventory locations. They must also be available for order pick or if the warehouse has assembly and packaging operations, kit manufacturing. Planning optimum routing implies the system has been designed to minimize transportation costs by using correct routing and freight classification rates based on the design of the distribution system. While equipment maintenance naturally falls under the category of total productive maintenance (TPM), there are many process breakdowns in maintenance operations. Lean Six Sigma projects can be deployed to eliminate these maintenance problems. Lean Six Sigma projects can also be deployed to improve low direct labor efficiencies. In these situations, the team can analyze the reasons for the low efficiencies and implement the correct solution to the problem. In each of these examples, the goal should be to reduce cost and improve quality. As an organization executes strategically aligned Lean Six Sigma projects, it will become a "Lean Six Sigma supply chain."

Carrier Operations

External carriers are an integral part of the supply chain. No matter how well an organization's supply chain operates, if external carriers perform poorly, customers will be disappointed with an organization's products and services. Carrier management, from an operational perspective, includes ensuring that materials are efficiently picked up and delivered on time to the customer without damage or loss of product. Given the complex nature of the tasks associated with carrier management including IT, equipment, and manual tasks, there will be process breakdowns. Lean Six Sigma projects are created to identify and eliminate these breakdowns. In a manner similar to other improvement projects, the Lean Six Sigma team verifies the current performance baseline and identifies the root causes of poor performance, but, unlike internal projects, in these projects the team must also include external people from the carrier organization and perhaps some specific customers to provide feedback and take part in system evaluations.

One classic project example where customer and carrier involvement is very important is "customer shortages." The shortage problem surfaces through customer complaints regarding shipments that are missing items.

After the organization receives the customer complaint, solution is difficult because the problem could be caused by many factors. These include the organization's internal picking operations, internal carrier operations, or even the customer's receiving process. Within these three basic areas of investigation, specific reasons for the shortages could include items inadvertently substituted for the correct item, items inadvertently left off the final shipment to the customer by the carrier (if the original shipment had to be broken down internally at the carrier's "break bulk" operation center), or items that the customer received incorrectly due to a process breakdown. There is also the possibility the item could have been stolen. The classical way to investigate shortage problems is to tighten up dock audits to ensure that orders are sent out correctly the first time. However, the distribution system is complex and the fact shortages may occur outside the organization's control, internal actions such as dock audits may not prevent product shortages. To eliminate shortages, an improvement team would be formed with key customers and carriers and, working backward from the customer, would audit selected shipments through the entire distribution system. Shortages could be occurring in either of three major areas of the system. These include internal warehousing operations, internal carrier operations, or the receiving operation (or ordering operation) at the external customer. Also, there could be several reasons (or root causes) for the shortages.

Another major function of carrier management includes effective load management. Load management is a complex task, which includes ensuring that the correct quantities of each ordered item are placed on the trailer, in the right location on the trailer, and stacked in a manner that will facilitate unloading by the customer, as well as transportation through the system without damage caused by load shifting or environmental conditions such as rain or snow. Another major function of load management is to ensure that the trailer carries the maximum weight, but no more due to the possibility of governmental fines. One interesting Lean Six Sigma project occurred several years ago at an organization that had several warehouses, which were used to inventory items manufactured at each warehouse location. These items were transferred back and forth between sister warehouses in other states. The problem was that the trailers were not being filled to their maximum weight capacity and left the trailer partially filled with product. In those days, a trailer cost approximately $2,500 to go across the United States, so partial loads were costing millions of dollars in incremental transportation costs. The initial solutions used to fix the problem were to write up the employees, train them, and so on. These solutions were all found to be ineffective because they were not tied to root cause analysis (not fact-based). When the project was finally handed off to a Lean Six Sigma black belt, the black belt quickly discovered the problem was caused by a flaw in the IT logic flow. This logic flaw caused the trailer to close at a lower than optimum weight. This electronically closed the trailer, preventing the employees from adding any more products to the trailer. Why did this problem persist for so long? Well, no one had focused on its root cause and its solution. It was easier for the corporate traffic department to assume the people were not doing their jobs properly rather than dig deeper into the real causes for the problem. Many Lean Six Sigma projects are similar to this situation. The resultant countermeasures to eliminate these problems are usually simple and inexpensive to implement.

Fleet Operations

The question often arises as to why systems as deceptively simple as the deployment and maintenance of trucks, that is, fleets, can possibly have systematic problems? The fact is there will be process breakdowns whenever there are systems that include materials, equipment, and people. Typical Lean Six Sigma project opportunities in fleet management and maintenance include optimum routing of vehicles, personnel hiring, employee scheduling and productivity, and vehicle purchase and maintenance as well as the purchase of all maintenance, repairs and operating (MRO) supplies including fuel. Specific Lean Six Sigma project applications include various aspects of the efficient delivery and pick-up of products at customer or supplier distribution centers. As an example, distribution centers are very busy places, which have assigned times when trucks may deliver or pick up products. If trucks arrive too early they must wait for their loading dock to be available. This situation wastes time and reduces employee productivity. On the other hand, if the truck arrives too late, then the driver may have to wait an exceedingly long time before the product can be unloaded or the customer may experience shortages in the order fulfillment process if their inventory levels are low. In both cases customer satisfaction and productivity is lowered from an optimum level. The root causes for arriving too early or late at the distribution center may be very complex given the truck and driver are both part of a complicated system. Examples of possible root causes for the problem might include poor scheduling of the delivery by either the supplier or customer, traffic delays, poor training of the driver or people receiving the product, inefficient unloading by the customer, or the unloading dock was not available to the supplier at the time originally promised by the customer. A Lean Six Sigma improvement team can systematically get at the root causes of these problems to eliminate them and ensure that the countermeasures are effective over time.

Warehousing

Warehousing is also a very complex process. Each supply chain has its special warehousing requirements depending on the types of products it handles. As an example, some supply chains may handle industrial products which require special packaging whereas other supply chains may handle perishable items like food or governmentally controlled items like alcohol, tobacco, and controlled substances which require special controls. Every supply chain has its own special operational issues with respect to the products it handles as well as the types of process breakdowns which occur within its supply chain. However, regardless of its inherent design, every supply chain can be improved using the Lean Six Sigma methodology, that is, use the 10-Step Solution Process to define the problem from the customer's viewpoint (working backward from the customer), and then systematically identify root causes and implement countermeasures to eliminate the problem.

Receiving and Shipping Operations

Receiving operations include ensuring that the truck is assigned a specific unloading dock, proper equipment is in place to unload the truck, people are trained to efficiently unload the product from the truck and record correct item descriptions and quantities, the unloaded items are properly placed in inventory storage, and the inventory transaction is properly closed. Other important activities are properly storing dunning, that is, pallets, totes and slip sheets and electronically closing the emptied trailer.

Shipping operations include the proper staging of orders on the shipping dock and efficiently preparing the order for shipment. Associated operations include ensuring that the order is staged at the correct shipping dock location with all the required paperwork and documentation. Once the order is staged on the shipping dock, it is loaded onto the trailer in an efficient manner. An important consideration is that the product not be damaged in transit and it arrives in a condition that will facilitate efficient unloading of the trailer by the customer. The product must also be protected from environmental conditions that may damage it.

These conditions include rain, humidity, temperature, and vibration. To protect against rain damage, various sealing techniques, that is, plastic shields, are used at the end of the trailer or in the case of perishable items, more elaborate environmental controls may be required such as refrigeration and shelf life controls. Vibration protection requires specialized container and package design. Process breakdowns in any of these areas can be eliminated using Lean Six Sigma projects.

Warehousing Operations

Table 7.1 lists "order accuracy" as a supply chain problem. Order accuracy problems are chronic in warehousing operations. Order accuracy is impacted by three major parts of the supply chain, that is, the distribution center, the carrier, and the external customer. But focusing just on the internal operations of the warehouse reveals many root causes for poor order accuracy. Order accuracy must be clearly defined from an external customer perspective. The customer expects the order to arrive defect-free, without damage, and on time. Focusing for a minute on just the concept of "defect-free," how is a defect defined from an external customer perspective? An order consists of "line items." Each line item represents one item and a quantity of the item. From an external customer viewpoint the line item is either correct or not. In addition to line item accuracy, the order must arrive on time and without damage. If there are process breakdowns which impact any of these requirements, the Lean Six Sigma improvement team working backward from the customer and using the 10-Step Solution Process can collect and analyze relevant data to drive to the root causes for the order accuracy problem. These root causes can take many forms. The original picking list could be incorrect due to errors in the ordering process, the person picking the order could misread the line item model or quantity information, or the item may not be in available inventory. There are of course many other issues which may be related to poor order accuracy.

Given the complexity of internal warehousing operations that include equipment, information, materials, and people there are numerous opportunities for process breakdowns. Table 7.2 lists several major operations associated with distribution center warehousing. These operations include material requirement planning (MRPII) or distribution resource planning (DRP), analysis of inventory location accuracy within the warehouse, material-handling strategies, packaging methods and technologies if applicable, light assembly and kitting operations, and direct labor workload management. Since MRPII was discussed in Chapter 5, we will discuss the other warehousing operations. DRP manages inventory-stocking strategies across the supply chain based on demand and supply constraints. Problems in inventory management are indicative of process breakdowns within the DRP function. Inaccurate inventory estimates will cause extra inventory to be ordered by material planners to meet shipment schedules, or inventory shortages may occur if orders cannot be shipped because inventory is not available. In the latter case, the items must be shipped separately to the external customer resulting in extra material-handling and shipment costs as well as customer dissatisfaction. Problems with damaged product are usually caused by poor packaging design or poor handling within the distribution center. Emergency orders result in increased handling and shipment costs. Also, emergency orders may take inventory from orders already promised to customers but not shipped. These situations will force redeployment of orders across the organization's distribution network creating backorders because the inventory was not available or was inefficiently picked by the warehouse operations people. Again, working backward from the external customer using the 10-Step Solution Process will eliminate these problems.

Facility Maintenance

Facility maintenance manages a variety of equipment used by distribution centers to move materials and products from receiving to inventory storage and on through to the shipping process. Those operations related to equipment include the purchase, maintenance, and safe handling of the equipment. Facility maintenance is also tasked with management of energy costs, buildings, and grounds. Energy usage includes lighting, heating, and air conditioning. Lean Six Sigma projects have been successfully applied in each of these areas. Using energy usage as one example of effective energy management, one organization wanted to optimize its lighting usage across three facilities. They had already installed automatic on-off sensors to conserve lighting, but initiated a Lean Six Sigma project to investigate further reductions in energy usage without creating an adverse impact on their employees. In this project, the Lean Six Sigma black belt made an initial baseline study of lighting throughout each facility using a light meter. The light meter measured the light in every room in each of the three facilities. The goal was to meet OSHA lighting standards for all employees in all locations and reduce lighting costs. The Lean Six Sigma black belt found that even after using automatic lighting sensors, some areas of each facility had too much light and others too little. Other issues associated with janitorial staffing surfaced; for example, one janitor read the paper everyday during his lunch break, but the lights in the entire office area were all turned on since he was in that large room. The final solution was to implement an energy savings program in which light would be right-sized in each room throughout the three facilities. Since minor amounts of capital were required, a long-term capital plan was implemented to gain the annualized business benefits net of capital. These benefits were significant. This project is a good example of how even the most studied process may be significantly improved by listening to the voice-of-the-customer (VOC), which in this case was the OSHA lighting standard, and conducting careful measurements of the key process output variables (KPOV).

Distribution Inventory Management

The management of distribution inventory requires visibility to all inventory locations throughout the system. Inventory policies are then established based on expected demand at a given facility versus the lead-time required to replenish the inventory at each location. Also, low demand items may be centrally warehoused to minimize cost but maintain customer service targets. In this context, distribution inventory management requires system-wide planning and control.

System Inventory

To optimize inventory investment, all warehousing locations should have visibility to system inventory, but inventory should be managed and controlled by individual warehouses or branches throughout the distribution system in real time according to established policies and procedures. In a distribution network, once the number of distribution branches (stocking locations) has been calculated (network system optimization), it is important to establish standardized policies and procedures to maintain optimum inventory levels across the system to meet customer-service-level targets at minimum cost. The optimum number of inventory stocking locations is determined based on trade-offs between inventory quantities at each item and location versus transportation and facility costs. Analysis of the system trade-offs allows the analysts, who design the system, to construct an optimization curve similar to Figure 7.4. Reviewing Figure 7.4, we see fewer inventory locations results in a lower overall investment across the system, but a higher total transportation cost. This fact is shown in Figure 7.5 where safety stock increases from 10,000 units using one stocking location to more

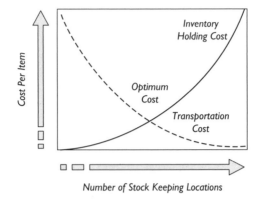

Number of Stock Keeping Locations

Figure 7.4 Optimum Stock-Keeping Locations

than 20,000 using four stocking locations. The optimum solution for a given network requires balancing off stocking locations which impact total system transportation cost versus inventory investment across the entire system. Using this simple analysis as a baseline, the system can be modified to determine an optimum stocking strategy based on anticipated product demand for every item and location. This type of analysis is also useful to reduce investment in slow-moving items, which can be stocked at a centralized location to minimize inventory investment but meet customer service targets. Process breakdowns in the original stocking strategy can be eliminated using Lean Six Sigma projects to identify their root causes and developing countermeasures to eliminate the root causes from the process.

In addition to poor design of the distribution system network, inefficiencies in routine warehouse operations can significantly decrease operational efficiencies and increase inventory levels. To improve

$$\text{Safety Stock For Each Location} = \frac{\text{Safety Stock @ One Location}}{\sqrt{\text{Number of Locations}}} = \frac{10,000 \text{ Units}}{\sqrt{4}} = 5,000 \text{ Units}$$

$$\text{Total Safety Stock} = 4 \times 5,000 \text{ Units} = 20,000 \text{ Units}$$

Figure 7.5 Inventory Impact

operational efficiencies, replenishment policies and procedures for every item and location should be established by their warehousing location. Replenishment policies and procedures should be based on actual item demand by location. High-volume items or items that are used to fill the majority of customer orders should be located for easy access for order filling and minimization and handling by the order pickers. Location accuracy and quantities should also be monitored to ensure that there will be no backorders or emergency replenishments. To the extent these conditions do not exist, productivity is reduced, order lead-time increases, and customer satisfaction is reduced. Lean Six Sigma projects are often deployed to improve operational efficiency in these processes.

Kits and Assemblies

Poor scheduling of kitting activities can result in high levels of work-in-process (WIP) inventory and poor on-time delivery of the finished kit to the external customer. This situation can be caused by numerous operations that are necessary to bring together various kitting subcomponents. A good practice used to ensure smooth build-up of kits is to maintain a separate material list for every end item (kit) and service part. Also, keeping accurate and separate demand records for kits and service parts ensures that subcomponents will be assigned correctly based on bill-of-material (BOM) structure. When process breakdowns occur in kit-building operations, the implementation of Lean methods including 5-S, mistake proofing, and a demand pull system are the most effective ways to eliminate the problems related to scheduling and building the kit assemblies. This is because they use visual control systems such as Kanban to minimize the impact of lead-time and demand variation on the system ensuring kits are started only when there is demand for them and subcomponents are available to complete the kitting order.

Consignment Inventory

Consignment inventory is held by a supplier based on customer request. Normally, the supplier agrees to hold the inventory at no cost until it is needed by the customer. There are pros and cons to this practice depending on specific circumstances. If the supplier has refused to improve productivity and quality, consignment can be used to force improvements to the supplier's operations. Also, if the same material or component is provided to several different locations of the same customer or customers in different industries, it may be more cost-effective for the supplier to manage the inventory for everyone. However, in most situations, consignment inventory is used to perpetuate poor inventory practices across the supply chain. This is particularly true when the customer does not want to make supply chain improvements. Problems with consigned inventory occur if a supplier has little control or visibility to external customer demand patterns. In these situations, overall supply chain cost will be higher since there is little incentive for participants to reduce lead-time or more effectively manage demand. This is in contrast to "vendor-managed inventory," in which a supplier actively manages inventory at a customer location using real-time information from the customer's customer. In this latter scenario, total supply chain cost can be significantly lowered for all supply chain participants.

Vendor-managed inventory in retail environments is a particularly interesting way for a supplier to manage inventory at a customer's location. In these systems, the customer assigns the supplier a certain amount of volume to inventory the supplier's products. The expectation is the supplier will manage the inventory types and quantities in such a manner as to maximize an item's profit flow across the store shelf. The customer is interested in a metric like an item's *profit-per-cubic-foot (PCF) per day.* However, to maximize the PCF metric a supplier must understand customer demand patterns for the item by store in order to stock the correct product. As an example, in one Lean Six Sigma black belt project, the customer was a major retailer and the supplier manufactured automobile replacement parts. The black belt analyzed the current performance level, that is, PCF and showed the customer PCF data for every retail store in the United States. Then, using marketing research data that correlated vehicle demographics to every one of the customer's stores by zip code, the black belt showed how modifications to stocking levels of each product by store could drastically increase each store's PCF metric. Needless to say the customer was very impressed by this detailed analysis and profitability improvement plan.

Cycle Counting

Unfortunately, cycle counting is a necessary auditing process because few organizations have effectively deployed Lean systems. Cycle counting systems estimate inventory quantities by item and location at regular intervals to detect differences between actual versus stated book value for every item. In these systems, the cycle counting can be either a 100% count or a sample taken from the population. If inventory accounts are to be 100% counted on a frequent basis, strategies to minimize direct labor are often utilized when conducting the cycle counting. As an example, in these cycle counting systems, the cycle counting auditor is instructed to count inventory when its estimated quantity is low or zero. In distribution centers which have a warehouse management system (WMS), order pickers are instructed to count inventory when its estimated quantity is low and they are new at the inventory location. If statistical sampling is used as the basis for the cycle counting system, a representative number of inventory accounts are taken from the inventory population (inventory records) and statistical estimates made of inventory value. There are several ways to obtain a statistical sample. In the simplest system, the account population is classified into three groups (ABC method) by value, volume, criticality, or some other criterion. Accuracy targets are set based on the relative importance of each category. Accuracy targets are usually set at 98%, 95%, and 90% for A, B, and C categories respectively (although these targets vary by organization). In more sophisticated systems, the inventory population is stratified by book value and stratified to minimize within-stratum valuation variance. In this latter sampling method valuation estimates can be obtained by sampling 80% fewer accounts than if samples were drawn from the entire population. These valuation estimates are extrapolated to estimate total inventory book value. However, the most efficient cycle-counting system is obtained through Lean implementation using a pull demand system and visual inventory control systems such as Kanban. In Lean systems inventory quantities are calculated and deployed throughout the process in the form of Kanban quantities.

The basic steps necessary to create an effective cycle-counting system are shown in Table 7.3. The first step is to process-map the

---

1. Process-map system transactions.

2. Develop procedures/policies and train people to work the system.

3. Conduct audits and determine baseline inventory accuracies.

4. Establish methodology (statistical sampling or other).

5. Assign responsibilities.

6. Analyze and eliminate errors.

7. Conduct ongoing audits.

8. Immediately correct inventory inaccuracies.

Table 7.3 Design of Cycle-Counting System

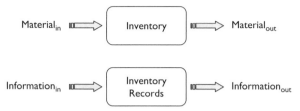

Figure 7.6 Understand Material Flow

inventory transaction flows including all steps and rework loops. The process map should be constructed based on material and information flows through the system as shown in Figure 7.6. In the second step, policies and procedures should be developed to control inventory transactions. Also, people should be trained to correctly follow policies and procedures, which should be easy to understand and include all relevant system information necessary to track and manage the flow of material and information. The third step in creating an effective cycle-counting system is development of the auditing system. Audits should be conducted periodically to identify areas where discrepancies currently exist. In the fourth step, the audit methodology may be either a 100% item count of the inventory accounts or the application of statistical sampling methods. As discussed earlier, the advantage of using statistical sampling methods is cycle counting audits can be conducted more frequently to facilitate immediate corrective action to eliminate inaccuracies. Also, auditing costs can be reduced by using Lean methods and visual controls. In addition to using Lean methods and visual controls, reducing product design and process complexity will decrease the number of items and subcomponents that must be managed by the cycle counting system. Reducing the system complexity also includes reducing the number of suppliers in the system. The balance of the steps required to implement a cycle-counting system include assigning responsibility for the auditing and systematically investigating and eliminating errors found during the cycle-counting operation to systematically improve system accuracy.

Inventory records should correspond to the physical movement of inventory as well as the information necessary to move the materials. This requires analysis of how materials and information move through the system including planned versus unplanned receipts, planned versus unplanned issuance of materials, stock returns, and their associated inventory adjustments. Although, cycle-counting systems require time and resources to create and deploy, major benefits are the identification and systematic elimination of the root causes that adversely impact inventory record accuracy and cause book value discrepancies. Poor inventory accuracy reduces productivity since workers must travel to more inventory locations to obtain items to build customer orders. If inventory is not available, backorders must be shipped to the customer later, increasing operational expense and reducing customer satisfaction. There are many reasons for low inventory record accuracy including poor auditor training, poor document control, inaccurate inventory databases, and poor physical control of assets including damage and theft. Lean Six Sigma improvement teams have been very effective in identifying and eliminating process breakdowns identified by cycle-counting systems.

Import/Export Operations

Import/export operations are another area in which to apply process improvements. Supply and demand are now global in nature because supply chains span huge geographical distances across many diverse culture and countries. Process breakdowns occur frequently in import and export processes due to the complexity of these global supply chains. These process breakdowns become the basis for Lean Six Sigma improvement projects. As an example, if an item is chronically held up by the customs office when it enters the United States the result will be delays in receiving it into an organization's inventory system, which will adversely impact overall lead-time. There could be many reasons for the item being held up at customs. Perhaps it requires special handling, or import duties may not have been paid, or paper work has not been properly completed by the supplier or customer. These chronic issues could be eliminated through deployment of Lean Six Sigma projects. As another example, several years ago export orders were taking a very long time to fill because the process consisted of many low-volume items having long lead-times. Complicating the situation, was the requirement that the entire order had to be sent out at once since export shipping costs were very high. In fact, since several other orders were being shipped to the same country, they had to be aggregated to completely fill shipping containers to save shipping costs. As a result, the order was staged on the shipping dock for several weeks during the time it was being assembled for shipment. However, in the interim, items were sometimes taken from the export order to fill domestic orders which needed the same items. As a result of all these complicating factors export orders were tied up for inordinate amounts of time. Customer dissatisfaction was also very high.

A Lean Six Sigma team studied the problem and eventually eliminated the root causes related to low-volume inventory levels. The solutions were related to improvements in policies and procedures and increased inventory availability for the items which were chronically in short supply. This greatly reduced the lead-time to complete the export orders since items were on hand to complete the export order. The shipping costs were reduced since the order was aggregated with other orders going to the same country.

Third-Party Logistics

Many organizations are outsourcing logistics' functions to third-party providers to concentrate on their core competencies, which are not focused logistics. As a result, outsourcing has accelerated over the past several years and become global. In parallel, organizations are also insourcing value adding goods and services. As an example, United Parcel Service (UPS) not only picks up computers for various manufacturing companies but actually repairs these computers at central locations to minimize supply chain transportation costs. However, breakdowns in the third party logistics' supply chain will definitely require Lean Six Sigma methods due to the inherent system complexity associated with the interacting organizations.

Outsourcing Operations to Improve Quality

Many Lean supply chains are outsourcing operations to third-party suppliers. This practice began with noncore operations within the supply chain, but over the past several years, some supply chains have outsourced their entire distribution system to third-party suppliers that can distribute products more efficiently. All logistics functions can be outsourced, with the greatest efficiencies occurring with small and mid-size businesses. The organizations which outsource do not consider distribution their core competency. Figure 7.7 shows basic services offered by third-party logistic providers, such as transportation, storage, and order processing. Additional services may include electronic data interchange (EDI), advanced shipping notification (ASN), order consolidation, and special packaging services. Typical reasons why organizations use third-party outsourcing are to compensate for inadequate internal resources (skills, capital), to quickly add capacity and operational capability, and to develop new operational capability.

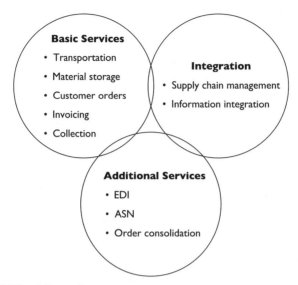

Figure 7.7 Third-Party Logistics

Integrating Third-Party Providers into Lean Six Sigma

Integration of third-party suppliers into the supply chain often starts with outsourcing of noncore tasks such as freight payment, customs clearance, carrier rate negotiation and routing analysis, leasing of equipment and temporary workers, and information technology and data processing, to name just few third-party activities. However, as mentioned earlier, the recent trend has been to outsource what used to be considered as core tasks such as order processing and fulfillment. In the past several years, third-party suppliers have taken over management of entire distribution networks from manufacturing organizations. In these more complicated outsourcing initiatives, the integration of third-party suppliers into the supply chain requires coordination of all demand and supply information. This practice is called "hybrid decentralization." Hybrid decentralization reflects the fact that given the high degree of automation and real-time information of the supply chain, it makes sense to have work performed where it can be done more cost-efficiently and with higher quality. Examples are design engineering, software development, and call centers and other work outsourced to India and China.

The integration of third-party suppliers is facilitated by process simplification and elimination of redundancies and rework loops, that is, the hidden factories. By capturing process information in real time, analyses of supply chain status and decision making are made easier, allowing coordination of all supply chain functions across vast geographical distance. Ideally the process status of all system elements will be visible to all supply chain participants through information technology. At a local work-team level, the major principles of Lean Six Sigma are clearly visible. Through elimination of process waste and creation of localized work cells, the users of the outputs from a process perform upstream operations (previously performed by other functions) to the greatest extent possible. In other words, the decision points should be located where the work is performed in the process or at a local level. The strategy is to organize around the process outputs rather than creating systems in which there are functional silos and divergent workflow streams which create inconsistent process performance metrics.

Reverse Logistics

Reverse logistics is used to receive, evaluate, record and process products that are returned from customers for various reasons. The entire reverse logistics system is a rework operation since there would be no product returned by the customer if the supply chain were operating optimally. Although it may not be currently feasible, the ultimate goal of a Lean supply chain should be the complete elimination of reverse logistics from the supply chain. Most reverse logistics functions are outsourced to third-party providers who receive items into their reclamation center. At the reclamation center, the workers evaluate and record the reason for each item's return. Lean Six Sigma projects can be easily deployed at reclamation centers since data on product returns is readily available. Another advantage of deploying Lean Six Sigma projects at reclamation centers is the "reasons for product return" are closely linked to the voice-of-the-customer (VOC). However, eliminating the "reasons for product returns" may be complicated since they may be tied to consumer or employee behavior that is difficult to analyze and control. This is especially the case where the customer does not know how to use the item or the item is returned for shelf-life issues.

In some supply chains Lean Six Sigma improvement projects are fairly straightforward to deploy and complete since these supply chains handle products that don't really have an expiration date. However, in retail or food supply chains return product issues can be very complex.

As an example, in food supply chains, foods are perishable, have a limited shelf-life and must be protected from the environment. In retail supply chains, especially those in clothing, customer preferences change by the season, necessitating product returns at the end of each season. These returned products might already be obsolete. In retail supply chains, seasonal products associated with holidays, especially those purchased from other countries such as India and China, are particularly problematic since the product's order cycle is several months in advance of the holiday. In the interim, customer forecasts for the product could have dramatically changed resulting in too much or too little available product. Reverse logistics is an area that can greatly benefit from the 10-Step Solution Process using Lean Six Sigma as well as operations research tools and methods.

The reasons for product returns also vary by industry. But the most common reasons are corporate return policies for items that did not sell, items beyond their shelf life, items that are damaged or contaminated, and items that the customer either did not know how to use or ordered incorrectly. Each of these return categories offers opportunities to apply Lean Six Sigma tools and methods since they require statistical analysis and elimination of the root causes for the problem. As an example, one corporation had a product return rate that exceeded 5% of annualized sales. However, corporate policy allowed just a 3% product return. When a Lean Six Sigma improvement team investigated the issue using the 10-Step Solution Process, they found situations in which customers were allowed to return product that was more than 10 years old. Some returned products were from competitors. As another example, items returned because of shelf-life issues are common in grocery supply chains. In these situations food may not be rotated on the shelf at a store level. This practice eventually creates a situation in which an item goes past its shelf-life and cannot be sold at its regular price. If the out-of-date items cannot be sold, they must be aggregated by store and sent to a reclamation center for subsequent processing. The reclamation center sends the out-of-date items to charities or throws them away. In either situation, the item standard cost and extra handling results in an economic loss to the organization. Typical solutions to these problems are often the application of 5-S techniques at a local store level. 5-S techniques ensure that the product shelves are orderly and contain just the minimum amount of the item necessary to meet expected customer demand, and the item's date code is easy to see at a glance. As a third example, items that are damaged must be repackaged, sold at a discount, or thrown away. In one Lean Six Sigma project, a Lean Six Sigma team investigated damage to cartons of a consumer product. The damage rate exceeded 2%. Eventually the cause of the damage was traced to one shift at just one facility and process controls were implemented by the improvement team to eliminate the damage problem.

The application of 5-S techniques is also useful in situations where the customer does not know how to assemble or operate the product correctly or orders the wrong product. In these situations, standardizing the product's assembly operations mistake proofing the assembly operations might reduce the product return incident rate. However, the most effective solution may be to redesign the product for easy assembly by the customer. Design of efficient product assembly techniques falls into the tool set called design for manufacturing (DFM). Using DFM, the number of components in the product is reduced and remaining components can be assembled in only one way (mistake proofed). There are other important characteristics of DFM which can be used to reduce the incident rate of product's return.

Food and Commodity Supply Chains

Lean Six Sigma projects are particularly important in distribution systems that transport food and controlled substances. These items require special handling and strict environmental controls. Given the short shelf life of these products, spoilage is always a real problem and is analogous to the concept of manufacturing "scrap." In addition, to the materials that are being moved through the distribution system, there is a heavy external customer interface at the retail level of the system. Breakdowns occur at these interfaces relative to item pricing (especially relative to price discounted and sales items) as well as product availability. Problems in these systems directly impact customer satisfaction. Complicating the situation is the large number of suppliers in retail systems. The number is typically in the thousands. Also, if controlled substances, alcohol products, and tobacco are sold at the retail level, regulatory issues may periodically become a problem to the organization. In all cases demand and lead-time management as well as inventory investment are critical metrics.

The complexity inherent in these systems causes numerous process breakdowns throughout the system. But these breakdowns can be systematically analyzed and eliminated by deploying Lean Six Sigma projects. As an example, incoming quality audits can be made at the back end of the process at receiving. These incoming audits verify the order has the correct line items and quantities and is not damaged on receipt. But the concept of "damage" takes on new meaning in food supply chains since shelf-life issues may be difficult to detect. Physical damage of the product is also a constant problem since the packaging of the items may be designed more for aesthetics than to protect against handling damage within the distribution system. These problems become exacerbated as the product is moved through the distribution system to local retail stores for sale to customers. As an example, complicating the product damage problem is that different types of products may be stacked together on pallets or loaded onto smaller trucks for delivery to retail stores. In these situations, heavier products may be stacked on top of lighter ones, causing damage to the lighter products. These situations may increase the likelihood of product damage and subsequent shelf-life issues at the retail stores.

The fact that product shelf life is subject to local work standards and practices related to rotation of the products on the shelf at the retail store level further complicates the situation. Customer handling damage within retail stores increases product damage when customers select and discard unwanted products throughout the store. After the customer discards the product, it must be retrieved by store employees and restocked on the shelf if it has not been damaged by handling. When the product is eventually taken to the check-out line by the customer, the purchase transaction will proceed well if item pricing is correct. However, given the large number of weekly price changes in these retail systems, pricing inaccuracies may also be a problem. Item pricing issues are exacerbated by pricing discounts and sales promotions.

At a macro level, food distribution systems may also offer customers specialized services such as cooking meals, selling non-food items, pharmacies, photography, and so on. These additional services complicate the system since they are characterized by many manual and specialized operations which contribute to process breakdowns. As an example, in stores in which meals are prepared, there will be spoilage issues since food must be discarded if not sold within its shelf-life window. Also, employees or customers may be subject to accidents and injuries since they interface with many of the store's operations. Lean Six Sigma tools and methods are useful in retail environments to analyze what are often huge amounts of available supplier and customer transaction data. Value stream mapping is also a necessary analysis tool to understand the material and information flows throughout these systems. As the Lean Six Sigma improvement team works towards solutions, Lean tools such as 5-S and mistake proofing as well as information technology (IT) changes may be necessary to ensure that solutions are sustained over time.

In contrast to food supply chains, commodity supply chains are relatively simpler. Although the overall complexity of these systems is lower than other supply chains, there is high leveragability of Lean Six Sigma tools and methods in areas where there are manual operations. These areas are often at the supplier customer interface relative to on-time delivery, lot sizing, and quality of the material relative to specification and weight accuracy or in the front office. At a more advanced level, supplier and customer cooperation relative to changes in product characteristics might allow both the supplier and customer to make product and process improvements that would increase internal efficiencies and quality. In these latter projects, DFSS and experimental design tools would be very useful. In all cases where Lean Six Sigma methods are used by the improvement team, the team works through the identified process issues, creates operational baselines, and evaluates measurement systems prior to data collection and analysis. Unless the problems require major system modifications, the eventual controls will rely heavily on Lean tools and methods including 5-S and mistake proofing. Modifications to the IT systems may also be required to implement the required solutions.

Summary

Lean supply chains are characterized by the ability to respond dynamically to changing customer demand as well as the efficient utilization of assets. These characteristics translate into high profitability and increased customer service for successful Lean supply chains. The effectiveness and maturity of a Lean supply chain can be measured using the 20 important supply chain metrics. Superior supply chain performance also depends on development of one or more core competencies. Core competencies include such capabilities as improved product development, implementation of Lean Six Sigma practices, elimination of promotional activity to smooth demand, and simplification of product and process designs as well as other systems.

The first step in the analysis of process breakdowns within a supply chain is creation of a value stream map showing material and information flows of the major work streams associated with the process. Creation of the value stream map is important to show the improvement team how the system works. This is especially important since a supply chain normally spans several functions and organizations. Also, breaking a supply chain into several natural work streams and analyzing their operational metrics against optimum targets will help to identify many Lean Six Sigma project opportunities. The analytical tools necessary to identify and analyze a problem may also differ from typical Lean Six Sigma projects. This is because supply chains are networks in which outputs must be maximized, minimized, or kept on target subject to resource constraints. Optimization analysis tools and methods include linear programming, queuing analysis, and related concepts. Unfortunately, operations research tools and methods are not normally taught in Lean Six Sigma classes. As project opportunities are identified by the Lean Six Sigma improvement team, it is important to ensure that all projects are aligned with senior management's high-level goals and objectives as well as external customer requirements.

Chapter 8

Root Cause Analysis Using Six Sigma Tools (With Operations Research Methods)

Key Objectives

After reading this chapter, you should understand how to use Lean Six Sigma tools and methods to drive to the root cause of the project problem, and understand the following key concepts:

1. Why measurement system accuracy and precision are important parts of every Lean Six Sigma project.
2. Why the improvement team should have listed the critical questions that needed to be answered by the data collection and analysis effort prior to doing these tasks.
3. Why creating a quantified system map using Lean Six Sigma yield metrics allows the improvement team to identify Lean Six Sigma improvement projects by looking for areas of low yield and obvious process waste.
4. Why in addition to Lean Six Sigma tools, other analytical tools such as queuing analysis, linear programming (LP), scheduling algorithms, simulation, and financial modeling, may be necessary to identify the root causes for the problem.

Root Cause Analysis Using the 10-Step Solution Process

Data collection and analysis are an integral part of the 10-Step Solution Process. This methodology, in a sequential manner, helps to precisely specify and quantify the project problem, measure the extent of the problem including contributing variables, analyze the data collected in relation to the problem, postulate improvements from the analysis, and test the best improvement scenarios under controlled conditions to ensure optimum solutions are integrated into the modified process. Integration and control of these process improvements will be discussed in Chapter 9. Additional considerations in root cause analysis are types of required data, their accuracy and precision, over what period the data will be collected (short or long) and the analysis methodology used to identify KPIVs. These concepts will be discussed in this chapter.

Operational breakdowns directly impact supply chain profitability and customer satisfaction. The root causes for poor process performance become apparent as we successively delayer a high-level operational problem. Taking inventory investment as an example, at a high level, lead-time and demand management issues directly impact inventory investment. At a lower level, longer lead-times are caused by operational issues related to lot size, poor quality, incorrect MRPII/MPS data constants, and on-time delivery, as well as others. In a similar manner, at a lower level, poor demand management is caused by issues related to poor forecasting models or a failure to accurately capture customer demand which are often seen as poor scheduling and inefficient capacity utilization. In Lean Six Sigma projects we analyze complicated supply chain problems to understand the relationships between high-level supply chain metrics and the operational root causes responsible for their poor performance.

Project Alignment (Review of Steps 1, 2, and 3)

The importance of project selection and strategic alignment were discussed in Chapter 1. Performance gaps in important supply chain metrics form the basis for improvement projects and are incorporated into the project charter's problem statement and objective. After the Lean Six Sigma improvement team has been formed, it should meet to review the project charter's problem statement and refine its project objective through subsequent data collection and analysis. When the project's objective is focused, the team should create a project plan using the

10-Step Solution Process. The project plan should use the ten steps as project milestones. Using the ten milestones, the team should identify major activities and their associated work tasks for each milestone to execute the project's plan and schedule.

The project charter's problem statement and objective are important since they drive the team's root cause analysis. The project objective should be focused on that portion of the supply chain associated with the specific operational issue under consideration, such as a scheduling problem, a lot size problem, and so on. This information is usually obtained by an operational assessment, analysis of a major work stream within the supply chain or analysis of key financial and operational metrics. This methodology ensures that the project will generate business benefits and is aligned with senior management's goals and objectives. A good place to start the root cause analysis is by using a high-level process map to analyze project metrics, which are also called the key process output variables (KPOVs) or "Y's" backward through the process. The Lean Six Sigma improvement team needs to understand how the KPOVs are impacted by the key process input variables (KPIVs) or "X's". KPIVs are identified using analytical tools that will be discussed in this chapter. This will enable the team to build the model $Y = f(X)$ which explains the KPOV in terms of the KPIVs.

Prove Causal Effects (Step 4)

The Lean Six Sigma improvement team fully characterizes the KPOV through data collection, measurement system analysis and capability estimates. Later through brainstorming and analysis the team will identify KPIVs which will be used to build the model of the work stream i.e. $Y = f(X)$ which is associated with the operational problem.

Data Collection on Process Outputs ("Y's")

It is important to fully measure the KPOV since the Lean Six Sigma improvement project is based on analysis of the gap between its performance level versus the customer requirement or target. In addition to fully characterizing the KPOV, data collection involves brainstorming all the potential input variables which may be impacting the KPOV. After the analysis phase, one or more of these input variables will be found to have a major impact on the KPOV and will be called key process input variables (KPIVs). It is also important data be accurate and precise and collected in a manner which is representative of the larger population.

Measuring the Extent of the Problem　Measurement systems are used by the organization every day to analyze and control ongoing operational and financial performance but people seldom consider that their measurement systems may contain errors. In the Lean Six Sigma program, measurement errors have been found of varying degree across many projects. Table 8.1 shows common examples of measurement error, which occur within supply chains. As an example, cycle-counting accuracy rates are seldom 100%. In fact, commonly accepted cycle counting error rates are usually set between 1% to 10% depending on inventory monetary value or other considerations. This is not surprising since cycle-counting systems consist of people and related data collection and analysis. Mistakes can occur if these systems have not been correctly set up, or people might make mistakes in collecting and entering data into the system. Also, because we rely on cycle counting systems to verify inventory quantities and locations, problems with these system will cause other process breakdowns, which will be difficult to detect until the cycle-counting system's accuracy and precision are analyzed and improved by the Lean Six Sigma team. As a second example, we use the MRPII system to make critical decisions regarding the status and control of inventory and other materials within the supply chain. The system constants used by MRPII systems, if incorrectly specified, will result in a systematic bias in their calculation, which may adversely impact estimates of lot size, lead-time, inventory quantity, and other constants. Also if MRPII algorithms incorrectly convert these constants, their calculations may have a significant offset bias. As an example, in one logistics project, the warehouse management system rounded a constant upward, effectively closing trailers prior to being completely loaded at the distribution center. This situation significantly increased transportation costs since trailers had to be sent across the United States partially full. As a third example, when using forecasting models measurement accuracy is

1. Cycle-counting accuracy
2. Accuracy of MRPII constants
3. Conversion constants in MRPII software
4. Forecasting accuracy
5. Incorrect software logic
6. Timing of report generation through file merge
7. Parallel generation of the data

Table 8.1　Measurement Error Examples

notoriously poor when making estimates of future demand. According to several benchmarking studies of forecasting accuracy, forecasting error rates, at the item and location level, can easily exceed +/− 25%. However, accurate forecasting is important since forecasts drive capacity planning, manufacturing schedules, inventory estimates, and many other critical decisions within a supply chain. For this reason the accuracy of forecasting measurement systems must be quantified and improved by your organization. Problems with incorrect software logic may take many forms including off-set biases (as mentioned above) as well as incorrectly calculating quantities related to inventory targets, lot sizing, lead-time as well as other parameters. Another example of measurement system error includes the timing of system reports. This can occur because databases are more accurate just after a system refresh of the database and degenerate during the work week as shipment and receiving transactions flows in and out of the system. As an example, shipments and receipts occurring within a distribution center may provide material planners with inaccurate inventory information as the shipment and receipt databases change until the system has been refreshed or brought up-to-date. As a last example, if system reports are created using different data streams, inaccuracies will result since the incoming data may be different for each data stream. These examples show that measurement system errors commonly occur within your supply chain. These are just a few examples where measurement errors may adversely impact supply chain performance.

It is not surprising that measurement errors are often problematic within a supply chain given the large number of functional and organizational interfaces collecting and analyzing data to create metrics. Since input and output metrics are often poorly defined and linked with each other, measurement system errors often occur at these functional interfaces. As a result, a project's data collection and analysis tasks cannot be initiated by the Lean Six Sigma improvement team until verification of the accuracy and precision of the inter-functional metric baselines i.e. verification of the measurement system. A measurement systems analysis is required to ensure that everyone agrees on the type and extent of the problem to be solved including its estimating system baseline metrics to build the business case for the project charter.

Measurement Systems Analysis (MSA) Supply chain measurement systems consist of people, equipment, and many other important components. Measurement problems with any one of these inputs will affect the team's ability to effectively measure both the initial process baseline and eventual improvements in process performance. For this reason, prior to data collection, the project team must analyze the measurement components of the MSA. Unlike a manufacturing *measurement systems analysis (MSA)*, in which a *gage reproducibility and repeatability (Gage R&R)* study can be conducted on an isolated system like a machine, supply chain MSAs must simultaneously measure several system inputs and outputs, which jointly impact each other. As an example, the evaluation of the required ending inventory investment is directly related to estimates of current on-hand inventory as well as demand estimates, which are used to build the MPS schedule. Other relevant information includes available capacity, lead-time, MRPII constants, and their relationship via the BOM to independent demand quantities. To understand the ending inventory investment target requires understanding its inter-related system components as well.

These system inputs must be verified as accurate and precise to the required degree by the team. But, how accurate and precise must they be? The MSA must enable the team to verify the process performance baseline and its impact on their project metric to eventually isolate the root causes of the problem. The Lean Six Sigma improvement team must also be able to verify, in the control phase of their project, their process improvements have been effective. The MSA should help determine measurement error sources as well as their magnitude so that they can be reduced or eliminated from the measurement system. The cause-and-effect (C&E) diagram shown in Figure 8.1 is a good tool to enable the team to "brainstorm" qualitative relationships between the project metric (effect) and system variables (causes), which may impact evaluation of the project metric. As an example, in Figure 8.1, the team believes the ability to accurately and precisely measure inventory investment depends on many other system variables including MRPII constants, their conversion factors, the presence of inaccurate data, how scrap and rework are accounted for by the system, and other variables. Errors in estimating any one of these system parameters will result in inaccurate estimates of required inventory investment. This example shows the complexity of MSA studies as they are applied to supply chain processes. Since supply chain systems are very complex, it is a good idea to train supply chain people as Lean Six Sigma green belts and black belts rather than only manufacturing people or engineers. Professional societies such as the *Association for Operations Management (APICS)* and related organizations are useful sources of supply chain information and methods for "belts" not familiar with supply chain systems.

The example in Figure 8.1 shows that the measurement of inventory investment (or any other output metric) consists of several components that can be studied by the Lean Six Sigma team for opportunities

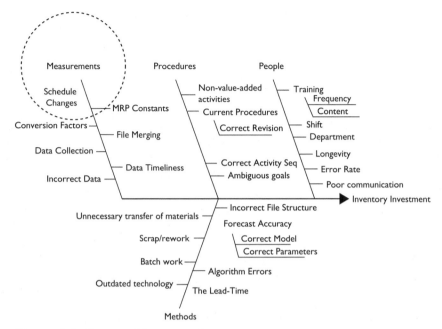

Figure 8.1 Causes of Inventory Investment

to improve their accuracy and precision. In other words, measurement consists of people, information, tools, and machines that have a cumulative impact on the KPOV.

Measurement System Components Analysis of measurement system components provides important information relative to establishing the baseline performance level of the project metric or KPOV. Measurement system components include resolution of the measuring device, accuracy (bias) of the measuring device used to conduct the MSA, repeatability of the device when used by one person under controlled conditions, as well as stability and linearity of the measurement system. MSA components have been very well-defined for manufacturing applications over the past several years. But there has been confusion relative to their application to supply chain and inventory systems. As each of the MSA components are discussed in the rest of this section, several applied examples will be provided to show how MSA concepts link to supply chain applications.

Resolution is the ability of the measurement system to discriminate changes in the characteristic being measured *(1/10 rule)*. An example would be lead-time analysis. In this scenario, our project objective might be to reduce lead-time; but, prior to data collection and analysis, we need to conduct an MSA on lead-time which is the KPOV. The first step in the MSA is to ensure that our measurement tool has sufficient resolution. If our project goal is to reduce lead-time by 10%, but our current measurement resolution is in units of "days", we may need to increase our resolution to "hours". In this scenario, the first step in our project analysis, would be to change the lead-time measurement system to an hourly basis. These situations commonly occur when a date stamp is used to record system transactions on a day-by-day basis but the Lean Six Sigma team needs to detect process improvements on an hourly basis. As an example we might have a situation in which we need to understand system transactions on an hourly basis, but the current data collection form uses discrete date stamped time increments such as A.M. or P.M. Other examples of poor resolution includes situations in which we need to identify all categories contributing to a problem, but the data has been collected in an aggregated manner in which several categories have been combined into just one category. In this situation, we would need to disaggregate the measurement system to include all categories of interest to the Lean Six Sigma team. As you can see, resolution problems are very common in supply chain systems.

An accurate measurement system has the ability to correctly measure a metric on average over many samples (no bias or offset in the average of the measurements). There are many examples in which supply chain measurements are inaccurate. As an example, bias occurs when algorithm conversion constants have been programmed to round to the next higher number, when trailers are closed at a certain weight or lead-time constants are inserted into on-time delivery reports. The latter situation occurs when distribution centers are allowed several days of extra lead-time to compensate for distance from one facility to another. Other examples include situations in which data collection forms have forced several evaluation categories to one end of a scale. This has the effect of biasing participant responses to one end of the scale. There are many other examples of biasing in supply chain measurement systems. The Lean Six Sigma team must ensure that their measurement system has been calibrated to ensure its accuracy so that the project metric baselines are not offset. An offset of the project metric baseline will have the effect of over- or understating operational performance and the associated project benefits.

Reproducibility measures the ability or skill of two or more people (or machines) to measure a metric with low variation between each person (machine). An example would be having people read the same inventory report, but make different interpretations of the data resulting in different actions. Another example would be people auditing the same inventory at the same time, but calculating different account balances for each item. These are examples of reproducibility errors associated with people measuring (or interpreting) the same thing differently from each other. Since many supply chain systems have a heavy manual component, reproducibility errors are prevalent. The best way to attack reproducibility errors is to use gage repeatability and reproducibility studies, and if the project metric is continuous, to break down the measurement system variation between components due to people and the measuring device they are using to evaluate the project metric. An attribute agreement analysis can be used to understand the impact of reproducibility on the project metric if the metric is discrete (percentage or ordinal).

Repeatability is the variation observed between measurements when made by the same person (or machine) when evaluating the same metric by measuring it several times under constant conditions. An example would be when a person counts an inventory population several times, but gets a different answer each time. The different answer obtained by the same person conducting the cycle count is due to the variation inherent within the cycle-counting system. This variation occurs because the person is part of the measurement system. Some reasons for poor repeatability might be caused when a person travels to an inventory storage location and fails to accurately record the inventory quantity at the location due to perception issues. Repeatability problems also occur whenever the measurement system consists of many manual operations when data is provided either to the system or people within the system, but the data is subject to variation. The result will be different decision outcomes due to minor variation within the data.

Stability is the ability of the measurement system to obtain the same measured value over time. An example would be when data collection forms (or historical baselines) are continuously revised, making it impossible to evaluate process changes relative to a historical standard. Another example would be degradation in measurement systems, which require people to be trained to evaluate the process metric. An example of this latter situation commonly occurs when employee evaluations degrade over time due to managers forgetting their original training instructions. In this latter situation, some managers may become inconsistent with respect to their employee evaluations.

Linearity is the ability of the measurement system to measure a metric over its entire range with equal variation (error). An example would be a data collection form having several redundant evaluation categories versus a few important ones. Too many categories cause noise at the extremes of the evaluation form. In addition to an excessive number of evaluation categories, perhaps the evaluation procedures are not clearly defined at the extremes of the measurement system. This may cause higher measurement system variation in this range of the evaluation system. Another example of linearity issues would be forcing employees to one side of a performance appraisal system or the other.

Estimating Process Capability on Y's

After the measurement system has been proven to be accurate and precise to the required degree, it is used to verify the performance baseline of the project metric(s). At this point in the project, the team should be able to measure the project metric. They should also know what makes its performance good or bad. In other words, the Lean Six Sigma team should know the "defect definition" of their project metric. The project metric is the operational metric or KPOV the team will improve through the root cause analysis. Closely tied to the operational metric is the actual financial benefits realized by the organization if the operational metric's average performance level is moved closer to the target or its variation is reduced through the eventual process improvements. As an example, if the KPOV is lead-time, the financial metric may be inventory investment. As we decrease lead-time, inventory investment should decrease, all other things remaining constant since it is highly dependent on lead-time. In addition to operational and financial metrics, the project may also include a second operational metric related to the root cause analysis. As an example, the project may become focused on one of the root causes for long lead-times. Perhaps large lot sizes are found to be a major contributing factor of long lead-time. In this case the project may include lot size as a second and lower-level operational metric. However, the improvement team must clearly establish the relationships between all metrics so that senior management will know how the financial metric improves as each of the operational metrics improve. As an example, the team should be able to state to management something like: "If lot size is reduced by 50%, then lead-time will decrease by 10%, reducing inventory investment by $50,000 for product category XYZ." Finally, the team may have to bring on a compensating metric to ensure that there are no adverse impacts on the external customer. In an inventory investment reduction project, the team would include the target level of customer service in the project charter with the statement that customer service would be maintained constant (or improved) regardless of the eventual recommended process improvements, which lower inventory investment.

Some operational metrics by their nature have an implied defect definition. As an example, scrap percentage or weight should be driven to zero. Rework percentage should be driven to zero. Forecast error should be driven to zero. But other metrics may have an inherent physical limitation. As an example, although lead-time drives inventory investment, in most systems it cannot be driven to zero, at least not without an entirely new process design, for example, moving to e-mail versus mailing a letter. In most systems, a defect relative to lead-time would be any time component that is not value-adding. The ultimate goal of the organization would be to reduce lead-time to its inherent value added level. However, from a Lean Six Sigma improvement team perspective, the current project might only bring the lead-time down a portion of the way, relative to the organization's ultimate goal, to the project's lead-time target. This target would be determined, after a preliminary operational analysis, by the Lean Six Sigma improvement team. In fact, it may require many projects to reduce the lead-time towards its 100% value added level.

Table 8.2 defines a defect from a Lean Six Sigma viewpoint. It also introduces the concept of "opportunity counting." Opportunities refer to the number of operations within a process (or components in a product design). Opportunities are a measure of process (or product) complexity. The basic concept is that the more complex the process, the greater its opportunity count, and the more defects it will produce over time. Dividing defects by opportunities provides a yield metric that can be used to compare the quality level of one process against another having higher or lower complexity. Opportunity counting is most valuable when two processes have the same defect percentage, but one has a much lower opportunity count reflecting the fact that it is a simpler process. This will

What is a defect?
A nonconformance that causes a product or service to fail to meet customer requirements.
What should be counted as defective ?
Administrative examples include inventory, long cycle time, rework, poor order accuracy or delivery performance, and so on. Manufacturing examples include scrap and rework.
How should the defects be counted ?
Defects should be counted at every value-adding and non-value-adding operation and summed over the entire process.

Table 8.2 Defect Definition

tend to focus the Lean Six Sigma improvement team's attention on the less complex process. The sigma calculation uses a normalized quality metric called *defects per million opportunities (DPMO)*. The DPMO statistic is calculated by dividing the process defect fraction by the opportunity count and multiplying by 1,000,000. This concept will be discussed in detail starting at Table 8.4, using "Order Fulfillment" as an applied example. The order fulfillment example will show how to improve the quality level *(sigma level)* of a supply chain process by first simplifying it and then reducing the defect levels of the remaining process steps including their internal operations.

What is process capability? Figure 8.2 shows that it can be viewed from two perspectives. The first is from the external customer's perspective as represented by internal specification limits on the KPOV. These specification limits should have been developed using the voice-of-the customer (VOC) information. VOC translation was discussed in Chapter 1. The second perspective is derived using process performance relative to the KPOV under analysis. This is called the voice-of-the-process (VOP). Comparing process variation (VOP) to customer specification limits (VOC) allows calculation of *capability ratio*. A high capability ratio indicates that a high percentage of the process variation is within the specification limits. This means the defective percentage is low (a small proportion of the process is outside the specification limits).

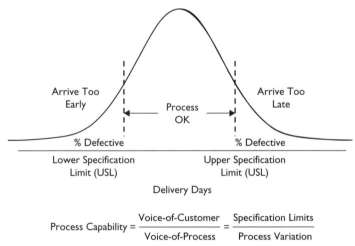

$$\text{Process Capability} = \frac{\text{Voice-of-Customer}}{\text{Voice-of-Process}} = \frac{\text{Specification Limits}}{\text{Process Variation}}$$

Figure 8.2 Estimating On-Time Delivery Performance

Samples of the KPOV taken from the process under analysis will exhibit natural variation (VOP). If process variability is very small, the customer will see a very low defective percentage and correspondingly high quality levels of the KPOV relative to customer requirements. On the other hand, if the process has high variability or shifts to one side of the upper (USL) or lower (LSL) specification limits, defect percentages will significantly increase. In the example shown in Figure 8.2, the project metric "delivery days" has been specified by the customer to be within the lower and upper specification limits (not specified in this example). If the delivery arrives too early, there will not be room at the unloading dock. This results in driver and equipment delays. In this scenario, the supplier delivering the materials has incurred a cost (due to the extra time required for the delivery) in excess of their operational standard. This condition results in a process defect. The operational metric is time. The defect is wasted time (or non-value-adding time). The defect is directly correlated to the cost of waiting for the unloading dock door to be free. The Lean Six Sigma improvement team would reduce wasted time to create a business benefit. On the other hand, if the delivery arrives too late, the customer will incur a cost either in time lost waiting to unload the trailer, or overtime required to unload the trailer, or perhaps in lost sales because the inventory was not available to fill orders. The late arrival is also a defect. The defect condition is wasted time, but the root causes and business impact may be different than the situation occurring due to early deliveries. Measuring delivery time in days and comparing the process performance to customer requirements allows the improvement team to measure delivery performance against the customer delivery standard. If there is a significant performance gap, then a Lean Six Sigma project can be created to close the gap. In this example, it is easy to see the difference between operational versus financial metrics and how they relate to VOC and VOP.

Table 8.3 cross-references the defect percentage of a process to Lean Six Sigma and classical capability metrics. There are many tables that will allow you to compare defect percentages to Lean Six Sigma metrics and cross-reference the metrics to classical quality metrics. Table 8.3 has to be carefully read since there are assumptions built into its construction. As an example, notice the table is divided into Sigma versus Z_{lt}, defect percentage, and PPM sections. The Z_{lt}, defect percentage, and PPM statistics are what most people would expect if they took an elementary course on statistics. Any standard normal table would have $Z = 0$ at a defect area = 50%. But notice that in the Sigma section, the Z number has been increased by a constant, 1.5. This is a vestige of the Six Sigma

Sigma	Z_{lt}	Defect %	PPM
1.50	0.00	50.00%	500,000
2.00	0.50	30.85%	308,500
3.00	1.50	6.68%	66,809
4.00	2.50	0.62%	6,210
5.00	3.50	0.02%	233
6.00	4.50	----	3.4

Table 8.3 Sigma Conversion Table

manufacturing program at Motorola. However, in our experience a shift of this magnitude in non-manufacturing is not commonly seen by black belts. The best way to estimate the true shift of your processes is to collect the data in subgrouped fashion over time and directly estimate the magnitude of shift for yourself. However, for the sake of ensuring completeness of our capability discussion, the *sigma shift* concept has been included in the order entry example.

There are two key concepts relative to calculating process capability. These are variation of the central location of the process KPOV (shifting of the process mean) and increasing dispersion of the KPOV over time (the variation around the mean increases). As a result, the KPOV capability will vary over time since it is calculated using the process standard deviation of the KPOV relative to its mean and the specification limits. The customer will evaluate process performance by comparing the cumulative variation of the KPOV against the specification limits over an extended time period, that is, both good and bad days. An interesting concept derived using the capability analysis is that an organization can compare day-to-day process performance against the "good" days. This is an internal benchmarking analysis called an "entitlement analysis." Imagine your best person meeting your performance standards with high capability, using the best method, the best materials, and the best equipment on a given day. This situation might encourage you to ask, "Why can't we do this all the time?" That is a basic concept of the Lean Six Sigma DMAIC methodology, that is, to bring a process up to its entitlement level, which is the best level at which it was designed to perform, in other words, the best day's performance level. However, identification of the conditions resulting in entitlement situations is not always obvious at the outset of the project. This is why the Lean Six Sigma improvement team must identify the root causes responsible for

poor process performance. This requires the Lean Six Sigma improvement team to work through a phased problem-solving methodology to
systematically identify and eliminate the root causes for poor process
performance as measured by the KPOV's initial capability level.

The "entitlement concept" is shown in Figure 8.3. Figure 8.3 contains
four quadrants. In the top left quadrant, the capability metric C_p measures
how well the process meets customer requirements when it is centered
on-target and the variation is equivalent to the smallest amount the
process is capable of producing under the current process design. The
smallest amount of variation in a system is within a rational subgroup.
An example of rational subgroups includes measuring clerical errors by
person on a shift to compare process variation between people, shifts, day
of the week, and different types of invoices. Another example of selecting rational subgroups would be measuring picking errors by person on a
shift in a distribution center to compare variation between pickers, shifts,
order types, days of the week, and so on. In both examples, the subgroups were selected in such a manner as to evaluate between subgroup
variation. The calculation of C_p uses pooled estimates of the subgroup's
standard deviations to estimate the overall process variation. C_p can be
converted to Sigma using the following formula: $Sigma = 3*C_p$.

In the lower left quadrant of Figure 8.3, the capability metric P_p measures a process that is centered on-target, but exhibits longer-term variation (subgroup-to-subgroup variation over an extended time period). In
the upper right quadrant of Figure 8.3, the capability metric C_{PK} measures
a process that is off-center, but exhibits short-term variation, that is, within subgroup variation. Finally, in the lower right quadrant, the capability
metric P_{PK} measures the worst-case scenario, a process that is shifted
from target and exhibits the maximum amount of process variation. The
customer measures the process capability using the P_{PK} metric. The Lean
Six Sigma strategy is to move the process on target and then analyze the
root causes for variation. Once the root causes have been identified, they
are systematically eliminated from the process to improve capability from
the customer viewpoint by decreasing process variation.

	On Target	Off Target
Short-term	C_p	C_{PK}
$Sigma = 3*C_p$		
Long-term	P_p	P_{PK}

Figure 8.3 Capability Metrics

1. An analysis was conducted on orders shipped from a distribution center. The analysis showed errors occurring at various stages of the order fulfillment process. Opportunities for error occurred at both non-value-adding and value-adding operations.
2. To understand the concept of opportunity counting and estimating capability statistics, three scenarios will be investigated against the baseline analysis.

Table 8.4 Order Fulfillment Example

To show how capability metrics are calculated and applied in practice, an "order fulfillment process" example is shown starting at Table 8.4. Figure 8.4 shows the baseline scenario called the "current process." The current process contains three process steps. These steps are "order entry," "distribution," and "carrier." At the order entry step, the distribution center clerks enter customer ordering information into the system. At the distribution step, the order is picked and staged for shipment to the customer. At the carrier step, the order is picked up from the supplier and transported to the customer. In scenario 1, an assumption is made that the process can be simplified by eliminating "order entry" using an electronic data interchange system (EDI). In scenario 2, the "pallet-wrapping operation within distribution" is eliminated along with the entire "order entry" process step. As we work through the various scenarios, it will become obvious that simplifying a process and reducing

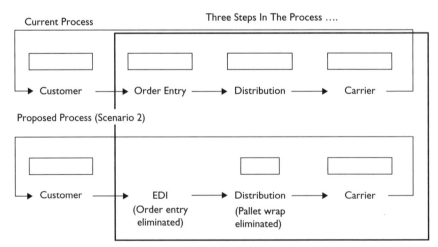

Figure 8.4 Order Fulfillment Process Map

defect levels will increase quality levels. However, in scenario 3, everything is purposely moved in the wrong direction to show what happens when processes are made unnecessarily complex. Scenario 3 shows quality decreases from its baseline level. For each scenario, defect types will be simulated at operations within the process, and yield metrics will be calculated for the baseline and three scenarios. Moving from one scenario to other, defects will also be systematically removed from the process to show how the quality metrics improve. This example will show how to quantify a supply chain process and calculate the yield metrics.

In Table 8.5, 100 orders are simulated through the order fulfillment process. The customer receives the 100 orders, but with 5 defects. The defects occurred in different steps of the order fulfillment process with 2 occurring at order entry, 2 at the distribution center, and 1 at the freight carrier. The 100 orders contained 1,000 line items. A line item is one product with an associated quantity. In addition to measuring the order by line-item quality, we also could have measured the order on other important characteristics such as on-time delivery, damage, and so on. We can calculate two key yield metrics. These are order yield and line-item yield. Order yield is important to the customer, but line-item yield is important to the supplier in the sense that the Lean Six Sigma improvement team will use this operational metric to analyze the root causes of the various defects found within each order. It should be noted that order yield tends to move in the same direction with line-item yield. The goal of the Lean Six Sigma improvement team will be to systematically improve process performance by reducing the defect levels throughout

Process Flow					
Process Step:	Customer	Order Entry	Distribution	Carrier	Customer
Orders:[1]	100	100	100	100	100
Lines:	1,000	1,000	1,000	1,000	1,000
Errors:	0	2	2	1	5
Types: Error	N/A	Two Incorrect Models	One Shortage One Incorrect Pallet Wrap	Damage	

[1]Orders is the unit of measurement in this example; but lines could also be used as an internal metric

Table 8.5 Process Flow and Simulated Defects

the order fulfillment process. The improvement goals will also be to "Lean out" the process by eliminating entire steps, that is, to eliminate "order entry" using EDI or operations within a given process step, for example, eliminate "pallet wrapping" within the distribution center and improve the process capability.

Table 8.6 shows the formulas necessary to calculate key yield statistics. *Defect-Per-Unit (DPU)* is calculated as the total defects found in the sample divided by the total units sampled. In Table 8.5, the defect total in the final order is 5 defects and from a customer perspective the sample is 100 orders. giving a 5% DPU. From an internal perspective, the DPU is 0.5% since we are dividing 5 defects by 1,000 line items. In either case, the DPU metric is converted to a *parts-per-million (PPM)* metric by multiplying the DPU metric by 1,000,000. The *rolled-throughput-yield (RTY)* measures the first-pass yield at every step in the process. RTY shows what percentage of units flowing through the process is defect-free. As an example, if we have two steps in a hypothetical process, each having a 90% yield, then the process RTY metric would be calculated as equal to 90% * 90% = 81%. In other words, 81% of the units passed through the two steps of the process without incurring any defects. "No defects" means the unit has not been reworked nor was a replacement unit substituted since there was no process scrap (no wasted material or labor). Under certain assumptions, such as many process steps, RTY can be approximated using the formula shown in Table 8.7. This formula provides the probability of zero defects though the process, that is, the yield faction. A summarization of the yield metrics for the order fulfillment process is shown in Table 8.8. To summarize, the external customer sees 100 orders containing 1,000 line items and 5 defects. The customer rates the supplier at a 5% order DPU. The internal Lean Six Sigma

$$\text{Defects-Per-Unit (DPU)} = \frac{\text{Total Defects Found in Sample}}{\text{Total Units in Sample}}$$

$$\text{Parts-Per-Million (PPM)} = \text{DPU} \times 1,000,000$$

$$\text{Rolled Throughput Yield (RTY)} = \prod_{i=1}^{n} \frac{\text{Defect Free Units at Each Step}_i}{\text{Total Units at Each Step}_i} \times 100$$

Table 8.6 DPU/PPM/RTY Calculations

RTY = (Yield 1) × (Yield 2) × (Yield 3)

Over a very large number of process steps the RTY approximation is:

RTY ~ $e^{-DPU_{Total}}$

Table 8.7 RTY Approximation

improvement team uses the order DPU and the line-item DPU to create improvement projects. In this baseline scenario, the process steps having the highest defect rates are order entry and distribution.

Most organizations would use the DPU and PPM metrics. The early Six Sigma program was based on the concept of opportunity counting. Opportunity counting was discussed earlier in this chapter, but will now be discussed in detail. If opportunity counting is used by the Lean Six Sigma team, the PPM metric will be converted to a defects-per-million opportunity (DPMO) metric. The DPMO metric is calculated by dividing the PPM metric by the number of opportunities in the process. What are opportunities? These are the total number of operations within the process. Each time an operation is performed, it can be right or wrong. This estimation process can be very difficult. A team must first define the smallest reasonable level to count operations and set counting standards so that everyone in the organization is counting

Process Flow				
Process Step:	**Customer**	**Order Entry**	**Distribution**	**Carrier**	**Customer**
Units:	100	100	100	100	100
Errors:	0	2	2	1	5
DPU:	0	0.02	0.02	0.01	0.05[1]
PPM:	0	20,000	20,000	10,000	50,000
RTY:	N/A	98%	98%	99%	95%[2]
[1]0.05 = (0.02) + (0.02) + (0.01)					
[2]95% = (98%) × (98%) × (99%)					
Note: These calculations agree because yields are very high.					

Table 8.8 Order Fulfillment Yield Statistics

opportunities the same way. If the standards are not properly defined, the capability metrics will not be consistent across the organization. Classifying operations as value-adding (VA) versus non-value-adding (NVA) is a second consideration when using opportunity counting. Table 8.9 shows the number of value-adding versus non-value-adding operations within each process step in the order fulfillment process. It should be noted if a process step is defined as non-value-adding, then so is every operation within the process step. This is because "order entry" does not have value-adding operations. Also, it should be noted, only opportunities from value-adding (VA) operations are counted to calculate the DPMO statistic. This practice will maintain a stable performance baseline throughout the process improvement cycle. As an example, in the process improvement cycle, we will be eliminating non-value operations as soon as practical. But we want to maintain a stable initial baseline against which to measure our improvement performance. If the opportunity counts of non-value-adding operations were included in the initial performance baseline, then the baseline would continually change as improvements were made to the process. Table 8.10 expands on this concept. The baseline for the opportunity count should only be changed when the process is redesigned by changing the number of value-adding operations, at which point a new opportunity count baseline is established using the newer process design. This practice brings stability to the yield evaluation system.

The simulated defect listing is shown in Table 8.11. There are five defects in the current baseline process. In scenario 1, when the Lean Six Sigma improvement team automated order entry using EDI, the result

	Order Entry	Distribution	Carrier
VA Operations	0	2	1
NVA Operations	2	2	1
Total Operations	2	4	2

- Opportunities are used to account for complexity
- Only value-adding operations are counted.
- Additional value-adding operations add opportunities.
- Non-value-adding operations do not add opportunities.
- Non-value-adding operations include storage, inspection, unnecessary movement, and so on.

Table 8.9 Opportunity Counting by Process Step

1. As non-value-added operations are eliminated through process redesign, the errors associated with those operations are also eliminated, resulting in a higher sigma level; but the original sigma opportunity count remains stable.
2. When value-added operations are eliminated through process redesign, the sigma baseline must be re-established to reflect the new opportunity count since the new process is different from the old process.

Table 8.10 Why Count Value-Adding Operations?

was that the two defects associated with the order entry, that is, "two incorrect models." were eliminated by the EDI automation. In scenario 2, the "pallet wrapping" operation was eliminated in distribution. This could have been done through a packaging design change. The quality levels successively increased from the current baseline process through scenarios 1 and 2. However, in scenario 3, additional NVA operations were added to the process, which introduced additional defects. This is a common occurrence when people unnecessarily complicate a supply chain process. In complicated processes containing non-value-adding operations, cost and lead-time increase and quality degenerates.

Table 8.12 breaks the simulation down for the baseline process as well as each of the three scenarios. The breakdown also includes classifying each operation as VA or NVA operations and their opportunity count. Notice the opportunity counts reflect the number of VA operations and

Current	Order Entry:	Two incorrect models
	Distribution:	One Shortage & Incorrect Wrap
	Carrier:	Damage
Scenario 1	Distribution:	One Shortage & Incorrect Wrap
	Carrier:	Damage
Scenario 2	Distribution:	One Shortage
	Carrier:	Damage
Scenario 3	Order Entry:	Two incorrect models & lost order & coding
	Distribution:	One Shortage & Incorrect Wrap
	Carrier:	Damage

Table 8.11 Simulated Defect Listing by Simulation Scenario

	Current Process	Eliminate Order Entry Operation (A)	New Process With Fewer VA Operations (B)	NVA Operations Added To Process
		Scenario 1	Scenario 2	Scenario 3
Number Steps:	3	2	2	3
VA Steps:	2	2	2	2
NVA Steps:	$1^{(1)}$	$0^{(4)}$	0	1
VA Operations:	$3^{(2)}$	3	$2^{(5)}$	3
NVA Operations:	$5^{(3)}$	3	3	$10^{(6)}$
Opportunities:	3	3	2	3
[1]EDI transfer could eliminate order entry				
[2]Order pick, wrap and carrier delivery				
[3]Inspection, handling, transport and so on				
[4]Product/process redesign using EDI				
[5]Pallet wrap was eliminated				
[6]New manager adds 5 NVA operations to order entry				

Table 8.12 Counting Opportunities by Simulation Scenario

not the total number of operations within each process step. Also notice the opportunity count went down in scenario 2. This was because we eliminated "pallet wrap," which was originally a VA operation required by the customer. So in a sense, scenario 2 represents a minor change from the original process design. If the opportunity counts change, the improvement team should establish a new performance baseline. But, practically speaking, the improvement team may want to wait until several minor process design changes have accumulated so as not to create too much confusion by changing the performance baselines on which capability metrics were originally calculated. In scenario 3, a new manager added 5 new NVA operations to the process. Increasing the number of NVA operations to a process only complicates it and reduces quality performance. Using the concept of opportunity counting, Table 8.13 shows how to calculate the sigma statistic of the simulation for the baseline process as well as each scenario using the DPMO metric. DPMO is an adjusted area under a normal curve (PPM divided by the process opportunity count). Sigma is calculated using a standard normal table and selecting the Z value associated with the DPMO defect area.

$$DPMO = \frac{PPM}{Opportunities/Unit}$$	
Z value from a normal table Sigma = corresponding to DPMO (must be converted to short-term Z)	

Table 8.13 Calculating Sigma using DPMO

Table 8.14 shows summarized yield statistics and capability metrics for both the current process and each scenario. Notice the original yield metrics such as DPU, PPM, and RTY are listed at the top of Table 8.14 and the Lean Six Sigma metrics are listed at the lower portion of the table. The DPMO metric was obtained by dividing the PPM number by the opportunity count. For the current process, the DPMO calculation would be 50,000 PPM divided by 3 opportunities to equal a DPMO of 16,667. The standard normal table Z_{lt} equivalent is 2.13. This means at $Z = 2.13$, 83.3% of the area under the curve (percentage good units) is to the left of $Z = 2.13$ and 16.7% of the area under the curve (16.7% bad units) is to the right of $Z = 2.13$. If the process analysis represented a long-term variation viewpoint of the process, then the calculated $Z = 2.13$ number would be adjusted upward by adding a 1.5 constant to the number as shown in Table 8.14. The mean shift could also be directly calculated if the data had been collected in subgroups.

	Current	Scenario 1	Scenario 2	Scenario 3	
Units:	100	100	100	100	
Defects:	5	3	2	7	
DPU:	0.05	0.03	0.02	0.07[1]	
PPM:	50,000	30,000	20,000	70,000	Customer Metrics
RTY:	95%	97%	98%	90%	
Opportunities:	3	3	2	3	
DPMO:	16,667	10,000	10,000	23,333	
Z_{lt}:	2.13	2.32	2.32	2.00	Complexity Metrics
Z_{st} (Sigma):	3.63	3.82	3.82	3.50	
[1]Assume two additional defects were created by addition of the NVA operation.					
Scenario 1: (.98)*(.99) = .97					
Scenario 2: RTY = (.99)*(.99) = .98					
Scenario 3: RTY = (.93)*(.98)*(.99) = .90					

Table 8.14 Summary Yield Statistics by Simulation Scenario

To summarize, there are two basic types of metrics. These are "customer" and "complexity" metrics. Since the external customers require orders to be correct, the metrics they use are percentage defective and PPM. The sigma metric is really irrelevant to the external customer since the customer cannot control its method of calculation. In the current process the customer sees a 5% defective process. In scenario 2, the external customer sees a 2% defective process. However, internally the Lean Six Sigma team uses the complexity metrics to measure their process improvement efforts, which are directly linked to external customer quality and satisfaction as well as operational complexity and performance.

In the current "order fulfillment" example, the sigma calculation was easy because the process steps were linearly arranged or *serial*. However, the Lean Six Sigma team will want to accurately characterize their process given it contains both steps in series and in parallel. Table 8.15 shows that in a serial process. The PPM numbers are added together and then divided by the total opportunity count of the process. Breaking the order entry step of the current process into its three parallel operations, we see there are three order entry clerks each handling a different incoming call volume and making different numbers of errors. As an example, clerk A handled 50 of the 100 orders and caused 1 defect. Clerk A's DPU is 0.02 or 2%. On the other hand, clerk C handled just 20 of the 100 orders and caused 1 defect for a DPU of 0.05 or 5% defective. If Clerk C had handled the same volume as Clerk A, then the defect count would have been between 2 and 3. To calculate the DPU for the entire process step, we weight the DPUs for each parallel branch (the clerk) by their transaction volume (number of orders). The result will be a volume-weighted DPU for the entire process step. The calculations are shown in Table 8.16. The PPM number is 20,000.

Serial Operations
PPM = PPM 1 + PPM 2 + PPM 3
PPM = 20,000 + 20,000 + 10,000 = 50,000
Parallel Operations
■ Suppose the order entry process had three clerks taking customer orders. This creates a parallel operation.

Table 8.15 Baseline Scenario PPM Consolidation (Serial Steps)

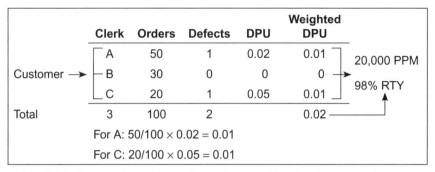

	Clerk	Orders	Defects	DPU	Weighted DPU	
	A	50	1	0.02	0.01	20,000 PPM
Customer →	B	30	0	0	0	
	C	20	1	0.05	0.01	98% RTY
Total	3	100	2		0.02	

For A: 50/100 × 0.02 = 0.01

For C: 20/100 × 0.05 = 0.01

Table 8.16 Process Step 1 PPM Consolidation (Parallel Operations)

This PPM calculation agrees with the aggregate DPU that was shown for the current process in previous tables.

The Lean Six Sigma improvement team uses this yield information to create improvement projects and focus on the root causes for the line-item error rates. The Pareto chart (first level) shown in Figure 8.5 is used to identify the process step having the highest PPM level. Using a second-level Pareto chart, the improvement team drills down to the specific clerk having the highest PPM number, that is, Clerk C. The team would then work to eliminate one or more defect types associated with Clerk C. Alternatively, the team could create a Pareto chart of all the defects in the order entry portion of the process and eliminate one or more defects for all the clerks. Finally, the team could automate the process to remove all manual operations and associated defects. This would permanently improve the quality level of the process. This was the improvement made in scenario 1.

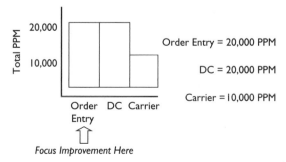

Order Entry = 20,000 PPM

DC = 20,000 PPM

Carrier = 10,000 PPM

Focus Improvement Here

Figure 8.5 Defect Analysis First-Level Pareto by Process Step

Data Collection on Process Inputs ("X's")

A key deliverable of the data collection process is characterization of the KPOV as well as the identification of all possible input variables, that is, the "X's," which may impact the KPOV. The KPOV could be one of the 12 key supply chain metrics or any other performance metric important to the organization or customer. The first step in data collection and analysis begins with bringing the team together to review the project charter. Part of the charter review includes verifying the KPOV is the correct one for the problem statement and project objective. A second objective of the meeting is to identify all the possible input variables that impact the project's KPOV(s). There are many ways to develop this list. The identification process begins with each member of the team making suggestions in turn. A facilitator records the suggested input variables. After the team has identified every possible input variable, the ideas are organized into major groupings or themes. The most common tool used to group the ideas into major themes is the cause-and-effect diagram (C&E). The C&E diagram was shown earlier in Figure 8.1. The causes are the various input variables that the team thinks might be impacting the KPOV (the effect). The C&E diagram allows the team to organize the input variables into major themes such as people, machines, materials, methods, measurements, and environmental inputs. If necessary, your teams can create major themes that are more useful to their projects or more relevant to your business or industry. The next step in the brainstorming exercise is for the team to prioritize the input variables for data collection. These are called *control* variables. The model Y = f(X) will eventually be built using key control variables or KPIVs. The second type of input variables are those which the team will maintain, using standardized procedures, at standard operating levels during data collection and analysis. These are called SOP variables. The third type of input variable is called a *noise* variable.

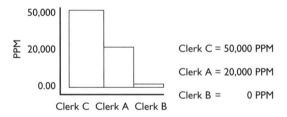

Continue the Root Cause Analysis with Clerk C

Figure 8.6 Defect Analysis Second-Level Pareto by Order Entry Clerk

Noise variables are thought to have little or no impact on the KPOV. During the data collection process the team may simply record the levels of the S&OP and noise variables.

The C&E diagram and other brainstorming tools are used to identify possible input variables. These inputs should be linked to the project's KPOV metrics as shown in Chapter 1, Figure 1.3, in which inventory investment is used as an example. If the project charter contains several KPOV variables, another prioritization tool called the cause-and-effect matrix (C&E matrix) is used to prioritize all the input variables relative to all the KPOVs. A C&E matrix is shown in Figure 8.10. In this analysis the KPOV variables are ranked in relative importance, and correlations are developed between each input variable and KPOV variable. This allows the team to prioritize data collection for those input variables which impact the most important KPOVs. C&E diagrams and the C&E matrix will be discussed in more detail later in this chapter.

Since we can seldom count everything, given resource limitations, a sampling plan is developed to ensure that useful information is obtained by the team during data collection activities at a minimum cost. Table 8.17 shows common data collection strategies used in supply chain improvement projects. Most of this information is collected on a continuing basis within supply chains. Important considerations in data collection and analysis are data representation of the population, data type and distribution, and sample size sufficiency.

Data representation refers to how well the data sample reflects the project problem statement and in particular the input variables and their ranges (levels). As an example, if the purpose of the project is to analyze inventory over three facilities, it makes sense to collect data over the three facilities as opposed to just one facility. This may not be obvious to everyone, and the concept gets more complex as the number of input variables increases. As a second example, if the intent of the analysis

- Cycle counting (simple or stratified sampling)
- Checking a sample of MRPII constants to verify accuracy
- Checking BOM items to verify database accuracy
- Analyzing supplier on-time delivery to determine differences
- And so on, and so on, and so on.

Table 8.17 Collecting Supply Chain Data

was to extrapolate the conclusions of the analysis over seven product categories and three facilities, we would need to collect data for all seven product categories by facility resulting in twenty-one different samples to ensure a representative sample of the project's problem statement. Effective data representation ensures the collection of relevant data that will be sufficient to answer all the questions posed by the project charter's problem statement and objective.

Data type refers to the fact some distribution forms of the KPOV contain more information than others. This will allow the team to reach a statistically significant conclusion faster for a given sample size. As an example, data that is continuously distributed will enable the team to very efficiently calculate sample statistics to make statistically valid conclusions. On the other hand, if the KPOV is characterized as pass/fail *(discrete distribution)*, the sample size necessary to reach statistically significant conclusions will be much higher than in the continuous distribution situation. As an example, if we measured the heights of a group of people using a continuous scale, for example, a tape measure, then we would know a lot about their height distribution and would be able to extrapolate out to the larger population of heights if the sample was representative of the population. On the other hand, if we collected the height information as simply the percentage over or under over 66 inches, we would know less about the heights of the people in the sample and the larger population of heights. Using this basic concept, your Lean Six Sigma improvement team should specify the KPOV and the input variables as continuous distributions if possible. As a second example, if we are measuring supplier on-time delivery, the KPOV (on-time delivery) should be specified in days or hours depending on the required resolution of the measurement system. It should not be specified as percentage, that is, percentage supplier on-time deliveries versus total deliveries. If it is currently specified as a percentage, then the Lean Six Sigma improvement team should immediately change the metric to make it continuous. This will provide the team with a more sensitive metric against which to measure their process improvements.

The concept of sample size sufficiency is more complex and depends on the statistical statement we would like to make based on the specific statistical analysis we need to conduct and how confident we would like to be in making the statistical statement obtained from the statistical analysis. Statistical statements include questions relative to central location, dispersion, and goodness-of-fit. Questions relative to central location include statements such as, "Has the process mean or median changed after the improvements were made by the team? Has the process variation decreased after the improvements were made by the team? Has the accounts receivable aging distribution been improved (goodness-of-fit test against the baseline aging distribution)?" These statistical statements should reflect the project's original goals and objectives. Answering the statistical statements allows the team to eventually answer the project questions in a fact based manner.

Statistical Sampling from a Population

A population is defined by the questions the analysis is designed to answer. As an example, if we are interested in average household income in Boston, our population would be the people of Boston (but we have to define Boston a little better). On the other hand, if our question is, "What is the average household income of people in the United States?" the population would be everyone living in the United States. In either example, there are important things to consider relative to better defining the population prior to data collection. To better define the population, we would stratify it by various demographic factors such as age or education level, and so on. The important point is we define our population relative to our analytical questions to ensure that it accurately represents the population of interest and answers all relevant questions. Relative to our project, we would stratify the population relative to our input variables. As an example, if we wanted to understand supplier on-time delivery for three major suppliers and five distribution centers, our stratification variables would be supplier at three levels and distribution center at five levels. This stratification would create a 3×5 matrix. The team would collect data for each combination of the stratification variables, that is, $3 \times 5 = 15$ combinations. This would ensure that data collection would be representative of the population of interest and allow the Lean Six Sigma improvement team to answer the relevant project questions. In summary, a random sample is collected from each combination of the stratification variables from the specific population of interest. This will ensure that the sample statistics represent the statistical characteristics of the population. These sample statistics will be the central tendency and variation of the population. Examples of central tendency include the sample average (or mean) and median. Variation is measured using either the sample range or sample standard deviation. The sample range is calculated as the maximum minus the minimum value of the sample observations. The sample standard deviation is calculated by the square root of the sum of squared distances of each observation from the sample mean divided by the sample size (minus one observation). Because the sample will not contain all possible observations of the population, there is

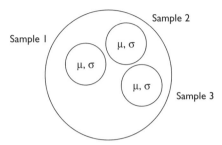

Figure 8.7 Population Sampling

some uncertainty around the estimated population parameters. In this sense, sample statistics, such as sample mean and standard deviation, are estimates of the true population parameters (i.e. μ, σ^2 for a continuous Normal distribution) as shown in Figure 8.7. As the sample size increases, population parameters can be estimated with increasing precision sample statistics.

Statistical sampling is important because we cannot count everything of interest in the population given the limited resources available to the Lean Six Sigma improvement team. As an example, if our project goal is to reduce lead-time, we measure lead-time for a representative sample of several time periods, but not all of them. As a second example, if we want to focus the project objective on lead-time analysis over many facilities and products, our sample should include a stratified sample for all the facilities and products of interest, but not for every possible time period. As a final example, if the project focus is to reduce lead-time for a particular product and facility, our population would be defined as all lead-times in that facility and product over several periods of time, but not all time periods. In conclusion, the defined population and subsequent data sample is determined by our project's problem statement and objective. These provide the focus for the analysis to answer the key questions relevant to the project's objective.

Common Sampling Methods   Table 8.18 shows four common sampling methods. These are simple random sampling, stratified sampling, systematic sampling, and cluster sampling. Simple random sampling draws a random sample of size n from the larger population. In *simple random sampling* a random sample is selected from the population in such a way that every combination of observations from all possible samples has the same probability of being selected in any subsequent sample. Random sampling is commonly applied to situations in which the population

Simple Random Sampling	Drawing a sample from a population at random. An example would be to take a random sample of weekly invoices and count the total number of errors for all invoices in the sample. Calculated statistics are average number of errors per invoice, standard deviation of errors per invoice, or percentage invoices in error.
Stratified Sampling	Using a stratification metric to divide a population into groups (strata), and then drawing a random sample from each group. An example is to divide an inventory population into groups (strata) based on stated book value, and then sampling within each stratum. Calculated statistics are overall inventory value as well as value within each stratum and standard deviation in overall inventory value.
Systematic Sampling	Drawing a sample using ratios, that is, on-time delivery of every sixth shipment from receipts records. Calculated statistics are average and standard deviation of delivery time.
Cluster Sampling	Drawing a sample using naturally occurring samples, that is, product groups. Calculated statistics are average and standard deviation for each cluster.

Table 8.18 Common Sampling Methods

is static. An example would be to take a random sample of inventory account values for a single month. *Stratified sampling* uses supplemental information relative to the input variables (stratification variables) to gain precision in statistical estimates. Stratification breaks the population into several strata (groups) in such a way as to reduce the within-subgroup variation relative to the overall variation of the population. Using stratified sampling, the overall sample size across the entire population is significantly smaller than if a random sample had been drawn from the same original population. Stratified sampling can be used in inventory cycle counts to obtain statistically valid estimates of the inventory book value and is a more efficient system than simply classifying accounts into arbitrary classes, that is, the ABC classification method. *Systematic sampling* draws samples from the population (or process) at equidistant time intervals as it varies over time. Systematic sampling is commonly used in manufacturing or other processes that vary over time. *Cluster sampling* draws samples from naturally occurring groups. An example of cluster sampling would be sampling by a particular customer demographic or geographical region. In each sampling strategy, the objective is to calculate, with predetermined statistical confidence and precision, efficient and unbiased sample statistics to make inferences relative to the population parameters. Statistical sampling is useful because we cannot physically count the entire population due to resource constraints.

Simple Data Analysis Tools

Several simple tools are very useful in the analysis of supply chain data. These analytical tools are also useful to communicate project status to the organization. A few of the most common analytical tools are Pareto charts, histograms, cause-and-effect diagrams (C&E), box plots, run charts (time-series charts), descriptive statistics, and process maps (of various types and degree of complexity). Most statistical packages offer these simple analytical tools (Minitab is the statistical software package used in the following examples). Note that the word "simple" does not mean these are not powerful tools to analyze process data to understand the root causes for the process breakdown. On the contrary, most supply chain problems are successfully analyzed using these simple data analysis tools. As an example, one of the most powerful analysis tools is the value stream map (VSM). A VSM allows the team to see the process in detail. This includes the spatial organization of the process including areas within the process that are broken, that is, rework loops and NVA activities. The Lean Six Sigma improvement team also uses the VSM to quantify key process output variables (KPOVs) related to time, material, cost, and quality. The specifics of value stream mapping were discussed in Chapter 4 and shown in Figure 4.4.

Pareto Charts   The original concept behind Pareto charts is the 80/20 rule, which states that 80% of the frequencies (counts) are associated with just 20% of the categories. Pareto charts display the count frequency of categorical data in descending order by frequency and category and are useful because they focus attention on those categories having the highest observational frequencies (counts). This is very useful in communicating project status as well as explaining why the team has focused the project in one direction rather than another. Pareto charts often allow the team to drive to the root causes of the problem or at least focus their attention in the direction of the major root causes. An example of a Pareto chart is shown in Figure 8.8. In this example, raw material inventory has a higher frequency (highest count) than either work-in-process (WIP) or finished goods inventory. If we want to reduce the number of inventory items (as opposed to the most expensive items), raw material inventory might be the initial focus of our project. Although the Pareto chart is an extremely effective tool to focus the team's attention on the categories having the largest occurrence frequency, many teams do not fully utilize this important tool. As a general rule, if the team has not scoped the project down to at least a third-level Pareto, then they are probably still too high in their project's problem statement to focus its objective and conduct an effective root cause analysis of the problem. Pareto charts are also useful analytical tools throughout the project. In the control phase of the project the team should be able to use Pareto charts and other analytical tools to show the process owner and local work team a reduction in defect occurrence. This concept is shown in Figure 8.8.

Histograms   A histogram is a graphical display of the data showing its central tendency and variation. Histograms are useful in the beginning of the Lean Six Sigma project to graphically show the KPOV average or variation against the improvement target. Later, in the solution phase of the project, the improvement team can show the process owner and local work team the improved shift in the process target or reduced variation due to the improvements. This concept is shown in Chapter 9, Figure 9.7.

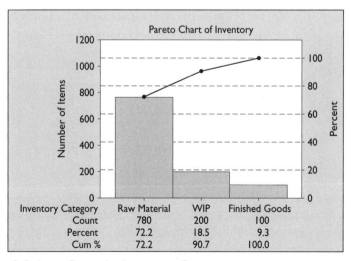

Figure 8.8 Item Count by Inventory Category

The histogram is constructed slightly differently than the Pareto chart. The X axis of the histogram is a continuous scale rather than consisting of several discrete categories. As an example, Figure 8.9 graphically displays "lot size" using histograms. In Figure 8.9, we can see the average lot size tends toward approximately 500 units and ranges between a minimum of 350 to maximum of 650 units before the process improvement, and after the process improvement, the average lot size shifts to the left of the original distribution, reflecting a lower average lot size. Assuming the decrease

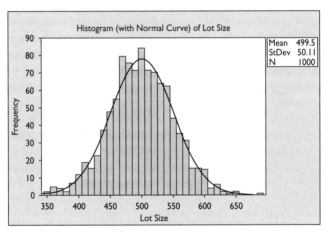

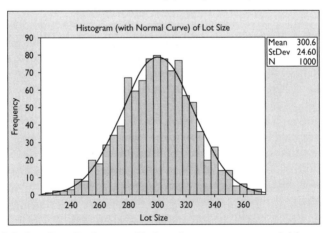

Figure 8.9 Lot Size Reduction (Before Improvements and After Improvements)

in lot size was due to process improvement efforts, a histogram allows the process owner and local work team to see the "before" improvement versus "after" improvement. Subsequent quantitative verification of process improvements using more advanced statistical methods can be used to "prove" the average difference. However, advanced statistical analysis usually begins with a simple graphical analysis of the sample data.

Cause-and-Effect Diagrams (C&E)   As mentioned earlier in this chapter, C&E diagrams are used to brainstorm possible inputs (causes) of a particular KPOV (effect). In Figure 8.1 the team brainstormed possible causes for high inventory investment. Inventory investment in this example could refer to inventory at a given facility or perhaps a product group or even a single item. The causes are initially categorized into major groups, that is, people, materials, methods, and so on to aid subsequent data collection and analysis. Lower-level inputs are progressively identified by the team until no more ideas are generated by the group. The "causes" of the "effect" form the basis for subsequent data collection since the important causes become stratification variables. In addition to the C&E tool, many other tools are used to brainstorm possible process inputs on which to collect data for inclusion into the initial inventory model. A team approach to identification of possible input "X's" minimizes the probability that unimportant inputs are inadvertently left out of the data collection phase.

If there are several KPOV variables that are part of the team's analysis, the input variables are ranked according to their relative importance to each KPOV. In addition, ranking the KPOV variables in relative importance to the customer allows the team to obtain a weighted ranking for each input variable to allow the improvement team to focus data collection efforts on input variables thought to be most important in explaining the variation in the KPOV variables. The Excel tool used to rank the input variables is called a cause-and-effects matrix. A C&E matrix is shown in Figure 8.10. The KPOV variables of the C&E matrix can be thought of as the "effects" from each C&E diagram. These effects or KPOV variables are listed horizontally along the top of the C&E matrix. Each KPOV is ranked relative to importance (usually with respect to the customer) on a scale of 1 to 10. A rating of 10 is very important to the customer. The causes, that is, possible input variables from all the C&E diagrams, are listed by row down the Excel spreadsheet. At the intersection of each input and KPOV variable is a correlation number of 1 to 10. A rating of 10 implies the input variable is thought to be highly correlated to the KPOV. The spreadsheet multiplies the "importance rating" of the KPOV by the

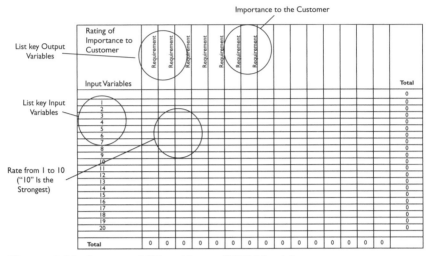

Figure 8.10 Cause-and-Effect Matrix (C&E Matrix)

"correlation" rating of each input variable to calculate a total rating for each KPOV and input variable combination. The total ratings are summed across each row for every input variable to calculate a weighted total for every input variable. The improvement team uses these weighted total ratings to prioritize data collection and analysis of the input variables.

Box Plots   Box plots describe the central location as well as the variation of sample data, but in more detail than a histogram. As an example, the box plot identifies the minimum and maximum values of the sample, the 1st quartile (below which is 25% of the sample), the median (below which is 50% of the sample) and the 75th quartile (below which is 75% of the sample). Figure 8.11 uses a box plot to graphically display the lot size data originally shown in Figure 8.9. The box plot provides more information on the distribution of the sample than a simple histogram of the data.

A multilevel box plot can be used to summarize KPOV sample data versus several discrete levels of an independent variable. As an example, the multiple-level box plot shown in Figure 8.12 breaks the inventory data down by inventory classification, that is, raw material, work-in-process (WIP), and finished goods. The multiple-level box plots compare discrete levels (categories) of the independent or stratification variable, that is, "inventory classification" relative to inventory investment showing the central location and dispersion, that is, median and range of inventory investment by inventory classification.

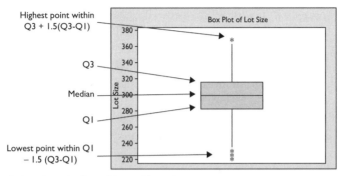

Figure 8.11 Simple Box Plot of Lot Size

Run Charts (Time Series) Run charts (also called time-series charts) graphically display sample data by its time-ordered sequence. They are useful analysis tools throughout many Lean Six Sigma projects. As an example, at the start of the Lean Six Sigma project, time-series charts are used to show the average level and variation of the KPOV over time when first defining the project. The project's improvement target can also be superimposed onto the time-series chart to show the performance gap between the current KPOV performance level and the improvement target. Time-series charts are also commonly used in inventory analysis

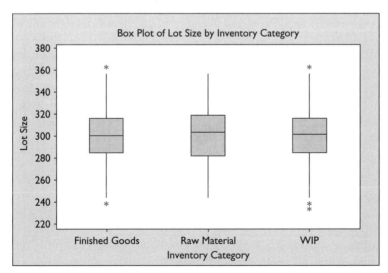

Figure 8.12 Multiple-Level Box Plot of Lot Size by Inventory Classification

to track changes in inventory investment or turns over time. If inventory turns are seasonal, the improvement team will be able to detect improvements in the turn's ratio from one time period to another, that is, independent of seasonality effects.

These charts, with the addition mathematical formulas, are also useful to forecast future demand. Figure 8.13 shows a sequential pattern of shipments versus demand over time as well as a forecast graph of the unit demand history over time. In the first graph, it is apparent that shipments are lower than unit demand for each time period. The differences

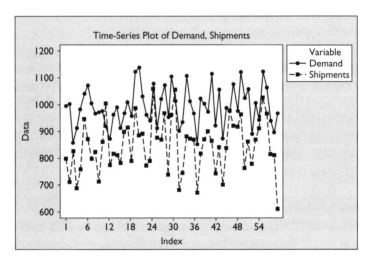

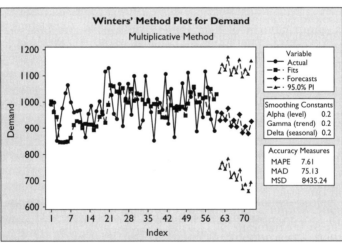

Figure 8.13 Shipments Versus Demand Over Time

between demand and actual shipments represent backorders which must be shipped to the customer at a later date. This performance gap between demand and shipments could serve as the basis for one or more Lean Six Sigma improvement projects. The Lean Six Sigma team might also want to use the unit demand data to better forecast future sales in order to improve shipment performance, that is, to have the correct inventory on-hand to reduce the backorder incident rate. Relative to Figure 8.13, the forecasting error statistics, that is, mean absolute error (MAPE) and root mean squared deviation (RMSD) were discussed in Chapter 3 and defined in Figure 3.15.

Time-series charts are also very useful in the control phase of the project to show improvement to KPOV's over time relative to their historical performance. This concept is discussed in Chapter 9 and shown in Figure 9.6. Figure 9.6 shows a control chart (a type of time-series chart) broken into before and after periods. In summary, using time-series charts to graphically show changes in the KPOV pattern is a powerful way to communicate the effectiveness of the project. The team could also show improvements using more rigorous statistical tools that would show differences between the average and variance of the KPOV over time.

Descriptive Statistics   Graphical techniques should be used in conjunction with quantitative analysis such as descriptive statistics to efficiently summarize large amounts of sample data. As an example, in Table 8.19, unit demand data is statistically summarized and a graphical summary is shown. Although the graphical summary provides very useful data about the distribution's central location and dispersion or variation, the statistical summary provides more detailed data, which can be used as the basis for more advanced analytical techniques. These advanced analytical techniques are used to prove differences in the population parameters. These concepts will be discussed later in this chapter.

Reviewing Table 8.19 in more detail, N is the total sample size. The *mean* is the average of the sample values. The *stddev* is the standard deviation of observations from the mean. The *minimum* is the smallest demand value within the data set. The *maximum* is the largest value in the data set. Q_1 is the first quartile of the data set. Twenty-five percent of the observations fall below this value (similar to box plot discussion above). Q_3 is the third quartile of the data set. Seventy-five percent of the observations fall below this value. Finally, the *median* is the value at which 50% of the observations are below and 50% of the observations are above. These are the basic summary statistics obtained from samples. However, depending on your analytical application, there may be others.

Variable	N	Mean	StdDev	Minimum
Demand	3146	1003.9	83.4	767.0
	Q1	Median	Q3	Maximum
	948.0	1002.0	1062.0	1394.0

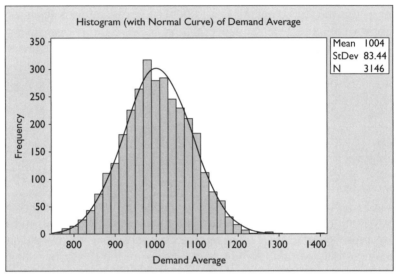

Table 8.19 Statistical Summary of Unit Demand

Process Mapping A process map is a diagram showing the flow of materials and information through the process. The focus of the process map is the project charter's problem statement and objective. The project's problem statement sets the boundaries of the process under analysis based on the KPOVs (metrics). Quantitatively mapping a process, even at a high level, relative to KPIV variables and their associated KPOV variables usually identifies numerous ways to improve the process by highlighting its strengths and weaknesses. The more highly quantified the process map, the more useful is its information relative to potential process improvements. A good example of a quantified process map is the value stream map (VSM) described in Chapter 6. The VSM quantities material and information flow through the process including equipment problems, inventory status, quality issues, and other process breakdowns. Mapping the process will show where these interruptions and inefficiencies occur. Often the process breakdowns will be associated with rework loops, increases in waiting time, work inspections, and other non-value-adding (NVA) tasks. Figure 8.14 shows a process as a logically grouped sequence of serial and parallel tasks

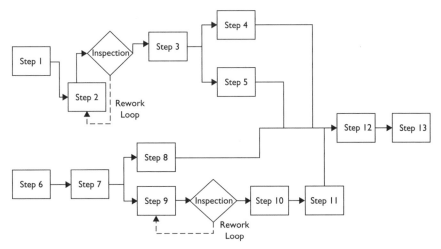

Figure 8.14 Simple Order Entry Process

consisting of people, machines, materials, and information. Also shown in this simple process map are two rework loops. Ideally, to effectively use the process map as an analytical tool, it must be quantified relative to the KPOV and KPIV variables. Process maps are especially useful in situations where a process is not achieving its operational targets or the work flow tasks are not well defined.

High-level maps show the major organizational relationships within a process. Functional process maps show the relationships between functions as well as the flow through each function over time. The basic steps necessary to create a process map are listed in Table 8.20. In the first step, the process map should reflect the original project problem statement and objective. This will ensure that the data collection and

- Problem statement should be well defined
- Include all required functions (team)
- Ensure the correct detail level
- Analyze the current process (metrics/interfaces)
- Create "current state" map and a "future state" map
- Compare both maps
- Migrate to the "future state" map

Table 8.20 Creating a Process Map

analysis efforts are correctly focused and the analysis will be relevant to the questions the Lean Six Sigma team has been asked to answer for senior management. Second, the process-mapping team should include key representatives from the impacted operations. These people should include functions just before the process (at the process input boundary) and just after the process (at the process output boundary). Third, the process map should be constructed top down with the correct level of quantification focused on the area of the process under analysis rather than the entire process. Fourth, once constructed, the process map should be analyzed to find improvement opportunities. If the map has been sufficiently quantified, this task will not be difficult. Fifth, working from this baseline map *(current state)* of the process, an improved process is designed *(future state)* and a project schedule is developed to make the necessary process changes. However, it might not be possible to migrate the process to its future state due to technological or resource constraints as shown in Figure 8.15. However, the analysis will usually identify many portions of the process which contain operational waste in the forms of unnecessary tasks, rework loops, waste, and KPOV performance gaps. These process breakdowns should be eliminated as soon as possible by deploying Lean Six Sigma projects.

Creating a System Model   In Chapter 1, Figure 1.3 shows, at a high level, how the quantified process map forms the basis for building a supply chain model and in particular an inventory model. Simulations of the model can be developed to analyze system performance if the process map is sufficiently quantified relative to key process input variables (KPIV) relative to key process output variables (KPOV) such as lead-time, demand, and other relevant metrics. The analysis of system performance will also identify Lean Six Sigma improvement projects that can be used

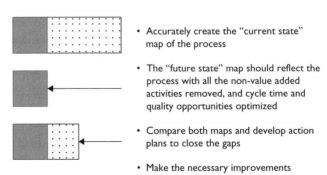

- Accurately create the "current state" map of the process

- The "future state" map should reflect the process with all the non-value added activities removed, and cycle time and quality opportunities optimized

- Compare both maps and develop action plans to close the gaps

- Make the necessary improvements

Figure 8.15 Decreasing Non-Value-Adding Time

to drive to the root causes for the process breakdowns. However, prior to building a quantified model, the relationships between KPIVs and KPOVs must be clearly understood by the improvement team.

The supply chain model that is built by the Lean Six Sigma improvement team will vary depending on the project's problem statement and objective. Well-established formulas are often available from the fields of operations management and operations research to help the team understand relationships between the KPOV and KPIV variables, although the major work stream and inventory models may differ by specific application. As an example, if inventory analysis is a major focus of the project, the calculation of an optimum inventory quantity is straightforward since the calculation is based on lead-time, demand variation, and service level as shown in Chapter 5, Figure 5.12. However, if staffing at a distribution call center is the project focus, then queuing models might be more appropriate to analyze the call center system. Finally, if the project focus is to identify and eliminate process waste, then a value stream map would be the appropriate analytical tool to answer the necessary questions regarding process implication and standardization.

Advanced Statistical Analysis Tools

Most Lean Six Sigma projects do not require advanced or complicated tools and methods for their solution. Our experience indicates that 50% or more of the Lean Six Sigma supply chain projects are effectively executed using Lean tools and methods. Another 25% of the projects use basic quality analysis tools to identify root causes. The remaining projects require varying degrees of analytical complexity depending on the root cause analysis necessary to achieve the project's objective. The following discussion is focused on projects requiring advanced tools and methods to identify and eliminate the root causes for the process breakdowns.

The discussion is broken into two sections. The first section discusses advanced statistical tools taught in every Lean Six Sigma black belt course. These include statistical inference, regression analysis, and experimental design methods. The second section reviews operations management and operations research tools and methods. Both discussions show how these tools might be applied in practice. These discussions end with a simulated example of a distribution call center showing how some of the advanced tools would be used in practice by Lean Six Sigma black belts.

Statistical Inference   The next level up, in analytical sophistication, from simple graphical analysis is statistical inference tools and methods. Using the tools and methods of statistical inference, we frame our problem statement and project objective in terms of statistical questions regarding the central location and the variance or distribution of the KPOV variables. This analysis also includes confidence and risk levels around the estimated population parameters. As an example, referring to Figure 8.12, that is, multiple-level box plots of lot size by inventory classification, we could take the analysis to a higher level using statistical inference. Qualitatively, Figure 8.12 shows that lot sizes by inventory classification appear to have the same mean and variance. We draw these conclusions because the median lot size of each inventory classification appears equal and the whiskers coming off the box plots by inventory classification appear to be the same length. The example shown in Figure 8.12 was easy to analyze graphically because the central location and variation of the subgroups were so similar. At a more advanced level, a *one-way analysis-of-variance (one-way ANOVA)* model could be built to answer the simple question, do the subgroups have the same mean (average) lot size? In the distribution call center example at the end of this chapter, Figures 8.22 and 8.23 show that the subgroup levels have statistically significantly increasingly higher average waiting times for different levels of incoming call volume. Statistical inference is useful in situations where there is some overlap between observations of the subgroups i.e. the distributions are not clearly separated or overlapping. Using a different statistical test, an equal variance test could also be used to answer the question of whether the subgroups have equal variances in lot size.

In Lean Six Sigma training, statistical inference is broadly divided into four major categories depending on the distribution of the key process output variable (KPOV) also called the "Y," "response variable," or the "dependent variable," versus the key process input variable (KPIV), also called the "independent variable" and "X." The first category is the situation in which the KPOV is continuous and the KPIV variables are discrete, that is, have one or more discrete levels. An example would be to measure the average delivery time of one or more discrete suppliers i.e. one distribution for each supplier with the output being delivery time in days. Statistical tests of central location associated with category 1, which use just one stratification variable, include one-sample t-tests, two-sample t-tests, and one-way analysis-of-variance (one-way ANOVA) tests. These statistical tests measure differences in the subgroup mean relative to a constant, another subgroup mean, or between several subgroups respectively. Additional tests on central location, but which use two or more stratification variables, each having a discrete number of subgroup levels, and a continuous response, are called *general full factorial models*. These models are also called two-way, three-way, and so on analysis-of-variance models or general linear models.

The second major category includes analytical tools used when both the KPOV and KPIVs are continuous. Correlation analysis, simple linear regression, and multiple linear regression models are in this category. Correlation analysis and simple linear regression are used to determine whether a linear relationship exists between two continuous variables. An example of simple linear regression would be measuring inventory investment as lead-time increases. Multiple linear regression models are used to build relationships between a KPOV and several KPIV variables. At least one of the KPIV variables is continuous in multiple linear regression models. An example would be building a linear model of inventory investment using lead-time and product type as its KPIV variables. Category 3 contains tools and methods that model situations where both the KPOV and KPIV variables are discrete. A key tool in this category is contingency tables. A typical contingency table compares defect counts cross-referenced using two variables. An example would be counting the number of defective invoices for three shifts and four facilities. The fourth category models the situation in which the KPOV is discrete and the KPIV variables are a mixture of either discrete or continuous variables. These tools are called logistic regression models. The KPOV of a logistic regression could have a "pass or fail" response *(binary logistics regression)*, have a discrete number of levels *(ordinal logistic regression)*, or the KPOV could be a label *(nominal logistic regression)*.

Regression Models   Several regression models are available to help determine a relationship between a KPOV and one or more KPIV variables, that is, $Y = f(X)$. These models depend on several assumptions concerning the KPOV and KPIV variables, which are discussed above at a high level but can be also be found in statistical textbooks. Multiple linear regression models are the most commonly used in the Lean Six Sigma program since linear relationships between KPOV and KPIV variables are common. But if the KPOV variable is measured as pass or fail, binary logistic regression models would be used by the Lean Six Sigma team. In addition to logistic regression models, there are other special regression models that have been developed for situations in which the assumptions on which linear regression models were developed are not satisfied. These special regression models can be very useful in the analysis of supply chain processes since many KPOV variables vary over time (regression-based time-series models). This is especially true for demand forecasting models. These special regression models are useful to handle situations in which there is time-related serial correlation of the model's residuals (differences between actual versus model-fitted values for each sample observation). As an example, multiple linear regression models can be modified by bringing in lagged KPOV terms, or the Cochran-Orcutt method could be used if the residuals are correlated. However, nonlinear regression would be used if the model contained lagged KPOV terms and the residuals were also correlated with each other. Unfortunately, most of these special regression models and time-series methods are not usually taught to Lean Six Sigma supply chain "belts".

Experimental Designs   Experimental design methods enable the Lean Six Sigma improvement team to quickly and systematically identify the KPIV variables and design experiments to evaluate, under controlled conditions, their combined impact on one or more KPOV variables. This allows the team to establish the $Y = f(X)$ relationship and make the necessary process improvements. Experimental designs efficiently model a process because the KPIVs were carefully selected and purposelessly setting their minimum and maximum levels to build a model that explains the KPOV variables over the full range of each KPIV (expected process performance). Another advantage of experimental designs is that they enable development of very accurate models using the smallest number of experiments. Lean Six Sigma black belts are trained in developing regression-based models using very efficiently designed experiments. As an example, if the KPIV variables are all continuous and their relationship with the KPOV is linear, then the analysis can be reduced to a two-level experimental design. In other words, the KPIV variables will be evaluated at just their low and high levels rather than intermediate levels (but counterpoints would be run to detect curvature if it in fact existed). There are many other aspects of experimental designs (blocking, fractional designs, response designs, and so on) that may be important to Lean Six Sigma projects but will not be discussed in this book.

Operations Research Tools   *Operations research (OR)* tools include linear programming models, transportation models (special case of linear programming models), assignment models, queuing models, network models, inventory models, forecasting models, simulation models, Markov processes, and decision models. Operations research tools and methods model the entire system being studied using relationships that are known beforehand to exist within the process. In other words, we don't have to do a designed experiment to understand the basic underlying functional relationships between the KPOV and KPIVs. As an example, the *average waiting time (AWT)* in a call center depends on the volume of incoming calls, the types of calls, the average length of each incoming call, the number of agents handling the incoming calls, and their skill level (training level relative to the information they are expected to provide the customer). Standard queuing models can help the team develop a general model for the call center system to optimize the number of agents to achieve target service levels at various incoming call volumes. The example at the end of this chapter uses both regression and queuing models to reduce customer waiting time in a distribution call center. Understanding that OR tools are the appropriate analytical method prior to data collection and analysis allows the team to quickly come up to speed on the problem statement by identifying relevant KPIV and KPOV variables necessary to build the OR model. Having this knowledge will greatly accelerate the Lean Six Sigma project completion cycle. This enables a faster implementation of the project's optimum solution rather than reliance on typical DMAIC tools and methods alone. This does not mean the Lean Six Sigma team would be jumping to conclusions, but rather it is advantageous to start the root cause analysis at a more sophisticated level. Why reinvent the wheel?

Linear Programming Models   Typical supply chain applications of operations research tools and methods are shown in Table 8.21. Several of these applications are based on *linear programming (LP)* methods. Examples include aggregate product planning, product routing, and some advanced inventory applications. Aggregate planning includes determining the best product mix to maximize profits given limited resources. Product-routing applications include problems related to how

▪ Aggregate planning	▪ Materials handling
▪ Product planning	▪ Waiting-line models
▪ Product routing	▪ Risk analysis
▪ Process control	▪ Simulation models
▪ Inventory control	▪ Decision analysis models
▪ Aggregate production planning	▪ Prioritization models
▪ Distribution scheduling	▪ Forecasting models
▪ Plant location	

Table 8.21 Operations Research Applications

products should be shipped from several manufacturing facilities to several distribution centers to minimize transportation costs. Inventory planning models determine when to manufacture products to minimize inventory and backorders to maximize customer service and minimize operational costs by time period.

LP models attempt to find an optimal solution when decisions must be made under constraints related to supply, demand, or other system conditions. These models consist of three major components including an objective function, the decision variables (which are modified to increase or decrease the level of the objective function), and several constraints. The objective function is usually specified in terms of "profit maximization" or "cost minimization" or similar maximization and minimization criteria. Decision variables could include questions related to "when and how much to order," "when and what to manufacture within a given time period," or "when and how much of a product to ship and to where." There are many other types of decision variables depending on the specific problem. Model constraints are limitations placed on the model. These limitations include resource limitations, capacity limitations, minimum demand levels, and other types of modeling conditions, which depend on the problem formulation. Once the relationships between decision variables, constraints, and problem objectives are specified, the goal is to achieve an optimum solution that specifies the levels for each decision variable and satisfies all problem constraints. These LP tools and methods should be taught to Lean Six Sigma "belts" who will be doing supply chain projects, but typically are not—also few master black belt *(MBB)* instructors have been trained in OR tools and methods.

Waiting-Line Models   Figure 8.16 shows a generic waiting-line *(queuing)* model. The model is based on several system components, which must be specified based on an analysis of the specific system being modeled by the team. The basic system components of waiting-line models include the arriving population, the arrival rate, the arrival time distribution, the behavior of the arrival population relative to remaining or leaving (balking) the queue, the allowed length of the queue, the number of parallel servers in the system, the service rate of each server, the service time distribution, and finally the service discipline, that is, first-come-first-served or some other service discipline. Table 8.22 lists a few other important waiting-line model characteristics. Waiting-line models are very complex and show how difficult it would be for an improvement team to create the model without having knowledge of the dynamic way

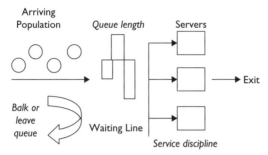

Figure 8.16 Basic Waiting-Line System

in which the system's components interact with each other. If a project involves waiting-time analysis in the context of resource optimization, the Lean Six Sigma team should understand waiting-line models prior to data collection and analysis, since the applicable system components and their dynamic relationships have been established through years of careful research. Although there may be other KPIV variables for the particular system under study, why not start the analysis at an advanced level to shorten the cycle time for project completion?

Other Operations Research Tools and Methods   Several other important classes of OR tools are very useful for supply chain analysis. The most common classes include specific transportation models, assignment models, network models, project scheduling models, simulation models, decision analysis models, and Markov processes. Transportation models are specially designed LP methods that model network flows. A common example is materials being shipped from several sources of supply to several destinations where the objective is to minimize transportation costs.

Population	Server
■ Infinite distribution	■ Constant service
■ Finite distribution	■ Variable service
■ Constant arrival	■ Single channel
■ Distributed arrival	■ Multichannel
■ Remain in queue	■ Single phase
■ Leave queue	■ Multiphase

Table 8.22 Queuing Model Assumptions

Common applications of assignment models include the assignment of people to machines and related scheduling applications. Network models and project scheduling models show how to optimally configure and manage a system having limited resources and several activities that must be completed in a predefined sequence. The critical path method (CPM) is an example of a network model used to control project resources and scheduling of the activities necessary to complete the project. Simulation models, which were discussed in Chapter 6, allow the user to change KPIV variable levels to analyze the resultant changes in the system's KPOV variables. Decision analysis models are used to evaluate several alternative solutions each with different benefits and risks of successful activity completion. Markov models allow the user to study how a system changes over time and estimate probabilities that the system will be in a certain state at a particular time. A common supply chain example using a Markov model would include predicting the probability of equipment breakdown in a distribution center.

Distribution Call Center Example

In the distribution call center example starting on Figure 8.17, the problem statement is: the average waiting time is averaging 100 seconds per incoming call versus the 90-second maximum target. The project goal

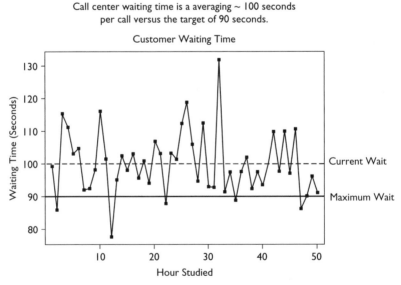

Figure 8.17 Project Problem Statement and Objective

is to reduce customer waiting time to 90 seconds or less for every call without deterioration in customer service levels. To reduce waiting time, the team must understand some of the input variables ("X's") that may be contributing to the long waiting time. Some possible "X's" identified by the team through simple discussion might include "the prioritization of calls by the automatic call routing system," "the region from which the call originates," "the type of customer since different customers will require different types of information," "the complaint type since different complaints will require different times to resolve by the call center agent," and "the correct routing of the call to an agent trained to answer customer questions relative to that market segment." KPOV variables include customer waiting time and satisfaction as measured by the agent's service-level report. Figure 8.18 shows how a SIPOC could be used to gather data on the KPOV and possible input variables. In addition to Figure 8.18, the C&E diagram shown in Figure 8.19 is also a useful tool to identify KPOV and possible input variables. Reviewing Figure 8.19 in more detail, we see the improvement team has identified, in addition to the queuing model inputs, several other possible "X's" including "call volume," "day-of-week," "environmental conditions," "the tools and materials the agent needs to handle each call," "agent staffing levels," and "the software tools that provide the agents with information to answer customer calls." These are the "X's" the team will need to collect data on to understand which "X's" are really the KPIV variables impacting the KPOV variables.

After the team verifies that it listed the correct KPOV variables and the "X's" most probably impacting the KPOV variables, a measurement

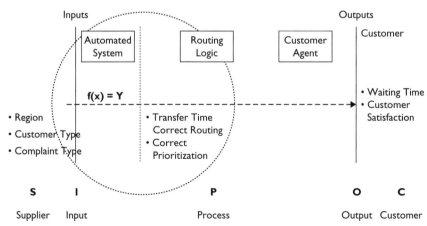

Figure 8.18 High-Level Process Map (SIPOC)

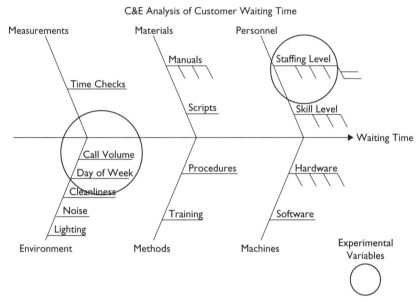

Figure 8.19 Analysis of Causes for Customer Waiting

system analysis (MSA) is conducted on each KPOV variable. The MSA will verify the KPOV variables baselines to ensure that the team has the correct baseline and will be able to detect KPOV changes when the eventual process improvements have been made. This concept is shown for the current example in Table 8.23 and Figure 8.20. The team will also need to review each of the input variables to verify they can be accurately and precisely measured throughout the data collection process. The team should also realize every project has its specific MSA strategies, tools, and methods. Figure 8.20 shows that a major MSA tool required for the MSA is an *attribute agreement analysis.* This is because the agent/customer interaction involves a subjective evaluation of customer information by an agent. The attribute agreement analysis is conducted to ensure that we correctly classify each complaint to evaluate whether the agent correctly answered customer questions relative to the complaint type. Remember, the project objective is to both reduce customer waiting time and correctly answer customer questions. If the number of incoming calls which must be rerouted is decreased, productivity will increase and customer satisfaction will improve. The attribute agreement analysis ensures that the team is correctly answering customer questions, and if not, where the MSA needs to be improved to achieve the necessary measurement

Measurement System Analysis in this project application takes on several forms...

1. We must be sure we are accurately recording customer waiting time... This requires we carefully define the waiting metric from the customer perspective... Waiting time begins when the customer places the call and ends when the customer's questions have been successfully answered by the system...

2. We must be sure the customer's questions have been accurately interpreted and answered by the agent... This requires an Attribute Gage R&R analysis since we must compare performance both to a standard and across many agents...

Table 8.23 Measurement Analysis of the Output "Y"

accuracy and precision. An attribute agreement analysis compares each agent relative to "agreement with themselves," "agreement with each other," and "agreement with a known standard." Lean Six Sigma ™ training programs discuss these measurement systems in detail as part of their MSA training.

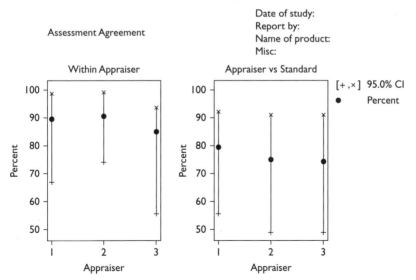

Figure 8.20 Measurement Analysis of Agent Service Level

After the Lean Six Sigma improvement team verifies that KPOV variables can be measured accurately and precisely, a capability analysis is conducted to determine the baseline performance level of the project's KPOV variables. A summarized version of the capability analysis is shown in Figure 8.21. The analysis shows 88% of the calls in the sample took longer than 90 seconds to resolve. There isn't a short-term (within-subgroup variation) metric since the data was not collected as subgroups and was gathered over several months. It would have been better if the team had collected data as subgroups since an entitlement analysis could have been made by the team. The concept of an entitlement analysis was discussed earlier in this chapter and was summarized in Figure 8.3. After the team makes its process improvements, a second capability analysis will be conducted on the improved process to see how well it meets the customer requirement of waiting less than 90 seconds.

The next step in the project analysis is to collect data on the process input variables and the corresponding levels of the KPOV variables. There are many tools and methods that can be used to conduct the root cause analysis depending on the analytical questions the team needs to answer. Figure 8.22 shows a "main effects plot" of a few input variables.

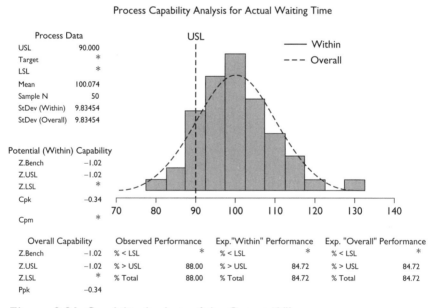

Process Capability Analysis for Actual Waiting Time

Process Data
USL   90.000
Target   *
LSL   *

Mean   100.074
Sample N   50
StDev (Within)   9.83454
StDev (Overall) 9.83454

Potential (Within) Capability
Z.Bench   −1.02
Z.USL   −1.02
Z.LSL   *
Cpk   −0.34
Cpm   *

— Within
--- Overall

Overall Capability		Observed Performance		Exp."Within" Performance		Exp. "Overall" Performance	
Z.Bench	−1.02	% < LSL	*	% < LSL	*	% < LSL	*
Z.USL	−1.02	% > USL	88.00	% > USL	84.72	% > USL	84.72
Z.LSL	*	% Total	88.00	% Total	84.72	% Total	84.72
Ppk	−0.34						

Figure 8.21 Capability Analysis of the Output "Y"

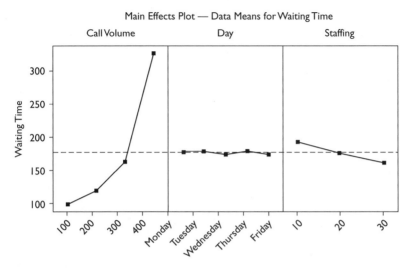

Figure 8.22 Analyzing the Inputs "X's" Using Main Effects Plots

The main effects plot displays the mean (average) waiting time for each level of every input variable independently of other input variables. As an example, it can be seen that as call volume increases, with the other input variables held constant, the customer waiting time increases. Also, the main effects plot shows that the input variables "day of the week" and "staffing" do not appear to significantly impact customer waiting time. However, a better way to see the impact of incoming call volume on customer waiting is to use the *multi-vari chart* shown in Figure 8.23. The multi-vari chart clearly shows the small impact of day-of-week on incoming call volume on customer waiting time, but the large impact of incoming call volume. The Lean Six Sigma improvement team can build a model explaining customer waiting time based on the KPIV levels. Table 8.24 shows a regression summary for the example. The regression model has identified incoming call volume and staffing levels as important predictors of customer waiting time assuming a linear relationship. Using R-Sq (adjusted), this model explains 71.7% of the variation in customer waiting time. But a more exact model is shown in Figure 8.24. This model assumes a curvilinear relationship that explains 84% of the variation in customer waiting time.

However, if the Lean Six Sigma improvement team understands queuing methods, then a more dynamic model of customer waiting time can be built based on staffing levels and incoming call volume. This is shown

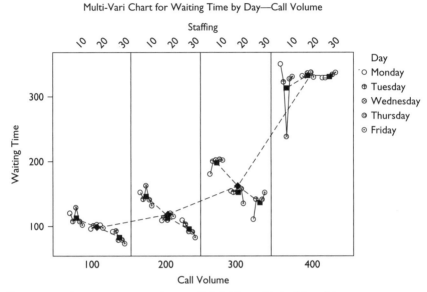

Figure 8.23 Analyzing the Inputs "X's" Using Multi-Vari Charts

in Table 8.25, in which customer arrival rates are compared against the number of agents as well as their service rates (average handling time in seconds). The queuing model will also help the shift supervisors manage customer waiting time by showing how to change staffing levels as incoming call volume fluctuates. To fully implement an optimum solution,

The regression equation is:					
Waiting Time = 26.8 + 0.726 Call Volume − 1.53 Staffing					
Predictor	Coefficient	SE Coefficient	T	P	VIF
Constant	26.79	16.33	1.64	0.103	
Call Volume	0.72560	0.04215	17.21	0.000	1.0
Staffing	−1.5338	0.5772	−2.66	0.009	1.0
S = 51.63	R-Sq = 72.2%		R-Sq (adjusted) = 71.7%		
PRESS = 326647	R-Sq (predicted) = 70.85%				

Table 8.24 Building a Regression Model for Waiting Time Y = f(X)

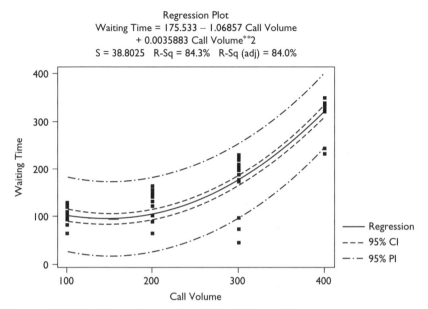

Figure 8.24 Building a Better Model for Waiting Time

Multi-Channel/Poison Arrival/Exponential Service Rate			
		Arrival Rate Per Hour	
		100	**125**	**150**
	10	#N/A	#N/A	#N/A
	11	360	#N/A	#N/A
	12	180	#N/A	#N/A
Call Center	**13**	119	720	#N/A
Agents	**14**	90	241	#N/A
	15	72	144	#N/A
	16	61	104	360
	17	50	79	180
	18	47	65	119
	19	40	54	90
	20	36	47	72
	21	32	43	61
Service Rate per Agent = 10 Customers per Hour				
*(Feasible Solutions Are Shaded)				

Table 8.25 Queuing Analysis of Waiting Time

the team must take additional actions, including evaluation of the proposed solution under controlled conditions and finalizing a project transition plan with the process owner and local work team. These actions will ensure that the process improvements will be maintained over time.

Summary

In this chapter, we discussed Step 4 of the 10-Step Solution Process, that is, prove the causal relationships, that is, $Y = f(X)$. Important analytical tools discussed in the chapter included both simple as well as more advanced tools that have been found very useful in the identification of root causes responsible for poor supply chain performance including high inventory investment. In addition to using standard Lean Six Sigma tools and methods it was shown advanced operations management and research tools and methods can be very useful in the analysis of supply chain processes. Through the root cause analysis, countermeasures can be developed to ensure that the system operates at its best level. All recommended improvements will eventually be tested under controlled conditions *(pilot studies)* as discussed in Chapter 9. Chapters 9 and 10 will also discuss the balance of the 10-Step Solution Process relative to improvements in the supply chain's measurement systems and key processes using inventory investment as an example.

Chapter 9

Lean Six Sigma Improvement and Control

Key Objectives

After reading this chapter, you should understand the importance of making process changes based on root cause analysis of the process breakdown and sustaining the process improvements over time, as well as understanding the following concepts:

1. Why evaluation of alternative solutions using a cost-benefit analysis is critical to ensuring that the organization realizes the maximum business benefits from its Lean Six Sigma projects with a minimum level of risk.
2. Why using a pilot study to confirm the final solution minimizes project implementation risk and gains support from the process owner and local work team since they can see the impact of the process changes under controlled conditions.
3. Why it is important to integrate process changes into current process improvement and control systems to ensure that the improvements become incorporated into daily work.
4. Why it is important to effectively communicate process changes using fact-based methods.

Control Strategies (Steps 5, 6, 7, 8, 9, and 10)

The 10-Step Solution Process, shown in Chapter 1, Table 1.3, should have been an integral part of the project identification, alignment, management and root cause analysis of the Lean Six Sigma project. At this point in the project, the process owner and work team should already agree on the basic steps necessary to close and transition the improvement project back to the process owner and work team. Also, the process should have been simplified and standardized using 5-S methods to eliminate obvious process waste prior to data collection and analysis to have enabled the Lean Six Sigma improvement team to systematically investigate and identify the root causes for process breakdowns. At this point in the project countermeasures (improvement actions) are developed to eliminate the root causes of the process breakdown. These countermeasures should demonstrate to the process owner and local work team the effectiveness of the proposed control strategies prior to closing and transitioning a project.

Reviewing Solutions

Since a multitude of factors may be responsible for process breakdowns, there may be several alternative solutions that would improve the process. A cost-benefit analysis is used to prioritize alternative solutions. This cost-benefit analysis depends on the specific root causes for the high-level problem. As an example, the root causes of high inventory investment usually fall into the two broad categories of long lead-times and high demand variation. Long lead-times could be caused by one of several types of operational breakdowns. These include large incoming raw material lot sizes, supplier delivery issues, or breakdowns in one or more operations within the supply chain. Different root causes may drive each situation. The countermeasures necessary to eliminate the identified root causes of the problem follow from the root cause analysis. As an example, if large incoming raw material lot sizes were found to be the problem, a cross-functional team would work with the external supplier to understand what drives the large lot sizes. The team would then develop Lean Six Sigma improvement projects to reduce their size. As another example, supplier on-time delivery problems may occur because of several different root causes. Each cause may require its own unique solution. If the on-time delivery problem is found to be associated with long lead-times due to an excessively complicated process, the process can be simplified using a value stream map (VSM). The VSM is used to identify and eliminate unnecessary operations and other operational waste.

Demand variation issues may be caused by several factors. Some of these include poor sales forecasts due to the loss or gain of major customers, poor forecasting models and other issues due to supply chain complexity. In the latter case, it should be noted, demand variations at each stage of the supply chain are additive. This creates a situation in which the cumulative variation across the supply chain becomes very large relative to the original external customer demand. Low inventory turns are caused by these as well as other reasons. Many examples have been provided throughout previous chapters of this book since there are numerous ways in which a supply chain can break down.

Process Control Strategies

As the Lean Six Sigma improvement team works through the 10-Step Solution Process, countermeasures are identified to attack the root causes of the process breakdown. However, some countermeasures are more effective than others. Some can be sustained by the process owner and local work team more easily. Others may be easier to implement because they use known technologies or require minimum resources. This concept is shown in Figure 9.1. The team will need to evaluate the proposed countermeasures and select the most efficient set of countermeasures to control the process over time. Figure 9.1 shows specific countermeasures on a continuum. This continuum is based on how easy it is to implement and sustain the countermeasures. The ease of implementing a process control depends on the nature of the control as well as the resources

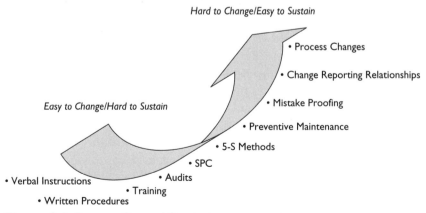

Figure 9.1 Process Control Strategies

required to put it in place. The ability to sustain an improvement is directly tied to the ability of the process owner and local work team to use the control tool as part of their daily work.

As an example, verbal instructions are inexpensive but are weak process controls since they must be implemented every day and can be misunderstood by the people doing the work. Written instructions are a little more difficult to develop than verbal instructions, but are more permanent and cannot be arbitrarily changed. However, people must read and remember them to control the process. On the other hand, mistake proofing and 5-S techniques are more difficult to implement, but they control a process more effectively than verbal or written instructions since some of the reasons for the original process breakdowns have either been eliminated or are prevented from occurring using various mistake-proofing strategies. However, 5-S controls are dependent on employee behavior because they consist of several manual components. One of the most effective controls is eliminating the portion of the process that causes the problem or redesigning the process to eliminate the problem. Designing out the root cause of a problem is always the best way to ensure that the process improvement will be sustained over time. However, it may not be technologically feasible or too costly to redesign a process. In summary, although a particular control tool may be weak by itself, it may be critical to implement it in the context of an overall control strategy. As an example, although verbal and written instructions may not be effective as mistake-proofing controls, they will still be required as part of the overall control strategy. This is also true for process audits and training of the local work team.

Critical considerations that must be considered by the Lean Six Sigma improvement team when developing their process control strategy include identification of the key process input variables (KPIVs), which must be controlled by the work team as well as how to control the KPIVs. In other words, the team must consider which control tools are required to maintain the KPIV at its optimum level to optimize the KPOV and reduce its variation. Another important control consideration is how the KPIVs will be measured, including their inspection and testing requirements, to ensure that they remain at their targeted performance levels. Other process control considerations include the frequency of their measurement, who is responsible for taking corrective action and the level of required training for the local work team. This information is incorporated into the process control plan as it becomes available to the Lean Six Sigma improvement team. The process control plan will be discussed later in this chapter.

Many organizations use verbal and written instructions, audits, and training to control their processes. Written procedures and training are also integral elements of the process control plan. However, unless there is a clearly documented audit frequency and assigned responsibility, these types of control tools will be difficult to maintain over time by the local work team. Another issue is that these controls must be coordinated with key functions across the organization including training professionals and human resources as well as the process owner. However, there are ways to increase the effectiveness of these control tools. The more visible, graphic, and simple the work instructions are, the easier it will be for the local work team to use them every day. Also, audits can be made simpler if they work on an exception basis or are designed to be easy to conduct and provide only important information. Lean tools and methods such as 5-S are very useful in these situations. Also, the concept of the visual factory including its tools and methods will greatly increase the effectiveness of verbal and written instructions.

Statistical process control (SPC) tools have been heavily used by manufacturing organizations for many years, but on a limited basis to control supply chain processes. Control charts are applied to the analysis of process variation. In these applications, they are used to monitor and distinguish common from special cause variation. Common cause process variation is composed of random variation due to many individual causes. It is difficult to economically eliminate common cause variation from a process. Also, a key concept of SPC is that a process can be in a state of statistical control, that is, having only common cause variation, but still not satisfy customer requirements. In these situations, Lean Six Sigma black belt projects are created to identify the root causes of the unacceptable process performance and set the process to a new performance level. The assumptions behind control charts are that if the process is stable, its behavior can be predicted and statistical conclusions can made concerning the system variation. Another key concept is that if control charts are used effectively, the process will operate with less variability since excessive tweaking of the process will be prevented due to the rules concerning control chart usage. The process owner and local work team use the control chart to identify and remove special causes of variation which adversely impact process performance. Figure 9.2 shows two commonly used control charts called the X-bar and range charts. These charts are used to monitor a continuously distributed process metric over time by plotting subgroup averages or the range (maximum minus minimum values of a subgroup). There are other

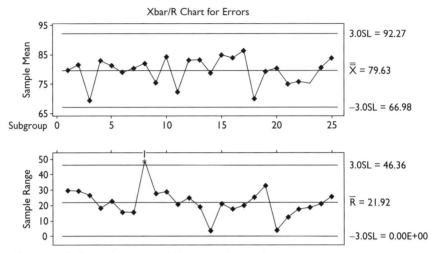

Figure 9.2 Typical Statistical Process Control Chart

types of control charts which differ based on the underlying probability distribution of the metric which is being charted over time.

If the original project problem statement and objective focused on machine or equipment breakdowns, then preventive maintenance (PM) control tools will be important countermeasures. In these projects, the specific PM countermeasures recommended by the improvement team must be tied to the root cause analysis of the problem. This is especially true in distribution centers having large vehicle fleets or other forms of rolling equipment such as forklifts and pallet trucks. As an example, if a machine is breaking down because of roller bearing problems, then the PM procedures impacting roller bearing maintenance must be reviewed and updated by the team. If forklifts are prematurely running out of power, then their maintenance and charging procedures need to be reviewed by the improvement team. In all situations the countermeasures must match the specific root cause for the machine or equipment breakdown. These recommended process changes must also be fully integrated into the current PM system.

The goal of the Lean Six Sigma project is to put the KPOV on target with minimum variation. This is done by developing the $Y = f(X)$ relationship through data collection analysis and verification of the $Y = f(X)$ relationship using process pilot studies. Unlike manufacturing environments, in supply chain operations the $Y = f(X)$ relationship may not be highly mathematical. As an example, the required process improvements, that is, countermeasures based on root cause analysis, may involve employee training, 5-S implementation, and mistake proofing controls rather than running experimental designs to develop regression models. But, regardless of the practical situation, Figure 9.3 shows that the basic concept remains to understand the relationships between the KPOV and KPIV variables, and then to set the KPIVs at levels that will ensure the KPOV variable remains at an optimum level with minimum variation with respect to customer requirements. These requirements are usually specified as lower specification limits (LSL) and upper specification limits (USL). The KPOV must remain within the LSL and USL limits to meet customer requirements. This may increase process capability up to the entitlement of the current process design. Figure 9.3 shows how the Lean Six Sigma improvement team works backward from their initial process capability analysis of the KPOV through the root cause analysis to identify the KPIVs. These KPIVs are then set at levels designed to improve the process capability of the KPOV to meet customer requirements.

Process maps are important tools used throughout the Lean Six Sigma project. Different versions of process maps were discussed over the past several chapters. In each process-mapping discussion, the key concept was to describe the material and information flows of the process and work down into the process only as required by the root cause analysis. In Figure 9.4 we see that the SIPOC is a high-level process map initially used to help link the process inputs (KPIVs) to customer requirements or key process outputs (KPOVs). At the next level of the analysis,

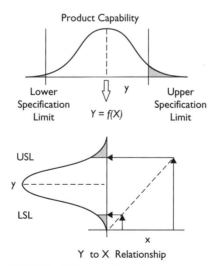

Figure 9.3 Setting KPIV Variables on Target

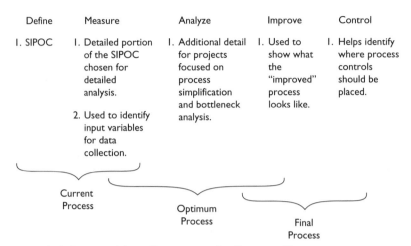

Figure 9.4 Process Maps Document the Process Changes

the process map is modified to create more detailed functional or value stream maps. In the control phase of the project, the process map associated with key portions of the modified process is used to document the process changes to enable the local work team to control the process. These more detailed process maps also become the basis for worker training, operational audits, and process control.

A failure mode and effects analysis (FMEA) is a detailed analysis of system failure points. Countermeasures are applied against these failure points to eliminate, control, or minimize the probability failure will occur at these points in the process in the future. The FMEA, shown in Figure 9.5, is one of the last control tools the team uses to ensure that nothing has been missed and all countermeasures will be effective over time. Also, if more work is required to make the new controls more effective, the FMEA can serve as a prioritized action list to ensure that every item not completed will be properly prioritized and tracked for future implementation by the improvement team and process owner. A "failure mode" is a description of a how a component (Design FMEA) or an operation (Process FMEA) can fail. We will use the term KPOV for the customer characteristic that can fail. A "failure effect" describes the effect of a particular failure mode on the KPOV. "Severity" (SEV) is an assessment of the seriousness of the failure effect of the KPOV on the external customer using a scale of 1 to 10. The higher the severity rating, the more severe the failure effect will be. A "failure cause" describes how the failure mode could have occurred in the process. The cause is the KPIV

| Process or
Product Name:
Responsible: | | | | | | Prepared by :
Date (Orig)   (Rev) | | |

Process Step/Part Number	Potential Failure Mode	Potential Failure Effects	S E V	Potential Causes	O C C	Current Controls	D E T	R P N
		Y's = X's						

Figure 9.5 Process Failure Mode and Effects and Analysis

contributing to the failure of the KPOV (these causes can be included in the data collection and analysis portion of the project if the FMEA is used as a brainstorming tool in the early stages of the project i.e. its measure phase). "Occurrence" is an assessment of the frequency with which the failure cause occurs using a scale of 1 to 10. The higher the rating, the more frequent is the occurrence of the failure mode. "Detection" (DET) is an assessment of the likelihood (or probability) that current controls will detect the failure mode using a scale of 1 to 10. The higher the rating, the poorer is the detection system of the current internal process controls. The *risk prioritization number (RPN)* is calculated using the formula: *RPN* = *(Severity Rating)* × *(Occurrence Rating)* × *(Detection Rating)*. It is used to prioritize recommended improvement actions. Attention should be focused on failure modes having high RPN numbers.

Cost-Benefit Analysis

Cost-benefit analysis compares one or more alternative solutions against current baseline performance. This analysis considers the business benefits if the project recommendations are implemented versus their implementation costs. Business benefits are both tangible as well as intangible. Examples of tangible benefits include higher sales, higher margins, and reduced operational costs and increased cash flow obtained through improvements in asset utilization. Intangible benefits include cost-avoidance benefits as well as improvements in customer satisfaction.

Cost-avoidance benefits include business benefits that are real, but whose costs can not be captured on the current financial statements. An example would be a requirement to implement a new cycle-counting system this year to prevent governmental fines next year. Another example would be to replace a material currently used to manufacture a product this year with a different material to prevent environmental fines next year. Project implementation costs include labor, material, and other costs necessary to change the process.

Cost-benefit analyses answer questions such as, "What are the time-phased costs and benefits of the project? What are the intangible benefits to the organization? How will the project impact key financial metrics?" Many other relevant questions depend on the specific project application and final solutions. Once the business case has been developed for each alternative solution, the alternative solutions are prioritized using one or more criteria including their specific business benefits. This methodical approach is likely to persuade key executives to commit resources and undertake the improvements necessary to eliminate the chronic process problem.

Piloting the Solution

Once a solution has been determined to be optimum, it is tested under very controlled conditions on a limited portion of the process. This methodology is called "piloting the solution." The piloting process is used to reduce the project's failure risk since the Lean Six Sigma improvement team can observe how their recommended process changes will work under actual process conditions. Pilots are particularly necessary when the proposed solution (process change) is large in scale or impact, costly to undertake, or there are significant external risks associated with the proposed solution. Another advantage of using a pilot is that the process owner and senior management are more likely to agree on a controlled evaluation of the proposed process changes rather than to a large-scale implementation, which may pose unnecessary risks to the organization.

Process pilots are also important since contradictory results may be obtained by not using a process pilot to carefully evaluate the proposed process changes. As an example, an unstructured evaluation of the proposed solution might be adversely impacted by an extraneous factor that the team is not aware of during the evaluation period of the pilot. Perhaps someone did not follow the recommended instructions or another process problem occurred, which negatively impacted the process evaluation. If these or similar types of situations occur, people might think the original analysis and solution were not feasible and should not be implemented by the organization. Piloting the solution under controlled conditions also increases process owner and local work team support. This is because they are intimately involved in the process changes and see their beneficial impact firsthand. In this sense, the pilot validates the original analysis and its assumptions and allows the Lean Six Sigma improvement team to further optimize the original solution prior to full-scale implementation of the process changes.

Planning the pilot is a methodical process requiring collaboration between the Lean Six Sigma improvement team and the process owner and their local work team. The information learned during the original analysis of root causes is an important part of the planning process. This information includes process maps, FMEAs, the analytical results gained from data collection, and all other information describing how the process functions. Checklists and templates should be also developed and used to ensure that nothing is missed during the planning or execution stages of the pilot. Contingency plans should also be developed to anticipate problems during the pilot evaluation. Other important planning activities include deciding where the pilot will be run, that is, which part of the process will be evaluated, what is to be measured, how it is to be measured, who will measure it, and how the subsequent analysis will be performed on the data that is collected from the pilot evaluation. The pilot should also provide relevant information on the key process input and output variables that were the basis of the original root cause analysis. Using this methodology to plan and run the pilot, the improvement team should be able to demonstrate significant KPOV performance improvements over the original process baseline as demonstrated using capability analysis and metrics.

After the pilot, the improvement team and process owner should review the pilot results. This evaluation includes reviewing the original project goals and objectives. As part of the evaluation process, participants should ask questions such as, "Did the pilot improve the key process output metrics relative to their targeted performance levels? What changes should be incorporated into the original solution? Have the business benefits originally estimated significantly changed?" In addition to this review of the pilot result, all changes to current process operations, critical process parameters, work and testing instructions, and other relevant process elements should be incorporated into the final process control plan.

Implementing the Solution

Implementation of the project's solution requires the identification and sequential listing of all work activities supporting the process changes. Identification of work activities is a straightforward process using information gathered during the preliminary phases of the project and the pilot. This process is called developing the "work breakdown structure." The work breakdown structure for the project is determined by analyzing the work activities and their associated work tasks, which must be completed to implement the project's countermeasures. These activities are the deliverables, which are grouped under the project's milestones. These milestones are the ten steps of the 10-Step Solution Process. Project milestones and activities (deliverables) are used to track and control the project to ensure that it stays on schedule and within budget.

The concepts of project planning and control, using the 10-Step Solution Process, were discussed earlier in Chapter 1. Figure 1.7 showed how a Gantt chart could be used to implement the 10-Step Solution Process using each of the ten steps as a milestone activity. After the milestones are identified, a Gantt chart is used to show the sequential relationships between all activities and their work tasks as well as their resource requirements. The Gantt chart is used to time-phase the work activities necessary to achieve the project's milestones. A responsible person should be assigned to ensure that all of the activities and work tasks necessary to complete the project are executed according to plan. The implementation plan should also include all required work and inspection instructions, tools, and equipment as well as required labor and materials.

Properly documenting the process changes is critical to implementing the final solutions and communicating them to the organization. It is essential to standardize the process as well as ensure that work and test instructions are updated. Process documentation includes identification of customer critical process elements, process maps showing how the process is structured (quantified process map), and identification of the key process input variables (KPIVs) including their target operational levels and how they could fail (Failure Mode and Effects Analysis [FMEA]). New or modified documentation must be integrated into the organization's quality control system.

Training

Although training is an integral part of the overall control strategy, it is often one of the most neglected aspects of the control phase. Having said this, training is only one of the many control tools that are necessary to ensure integration of the project's solutions. If training is to be an integral part of the control plan, it must be tied to the root cause analysis. In other words, failures in training should have been shown to cause poor process performance. To eliminate the identified process breakdowns, formal modifications to the current employee training program should be made through the correct organizational function. This function usually resides within human resources. The specific details regarding the proposed changes to the training elements should be listed as well as their training frequency.

Assessing Change Effectiveness

The improvement team should be able to show the "before" versus "after" scenarios associated with the process changes. There are many ways to communicate the process changes. Figures 9.6, 9.7, and 9.8 describe three simple graphical ways to show the improvements of the KPOV variables. In addition to graphical presentation, rigorous statistical analysis could also be used to demonstrate significant changes in the process KPOV relative to its average and variation. The process owner and work team should see proof the process has been effectively changed before

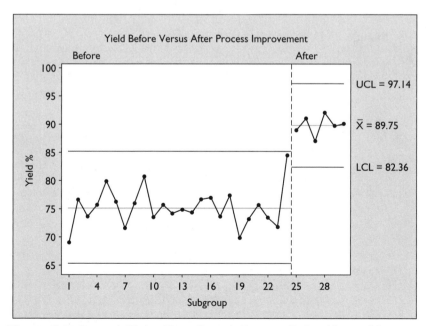

Figure 9.6 Control Chart (Time Series) Showing Before Versus After Yield Percentage

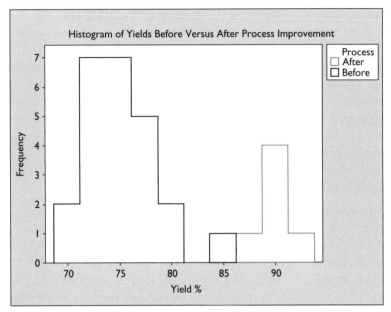

Figure 9.7 Histograms (Distribution) Showing Before Versus After Yield Percentage

the project transition is considered complete. At a higher organizational level, finance and senior management should see the business benefits from the project in the form of reductions in direct labor and material usage, increased revenue, or increases in cash flow typified by inventory

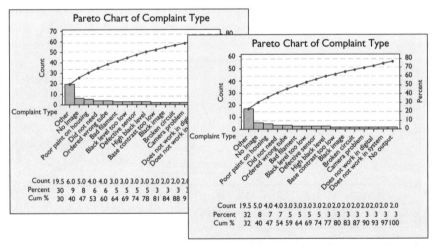

Figure 9.8 Pareto Charts Showing Before Versus After Categories Having the Highest Counts

reduction or other asset conversion activities. To verify that the process improvements are sustainable, the business benefits should be evident for an extended period of time and under a variety of processing conditions. The specific time period and conditions necessary to show the improvements can be sustained will vary by project and organization.

It has been shown that sustaining change is very difficult through numerous studies of change in diverse organizations and cultures. However, these same studies have consistently shown that successful change initiatives have similar characteristics. These characteristics include creation of real business benefits to the organization, strong support from senior management, effective communication of the goals and objectives of the initiative throughout the organization, team empowerment, alignment of the project selection process with business goals and objectives. To facilitate successful deployment of the Lean Six Sigma initiative, the improvement team should continuously communicate the project's status to the organization in clear and unambiguous terms. The improvement team and process owner should also be empowered to make the necessary process changes and improvements. These basic organization characteristics will ensure that the improvement team will be successful.

Assessing Rewards

After the improvement team completes the Lean Six Sigma project, team rewards should follow the organization's standard policy. Rewards should recognize individual as well as team performance. Individual rewards are important to encourage performance exceeding that which was expected by the team leader. However, team rewards are even more important since cooperation is one of the most critical aspects of a project team. How this balance is achieved varies by organization, but team reward planning is important to encourage successful project completion and organizational collaboration.

There are several strategies for rewarding team members. First, it is a good idea, just after the project is kicked off, to bring everyone together for an informal meeting. This informal meeting could be held around lunch or over dinner or another group activity. Second, a plan should be put into place to ensure effective communication. Communication is important to show the organization the project's progress over time and to build momentum for the Lean Six Sigma deployment. This will greatly increase the probability the team will be successful. Of course, being part of a successful team is also a reward. Third, everyone on the team should be involved with the project planning process so that they can influence the project direction and have ownership over the results. Fourth, being on a team is a great way to learn new skills and meet other people in the organization. Finally, the organization has many options relative to financial rewards. Financial rewards may or may not be part of your organizational culture. But it is difficult to ask people to execute very complex and difficult projects saving hundreds of thousands of dollars without some financial reward. We have seen organizations that "take and take" without giving back to their improvement team, with the result that people do not want to participate in the Lean Six Sigma program after a short period of time. On the other hand, "throwing" money at an improvement team is not a good answer either. Each organization must consider the correct reward mix for their improvement teams. These considerations may have a critical impact on the project's success and sustainability over time. They will certainly impact the success of the Lean Six Sigma deployment over time.

Transitioning the Project

The project transition work should begin in the early phases of the project. What people do not always realize is that even the best solutions require gaining support from very busy people throughout the organization. Most professional people want to do the right thing for the organization. But, because the root causes are not initially obvious, everyone has a different opinion of what the best course of action is for the problem at hand. To help facilitate the situation, there are many tools and techniques that can help get people around a solution. First, involving people who have a vested interest in the project and its solution will allow the team to obtain many very critical insights as to how the root cause investigation should proceed. Not having this external advice may cause the team to focus the project investigation in the wrong direction. Obtaining advice in the early stages of the project is also critical to gaining organizational support. Data collection can be focused on answering questions others in the organization have regarding the problem, using advice from people familiar with the project's problem statement. Without this external feedback, the project team could develop "groupthink."

At the end of the project, the project's solutions will have a direct impact on the process owner, the local work team as well as other organizational functions. In fact, there may actually be several ways to implement a given solution with some being more acceptable to the organization than others. Involving other interested parties with the solution evaluation process will increase the probability they will support the project implementation. However, involving others does not mean the improvement team should not act on fact nor ignore optimum solutions. Everyone does not have to be happy with the implemented solution, but everyone should have been able to represent their viewpoints in an unbiased manner and understand the factual basis for the recommendations. All relevant opinions, advice, and objections should be seriously considered by the team. At the end of the project, the goal is to implement a fact based win-win solution for the organization.

Communication

Communication of the project results is an ongoing process in which people within the organization are kept apprised of project milestones and results. Although there are many possible forms of project communication, the most effective channels tailor the communication vehicle to the emotional content and complexity of the message. If the emotional content is high and the message complex, face-to-face meetings would be the best communication vehicle. On the other hand, if the message is straightforward and fact-based, then e-mails might suffice to convey the required information. However, it is important to effectively communicate the project status to the people within the organization who need the information.

Another important part of the project's communication process is updating the organization as the improvement team makes progress through the 10-Step Solution Process. A good way to communicate the project is to develop a PowerPoint presentation template using the ten steps. Each of the ten steps conveys a critical message. The first step, project aligned with business goals, clearly communicates the importance of the project. The second step, buy-in from the process owner, finance, and others, shows how the project is linked throughout the organization. Finally, in Step 10, apply control strategy, the organization sees how the solutions have been implemented based on root cause analysis. Posting the project results, that is, the PowerPoint presentation of the 10-Step Solution Process, in public places will go a long way toward communicating the team's work activities to the organization.

Development of the Control Strategy

Development of the control strategy is directly based on knowledge gained from data collection, analysis of root causes, and evaluation of initial process improvements, that is, pilot studies. These activities and analyses should confirm major drivers of the process breakdown as well as the actions necessary to control the improved process to meet the project's goals and objectives. It is at this point in the project that the control strategies should be incorporated into a formal control plan. This control plan includes several important elements each containing specific tools and methods to ensure that business benefits are sustainable over time. As shown in Figure 9.1, typical control elements include changes to standards and procedures, upgraded training procedures to ensure that people have the relevant knowledge of the modified process, implementation of 5-S strategies, mistake-proofing operations to prevent operational failure, and elimination of unnecessary operations. 5-S methods include visual display of improvement metrics as well as their status to ensure real-time communication and decision making within the local work team. The improvement team may also have implemented changes to the organization's information technology (IT) systems. In parallel, an audit system should be created to conduct periodic operational reviews of the implemented countermeasures. Effective implementation of the necessary controls will also ensure project alignment and support of the organization.

When considering a control strategy, it is also important to realize some control methods are more effective than others. Using inventory turns as an example, if lead-time was shown, through the root cause analysis, to be a major issue impacting inventory turns for a particular product group, eliminating portions of the process to reduce cumulative lead would be a more effective and sustainable process change than reliance on simple procedural changes involving written instructions. This is because written procedures can be easy to ignore over time. Moderate changes to IT systems are also effective long-term controls. These changes may include modifications to MRPII and MPS parameters, software changes, changes to the forecasting models, and changes to purchasing and distribution systems. MRPII parameters include such things as lot size and lead-time. MPS parameters include item, location, and scheduling information. Forecasting model parameters include level, trend, and seasonal components as well as the historical time horizon on which the forecasting model is built. Major process improvements may also be possible through application of 5-S methods to simplify and standardize the workflow of the remaining work tasks. Mistake proofing the remaining operations will further enhance process control. Simple examples of mistake proofing would be making minor changes to software code to prevent data entry errors or automating forecasting-parameter selection to minimize forecasting error rather than having analysts set the parameter levels.

After the process has been simplified and the remaining work tasks have been standardized by the improvement team, the local work team can assume responsibility for ongoing process control. One of the most effective ways to ensure ongoing process control is to establish realistic metric targets on both the KPIVs and KPOVs and hold the local work team accountable for their sustained performance. The implementation of simple visual control systems is also very effective. These visual systems display meaningful performance metrics in real time. The ultimate goal of the Lean Six Sigma improvement team should be to create a simple process that will precisely, accurately, and consistently deliver customer value over time and require minimum manual intervention.

Control Plans

At this point of the project, the necessary process controls have been incorporated into the process control plan using all the information gained during the pilot studies as well as the original root cause analysis. It is also important to document the anticipated business benefits that will be gained from each improvement action as well as their completion dates, and the people responsible for each action as well as the specific countermeasures required to eliminate the root causes of the problem. Effective execution of all the improvement actions is also necessary to ensure that the project remains on schedule. The control strategy should also ensure an effective project transition from the improvement team to the process owner and local work team. This implies that the local work team, which is impacted by the improvements, has the necessary training and tools to ensure that process improvements are cost effectively sustained over time.

After conducting pilot tests of the optimum improvement scenario, the Lean Six Sigma improvement team finalizes all the required improvement actions necessary to implement the process changes. The process control plan is the key tool required to implement the necessary process changes. Figure 9.9 shows a generic process control plan. In addition

KPOV	Target	Defect Definition	Root Cause	Counter Measure	MSA Method	Sample Size	Frequency	Person	Reference Procedure

Figure 9.9 Key Elements of the Process Control Plan

to process control plans, many organizations also use quality control plans, preventive maintenance plans, and other systems that are useful to communicate and control their processes. The control plan lists each KPOV and its customer requirements. The KPOV defect definition is also specified so that the local work team will know what makes the KPOV good or bad. The root causes of the problem, which are also embodied within the process FMEA, show the local work team how the KPOV can fail as well as how the countermeasures will eliminate the root causes contributing to KPOV failures. Other relevant information contained within the control plan includes how the KPOV is measured as well as its measurement frequency, measurement sample size, and the people responsible for the control of the KPOV. All additional information regarding work and inspection procedures including preventive maintenance procedures is contained within the reference procedure section. Process controls remain effective over time because the control plan sets specific success criteria for every KPOV. In addition to administrative details describing the process, specific elements of the control plan specify operational targets for key process inputs including their allowed tolerance as well as the measurement methods required to monitor these critical inputs including required sample sizes and frequencies. Integral to the control plan are the specific actions necessary to react to out-of-control conditions. These actions should be incorporated within reference procedures.

Integration of Other Initiatives

At any given time, organizations have numerous ongoing initiatives. At an operational level, there are usually several important initiatives in place at once. Some of these are Six Sigma, Lean, and International Standards Organization (ISO) certification. However, there are others such as supply chain excellence and preventive maintenance. Figure 9.10 shows how Six Sigma, Lean, and ISO integrate within an organization to ensure that customer requirements are consistently met over time. The Six Sigma initiative uses an applied sequential problem-solving methodology to ensure that processes meet customer requirements with minimum variation. Six Sigma uses a five-phase methodology characterized as: *define* the problem relative to customer important outputs ("Y" or KPOV), *measure* the extent of the problem and determine the performance gap of the KPOV relative to customer requirements, collect and *analyze* data on the process variables (the "X's") which drive the KPOVs, *improve* the process by setting the KPOVs at their optimum target levels by conducting controlled tests to evaluate the relationship between the KPOVs

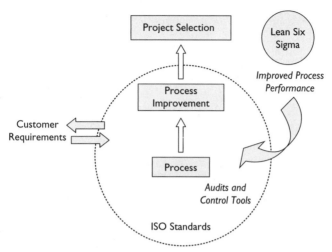

Figure 9.10 Improvement Control Strategy

and the "X's." Set the process at its new optimum level with minimum variation in the *control* phase and standardize the modified process. Lean initiatives reduce process waste through work analysis, simplification, and standardization. Integral to waste elimination is the concept of takt time. Takt time is used to match the process output to external customer demand. Waste is easily identified as those work tasks requiring resource levels greater than optimum to maintain takt time. An effective integration of the two systems is accomplished through Lean deployment followed by Six Sigma optimization of those remaining operations requiring statistical analysis. From this perspective, Six Sigma and Lean initiatives are complementary. ISO systems and certification consist of a series of mandated standards documenting the actual workflow to ensure consistency of process performance. It is important to apply ISO after Six Sigma and Lean improvements have been made within the process so that the final ISO requirements reflect an optimized process design.

Summary

The business benefits will not be realized by the organization unless the project's solutions have been implemented and can be sustained over time. There are many things to consider at this point in the project. Given that the root causes of the process breakdown were identified and several alternative solutions identified by the data analysis and pilot studies, the team should conduct a cost-benefit analysis of each of the alternate solutions. The cost-benefit analysis will be finalized after the optimum solution has been tested, under controlled conditions, within the modified process. The optimum solution is then implemented with the process owner and local work team.

Development of the control strategy is best done with the process owner and local work team incorporating all the information the team learned during the 10-Step Solution Process. This information is embodied in the control plan and integrated into the organization's standard control systems to ensure integration with other organizational initiatives. Integral to the implementation will be the effective use of control tools and strategies to place countermeasures against the root causes of the process breakdowns. The Lean Six Sigma improvement team must be careful to assess the effectiveness of the proposed solution including the ability of the local work team to sustain the process changes. At this point in the project, it is also important to ensure the team is rewarded and lessons learned are communicated to the organization. Ensuring that the improvements are firmly based on data analysis will also ensure a smooth project transition. Communication of the project's benefits across the organization is also important.

Chapter 10

Applying the 10-Step Solution Process

Key Objectives

After reading this chapter, you should understand the importance of building models of major work streams within supply chains to identify Lean Six Sigma projects that are strategically aligned and will improve key operational metrics, as well as understanding the following concepts:

1. Why it is useful to build and analyze simple models of major work streams using inventory optimization as an example.
2. How to use an inventory model to identify Lean Six Sigma improvement projects.
3. How to apply Lean Six Sigma tools and methods to eliminate some of the common process breakdowns encountered in supply chains and their distribution centers.

Lean Six Sigma Applied to Inventory

Our objective in Chapter 10 is to show the advantages of developing simple supply models using inventory investment as an example. Using the model, we will be able to develop internal benchmarks for the key financial and operational metrics shown in Chapter 1, Figures 1.1 and 1.2. Also, we will also be able to identify Lean Six Sigma supply chain projects using high inventory investment as an indicator of process problems. Microsoft Excel will be used to understand operational relationships and their impact on inventory investment and turns. This methodical approach ensures that the correct data is collected for analysis to use in the optimized model and conduct sensitivity analyses. These sensitivity analyses will also enable the team to identify and implement "quick hit" improvements through a simple rebalancing of the inventory by item and location based on per-unit service-level targets, demand variation, and lead-time. This balance analysis will also identify obsolete and excess items through comparison of current versus required inventory quantities. Further reductions in investment, over the initial balance quantities, will require identification and execution of Lean Six Sigma improvement projects to eliminate the root causes of the process problems which create inventory investment problems.

Preparing for the Analysis

Prior to development of the inventory model (or any model), the team should identify; with the aid of Figures 1.1 and 1.2, the specific questions the analysis is required to answer. Not every question requires development of a formal inventory model, but rather, can be answered by the team as part of a vigorous data collection plan followed by appropriate analysis. Development of internal benchmarks may also require answering questions relevant to the analysis such as those listed in Table 1.4. It is important to specify major analytical questions up-front in the project analysis to keep the improvement team grounded and focused on the project's problem statement and objective. Also, these questions will direct the team to collect data that is relevant to the goals and objectives of the project and pull together diverse data files and required data fields. In summary, knowing the correct questions to answer will show the Lean Six Sigma improvement team the specific data fields required for the inventory analysis. It should be also noted that collection of the data fields requires a multifunctional improvement team, which should include operations integral to the analysis. As examples, if received inventory is one of the major problems, suppliers should become part of the Lean Six Sigma improvement team. If work-in-process (WIP) materials are not available at start of production, then manufacturing personnel should be part of the team. If forecasting is an issue, then forecasting analysts should join the team.

Developing a PIM Model

Chapter 6 discussed several types of inventory models which varied depending on the underlying assumptions relative to how the process functioned. One of the most common types of inventory models is the perpetual inventory model (PIM). The PIM is discussed in Chapter 6 and shown in Figure 6.1. The PIM also is the basis for several more specific inventory models and is directly useful in the analysis of distribution center finished goods inventory. In the PIM model, the optimum inventory quantity (investment) depends on lead-time, expected demand over the order cycle, demand variation, and the target per-unit service level. At a lower level, there are several complicating factors or variables, which drive lead-time and demand. These variables can be built into the model by the improvement team. As an example, incoming raw material lot size can have a major impact on lead-time and ultimately inventory investment. If the lot size quantity is a multiple of the established order cycle quantity, then average on-hand inventory will be higher than established by the order cycle as discussed in Chapter 4. The graphical representation of inventory investment shown in Figure 6.1 is based on this concept and incorporated into the PIM/ Excel model. Using the PIM/Excel model, inventory is analyzed at each item and location and various operational and financial metrics are aggregated upward to the business unit level to create an optimum unit balance from which to begin the sensitivity analysis.

Initial improvement opportunities are identified for inventory items having excess inventory or lower than required inventory quantities using the balance analysis. Subsequent analyses identify projects at a lower level that contribute to excess, obsolete, or suboptimum distribution center inventories. These projects also identify contributing factors of long lead-times and demand variation. Using the PIM/Excel model, various "what-if" analyses are used to develop Lean Six Sigma projects. Business benefits are determined for each project based on detailed analysis of the modeling parameters and metrics. As an example, inventory turns can be calculated for every item and location. A sensitivity analysis can be conducted on the inventory turns ratio by changing lead-time and demand variation in the model. The team will be able to calculate reasonable improvement in the turns ratio (create an internal benchmark). It will also be relatively easy to identify the areas where projects should be deployed to reduce lead-time or demand variation, as well as their contributing root causes, to improve the inventory turns ratio.

As an example, if large incoming raw material lot sizes are found to drive inventory investment, then projects can be deployed in this area. On the other hand, if poor forecasting is found to increase inventory for a particular product group, then projects can be deployed in the forecasting area. This methodical approach will allow the team to optimize operational and financial metrics against internal rather than external benchmarks. External benchmarks may be misleading since process designs may differ between organizations distorting a realistic comparison between their operational metrics.

Figure 6.1 qualitatively shows how the PIM model works. It is easy to see graphically how potential reductions in distribution center inventory investment at a constant service level could be achieved through reductions in lead-time or demand variation. This concept is applied, using the model, to every item and location in inventory. Figure 6.1 shows that the actual inventory quantity could be lower, higher, or at optimum. If the actual quantity is higher than optimum, reductions in investment might be possible if excess quantities can be reduced in the near future. However, if the actual inventory quantity is too low, several courses of action may be required to achieve target unit service levels. If lead-time is very long, Lean methods can be applied to reduce long lead-times. On the other hand, poor item ordering, supplier delivery problems, or other problems would require alternative improvement actions. Figure 6.1 also shows that when lead-time is reduced, optimum inventory investment decreases along the current demand variation curve. Alternatively, if demand variation is reduced or forecasting accuracy improves, required inventory investment steps down from one demand curve to the next lowest curve. These relationships between inventory investment, lead-time, and demand versus unit service levels allow the team to aggregate information up to product group levels to identify Lean Six Sigma improvement projects having significant business benefits. These projects also allow an organization to systematically reduce inventory investment across the supply chain. The applied models can be created in Excel or using off-the-shelf modeling software.

The PIM model shown in Figure 6.5 can be found in various forms throughout the field of operations management. In this model, an economic order quantity for Q_1 units is placed at $Time = 0$. Q_1 is determined from standard economic order quantity (EOQ) or lot-sizing (ELS) formulas. It is the amount of inventory expected to be used during the order cycle time T. The usage rate (expected unit demand) is assumed to be linear over T. However, actual inventory usage could be higher or lower than expected during the order cycle time T. Given the lead-time offset LT or "reorder point," a second quantity Q_2 or "reorder quantity" is ordered at the reorder point to ensure that the inventory is brought back to the original target quantity (Q_1) at the end of the order cycle. In this system, the inventory depletion rate is seldom strictly linear due to unit demand over time being either lower or higher than expected. Given these assumptions, the average inventory for an item is set according to the classic formula shown in Figure 6.5. This classic formula is applied to every item and location in the inventory using the PIM model. Remember, the formula is just an approximation to the true inventory optimum quantity for the item and location since the original lead-time and demand distributions may not be normally distributed. Also, it is very important to remember the "actual inventory" quantity must be estimated carefully. This is because the difference between "actual" and "optimum" inventory quantities, for each item and location, will become the "reduction/addition targets" calculated by the model. However, minor errors in the recommended optimum inventory quantity can be subsequently adjusted after the analysis, item by item, to ensure that targeted customer per-unit service levels are met in practice. Also, it is easy to verify modeling results since items with service levels that exceed target probably have excess investment, while items with lower service levels may not have enough inventory (although there could be operational problems causing low customer service in these situations). Over an entire inventory, the modeling technique has proven very useful in identification of areas to both reduce inventory investment and improve customer service.

Table 6.1 shows the algorithm for the PIM analysis. Information and data necessary to create the inventory model include lead-time, expected demand and variation, unit service level, and descriptive information, for example, the original questions identified by the team and shown in Table 1.4. Common descriptive information includes product category, supplier name, types of inventory, forecasting accuracy, shipment information, and on-time delivery information as well as other categories. The safety-stock portion of the model is shown in Figures 6.6 and 6.7. Figure 6.6 shows that safety stock is necessary unless expected levels and variations in demand and lead-time are both zero. The only operational situation in which this will occur is within a continuous process. In make-to-order environments, demand variation is assumed to be zero (if the customer adheres to the established order schedule); but lead-time variation can be significant. Inventory investment levels are determined by item and location across the inventory population using these basic concepts.

The Excel Worksheets

The following discussion is designed to help green belts and black belts quickly model their inventory system, show where improvements would be most beneficial to the organization, and use the "lessons learned" to develop more sophisticated models. The ideal scenario is to purchase off-the-shelf software to store and analyze relevant supply chain data on a continuous basis to provide information necessary to make process improvements. On the other hand, the organization's IT department could also develop a simple Excel-based or similar spreadsheet based software program to do the necessary calculations or purchase specially designed software that models supply chain processes and their associated inventory. Although the specific inventory model used by the team can take many forms depending on system characteristics, a simple Excel model consisting of several worksheets has been developed to show the basic modeling concepts and their advantages. These concepts can be immediately applied by any Lean Six Sigma improvement team to analyze inventory investment and other key process output variables (KPOVs) to improve operational and financial performance.

The Excel data worksheets shown in Figures 10.1 to 10.10 inclusive are designed to periodically receive data downloads from IT system files each month using refreshed data. The worksheet entitled "Make-To-Stock Model," shown in Figures 10.11 to 10.14 inclusive, is constructed to calculate summary statistics using the information in the data worksheets. But not every organization may need to download all the data fields shown in the make-to-stock inventory model. As an example, if you do not forecast, then the Forecasting History worksheet will be left blank. On the other hand, if you need to analyze seasonal demand, you will need a minimum of 36 months of actual unit demand, that is, three years of history as well as the associated forecasts by item and location for each forecasted time period. In the case of seasonal forecasts, the Forecasting History worksheet must be modified to include three years of monthly forecasts rather than just 12 months. The history of actual unit demand should be placed in the Demand History worksheet. The history of actual lead-times should be placed in the Lead-Time worksheet. This information would be useful to increase the accuracy of the lead-time estimates. However, many organizations have system estimates of lead-time. In these cases, the Lead-Time worksheet may not be required in the model. However, since inventory investment is driven by lead-time, it is important that lead-time estimates be as accurate as possible. In the Backorder worksheet, the "actual shipment history" is compared to "actual demand."

The Inventory worksheet contains inventory quantities. This is a very important worksheet since the optimum inventory quantities calculated by the model depend on accurate estimates of average on-hand inventory quantities. For this reason, the time at which inventory is counted during the order cycle is very important in inventory estimation. As an example, if on-hand inventory quantities are estimated at the beginning of the month, then the model will calculate a higher-than-actual excess inventory level. On the other hand, if inventory quantities are estimated at the end of the order cycle, then the model will calculate a lower-than-actual excess inventory level. Assuming a linear demand rate, the best course of action may be to use the mid-order cycle point as the best estimate of on-hand average inventory level. But this assumption may vary by organization. The Standard Data worksheets contain information required by most manufacturing organizations to categorize the metrics by product group and facility as well as other demographic factors. This standard data worksheet reflects the Lean Six Sigma team's original project charter questions. Common questions are listed in Table 1.4.

It should be noted that the data used in the model is estimated on a unit basis by month. It may also be necessary to convert standard system data upward from one time scale, that is, weeks to months, using Excel formulas; but the reverse scenario is not feasible since variation will be artificially reduced by the downward unit conversion from months to weeks. Also, more complicated models, such as the "make-to-order" model, can be created by coding columns to capture workflow information. This workflow information describes an item's manufacturing sequence through the work-in-process (WIP) stages of the manufacturing system, that is, the system network. In make-to-order models, identifying the cumulative lead-time along the network's critical path is important since it drives raw material and work-in-process (WIP) inventory investment. In either case, a sensitivity analysis on the model's inputs and outputs will help focus the Lean Six Sigma improvement team on KPIVs having a significant impact on inventory investment and customer service.

Demand History Worksheet

Figure 10.1 shows the Demand History worksheet. Demand is an important driver of inventory investment. In fact, expected demand given limited productive capacity is one of the reasons we must invest in inventory in the first place. In the Demand History worksheet, we are interested in calculating unit "Average Demand" and "Demand Standard Deviation" to feed into the final model. But we are also interested in calculating

	A	B	K	L	M	N	O	P
1	Part Number	Type	September	October	November	December	Demand Average	Demand StdDev
2	1	Raw Mat'l	1513	865	842	1464	994	353
3	2	Raw Mat'l	993	1025	963	1032	999	136
4	3	Raw Mat'l	1000	1006	1141	17	854	316
5	4	Raw Mat'l	490	911	855	1157	911	187
6	5	Raw Mat'l	1155	1010	998	379	978	324
7	6	Raw Mat'l	1535	1041	1069	911	1034	172
8	7	Raw Mat'l	1080	966	976	679	1065	217
9	8	Raw Mat'l	656	938	994	544	1001	313
10	9	Raw Mat'l	1080	1034	955	731	963	106
11	10	WIP	475	973	991	636	967	259
12	11	WIP	1362	1096	1122	681	971	317
13	12	WIP	267	1051	1077	1004	915	239
14	13	WIP	720	960	1006	280	868	334
15	14	WIP	614	1011	1021	868	957	213
16	15	WIP	1256	1084	988	1010	987	159
17	16	WIP	901	1008	1009	712	906	131
18	17	Finished Goods	881	784	943	829	965	248
19	18	Finished Goods	1181	1113	1025	772	1006	191
20	19	Finished Goods	855	970	1064	759	951	276
21	20	Finished Goods	833	1127	917	2023	1117	333
22	21	Finished Goods	754	926	1098	1432	1131	234
23	22	Raw Mat'l	904	1221	901	1433	1025	171
24	23	Raw Mat'l	1169	863	1031	248	956	462
25	24	Raw Mat'l	363	1108	996	477	938	273
26	25	Raw Mat'l	1018	959	944	1510	1071	158
27	26	WIP	898	940	993	821	907	177
28	27	WIP	1209	955	1113	1177	1015	132
29	28	WIP	1566	985	1010	769	1067	328
30	29	WIP	305	998	1042	1348	951	235
31	30	WIP	459	963	796	1210	1099	527
32	31	WIP	201	1030	1148	201	1009	520

ᴵ◀ ◀ ▶ ▶ᴵ \ Definitions \ **Demand History** ⟨ Shipment History ⟨ Backorder ⟨ Forecasting History ⟨ Fo ◀ ▶ ᴵ

Figure 10.1 Demand History Worksheet

forecasting error. Forecast error calculations are made by comparing fore-casted to actual quantities using the formulas described in Figure 3.15. The mean and standard deviation of unit demand are calculated using standard statistical formulas of the item's unit sample mean and standard deviation, calculated on a month-to-month basis. Either the standard deviation of actual unit demand or the root mean squared deviation (RMSD) of their forecasted quantities can be used in the make-to-stock model. If the forecasting error, as measured by the root mean squared deviation (RMSD), is less than the calculated sample standard deviation, then it will be used with lead-time to calculate safety-stock levels for an item and location. This concept makes sense from the viewpoint that if we accurately forecast unit demand, then the required safety-stock levels should be lower since they exist as a protection against variations in lead-time and demand. In the final model, optimum inventory investment will depend on item unit demand, and improvement projects will be deployed to eliminate any root causes that adversely impact our ability to satisfy customer demand at the targeted per-unit service level.

Shipment History Worksheet

Figure 10.2 shows the Shipment History worksheet. Shipment history information is important because failure to ship an item on-time could result in lost sales *(lost margin)* or additional costs necessary to make second or third shipments to complete customer orders (backorders). Differences between customer demand and the resultant shipment quantities will also help to identify Lean Six Sigma project opportunities in situations where the shipping process breaks down. There may be many root causes of the failure to ship according to customer ordering requirements. Lean Six Sigma projects can be developed around these problems.

In the Shipment History worksheet, we are interested in calculating unit Average Shipment and Shipment Standard Deviation as inputs into the final make-to-stock model. As a side comment, some organizations incorrectly use actual shipments rather than the original customer demand as the basis for their forecasting models. This is a poor operational practice since it has the effect of creating a chronic inventory problem in which inventory may never be available to satisfy actual customer

	A	B	K	L	M	N	O	P
1	Part Number	Type	Septembe	October	Novembe	Decembe	Shipment Average	Shipment StdDev
2	1	Raw Mat'l	1222	590	842	1464	797	396
3	2	Raw Mat'l	445	362	963	1029	710	341
4	3	Raw Mat'l	1000	953	1141	17	824	386
5	4	Raw Mat'l	442	911	855	886	687	327
6	5	Raw Mat'l	636	1010	998	359	759	288
7	6	Raw Mat'l	1535	1041	1069	647	939	253
8	7	Raw Mat'l	1080	966	976	582	866	315
9	8	Raw Mat'l	174	847	994	544	796	500
10	9	Raw Mat'l	118	790	955	539	820	354
11	10	WIP	475	973	991	583	711	220
12	11	WIP	591	1096	1122	681	857	340
13	12	WIP	208	1051	1077	984	995	494
14	13	WIP	720	960	1006	133	773	320
15	14	WIP	614	1011	1021	868	814	253
16	15	WIP	1084	842	988	307	811	272
17	16	WIP	901	938	1009	569	782	385
18	17	Finished Goods	603	784	943	829	895	190
19	18	Finished Goods	1181	1113	1025	727	909	306
20	19	Finished Goods	855	970	1064	715	787	356
21	20	Finished Goods	833	1127	917	1629	980	415
22	21	Finished Goods	754	608	1098	1196	881	246
23	22	Raw Mat'l	867	1221	901	1215	886	331
24	23	Raw Mat'l	1169	863	1031	248	773	392
25	24	Raw Mat'l	341	1108	996	318	787	317
26	25	Raw Mat'l	967	922	944	1338	1051	309
27	26	WIP	898	940	993	821	872	395
28	27	WIP	887	533	1113	886	867	315
29	28	WIP	1566	680	1010	769	963	278
30	29	WIP	247	831	1042	513	739	287
31	30	WIP	228	768	796	922	955	500
32	31	WIP	201	1030	1148	201	1047	880

◄ ◄ ► ►\ Definitions ╱ Demand History \ **Shipment History** ╱ Backorder ╱ Forecasting History ╱ Fo ◄

Figure 10.2 Shipment History Worksheet

demand since forecasts will not capture the original demand for an item. This will put the item in a chronic backorder situation since shipments will not reflect original customer demand.

Backorder History Worksheet

Figure 10.3 shows the Backorder History worksheet. In the Backorder History worksheet, backorders are calculated by item and location using the formula *Backorder = Shipment Quantity − Demand Quantity*. We are interested in calculating unit Backorder Average and Backorder Standard Deviation as inputs into the final make-to-stock model. The existence of backorders may imply that item-level inventory investment is too low (although there may be other reasons). As the improvement team analyzes the inventory model, backorder problems in one or more product categories may help to identify Lean Six Sigma projects having significant business opportunity. Backorders are a problem because they lower customer service levels and increase shipment costs. Shipment costs increase due to order expediting, that is, overnight shipments, and increases in distribution center direct labor caused by extra product handling.

	A	B	K	L	M	N	O	P
1	Part Number	Type	September	October	November	December	Backorder Average	Backorder StdDev
2	1	Raw Mat'l	-291	-275	0	0	-197	385
3	2	Raw Mat'l	-548	-662	0	-4	-289	355
4	3	Raw Mat'l	0	-53	0	0	-30	389
5	4	Raw Mat'l	-48	0	0	-271	-224	318
6	5	Raw Mat'l	-519	0	0	-20	-219	255
7	6	Raw Mat'l	0	0	0	-264	-95	127
8	7	Raw Mat'l	0	0	0	-97	-199	292
9	8	Raw Mat'l	-482	-90	0	0	-205	325
10	9	Raw Mat'l	-962	-244	0	-192	-144	349
11	10	WIP	0	0	0	-53	-256	331
12	11	WIP	-771	0	0	0	-113	393
13	12	WIP	-59	0	0	-19	80	517
14	13	WIP	0	0	0	-147	-95	383
15	14	WIP	0	0	0	0	-142	318
16	15	WIP	-172	-242	0	-703	-176	256
17	16	WIP	0	-70	0	-143	-124	446
18	17	Finished Goods	-277	0	0	0	-70	151
19	18	Finished Goods	0	0	0	-45	-97	301
20	19	Finished Goods	0	0	0	-45	-164	320
21	20	Finished Goods	0	0	0	-394	-137	300
22	21	Finished Goods	0	-318	0	-236	-250	316
23	22	Raw Mat'l	-37	0	0	-218	-139	371
24	23	Raw Mat'l	0	0	0	0	-183	502
25	24	Raw Mat'l	-22	0	0	-159	-152	322
26	25	Raw Mat'l	-51	-37	0	-172	-20	249
27	26	WIP	0	0	0	0	-34	546
28	27	WIP	-323	-422	0	-291	-148	365
29	28	WIP	0	-305	0	0	-104	197
30	29	WIP	-58	-167	0	-835	-212	310
31	30	WIP	-231	-195	0	-288	-144	167
32	31	WIP	0	0	0	0	38	455

H ◄ ► H \ Definitions / Demand History / Shipment History \ **Backorder** / Forecasting History / Fo |◄|

Figure 10.3 Backorder History Worksheet

Forecasting History Worksheet

Figure 10.4 shows the Forecasting History worksheet. Forecasts by item and location as well as inventory type are imported into this worksheet to be used in the Forecasting Error worksheet to calculate forecasting error statistics using the formulas described in Chapter 3 in Figure 3.15. Although most organizations use simple time-series models to forecast unit demand the specific forecasting model used to make the original forecast may vary by organization. Simple time-series models such as Winter's method were discussed in Chapter 3. Winter's method is commonly used as a forecasting model since it can be automated and used to easily forecast thousands of items each month. It uses three parameters to create a monthly forecast for an item. These parameters build the model using the average level of the time series, its trend, and seasonal components. However, Winter's method requires at least 36 months of forecasting history to model seasonal demand patterns. In the worksheet, the Forecasting History is matched in each time period to the originally forecasted quantity so patterns can be identified by the team. The forecasting error statistics can be incorporated into the

	A	B	G	H	I	J	K	L	M	N
1	Part Number	Type	May	June	July	August	September	October	November	December
2	1	Raw Mat'l	314	1009	1012	760	1222	590	1394	1500
3	2	Raw Mat'l	1203	1072	338	655	445	362	1627	1029
4	3	Raw Mat'l	610	1603	825	862	1640	953	1927	28
5	4	Raw Mat'l	905	1827	734	74	442	1546	1071	886
6	5	Raw Mat'l	641	240	1463	1161	636	1085	1505	359
7	6	Raw Mat'l	1024	909	1575	1400	2212	1262	1315	647
8	7	Raw Mat'l	1590	1039	1172	492	1137	1042	1575	582
9	8	Raw Mat'l	1376	676	806	1656	174	847	1549	591
10	9	Raw Mat'l	986	1338	1498	1037	118	790	1275	539
11	10	WIP	1116	587	375	574	600	1249	1150	583
12	11	WIP	182	1242	966	554	591	1166	1589	916
13	12	WIP	662	3402	799	710	208	1853	1765	984
14	13	WIP	180	806	1138	1688	1174	1119	1353	133
15	14	WIP	509	950	715	767	671	1230	1573	890
16	15	WIP	1219	719	1029	1442	1084	842	1803	307
17	16	WIP	710	1796	1319	442	1101	938	1935	569
18	17	Finished Goods	1332	1263	856	1255	603	808	1213	1080
19	18	Finished Goods	603	926	1618	760	1313	1494	1950	727
20	19	Finished Goods	1527	1245	1156	808	1053	1408	1210	715
21	20	Finished Goods	1424	2922	827	393	1146	1402	1700	1629
22	21	Finished Goods	1547	752	1804	1770	905	608	1288	1196
23	22	Raw Mat'l	757	2368	449	681	867	1473	1635	1215
24	23	Raw Mat'l	1906	212	1246	1152	1903	915	1551	252
25	24	Raw Mat'l	1382	260	1308	666	341	1235	1711	318
26	25	Raw Mat'l	972	1986	593	879	967	922	1471	1338
27	26	WIP	438	2096	867	953	908	1085	1037	966
28	27	WIP	1105	88	1494	1229	887	533	1896	886
29	28	WIP	1846	1649	1470	1593	1812	680	1634	1035
30	29	WIP	1011	166	1414	808	247	831	1398	513
31	30	WIP	1468	1384	1732	1126	228	768	1366	922
32	31	WIP	1192		666	1525	823	314	1618	208

ⅠⅣ ◀ ▶ ▶Ⅰ ╱ Shipment History ╱ Backorder ╲ **Forecasting History** ╱ Forecasting Error ╱ Lead-Time ╱ Ir ◀

Figure 10.4 Forecasting History Worksheet

inventory analysis to help with the root cause analysis. If the organization does not use forecasting models, the Forecasting History worksheet will be left blank.

Forecasting Error Worksheet

Figure 10.5 shows the Forecasting Error worksheet. In this worksheet, three types of forecasting errors are calculated by item and location. The first forecasting error statistic is calculated using the formula *Forecasting Error = Forecasted Quantity − Demand Quantity*. The second statistic, "percent error," is volume adjusted, and is calculated to allow forecasting error comparisons between different items. The third statistic, root mean squared error (RMSD), is calculated as a unit deviation from forecast. It will be used in the safety-stock formula if its magnitude is less than the standard deviation of unit demand. The formulas for percentage error and RMSD are shown in Figure 3.15. Forecasting errors directly impact inventory investment as well as customer service levels.

	A	B	J	K	L	M	N	O	P
1	Part Number	Type	August	September	October	November	December	Percent Error	RMSD
2		1 Raw Mat'l	-297	-291	-275	552	35	-33.54%	480
3		2 Raw Mat'l	-383	-548	-662	664	-4	-56.81%	497
4		3 Raw Mat'l	-124	640	-53	787	11	-76.66%	533
5		4 Raw Mat'l	-753	-48	636	216	-271	-61.03%	522
6		5 Raw Mat'l	188	-519	75	507	-20	-13.22%	388
7		6 Raw Mat'l	339	677	220	246	-264	-19.88%	324
8		7 Raw Mat'l	-576	57	75	599	-97	-4.33%	423
9		8 Raw Mat'l	634	-482	-90	555	47	90.81%	517
10		9 Raw Mat'l	236	-962	-244	320	-192	-75.40%	487
11		10 WIP	-338	125	276	159	-53	-14.14%	425
12		11 WIP	-371	-771	70	467	234	14.98%	489
13		12 WIP	-261	-59	802	688	-19	66.36%	890
14		13 WIP	804	455	160	347	-147	11.57%	512
15		14 WIP	-344	56	219	552	22	54.68%	481
16		15 WIP	485	-172	-242	815	-703	55.57%	523
17		16 WIP	-536	199	-70	926	-143	-43.35%	558
18		17 Finished Goods	10	-277	24	270	251	18.39%	224
19		18 Finished Goods	-98	133	381	925	-45	11.13%	462
20		19 Finished Goods	78	198	438	146	-45	19.48%	405
21		20 Finished Goods	-760	313	275	783	-394	44.33%	597
22		21 Finished Goods	542	151	-318	191	-236	80.57%	552
23		22 Raw Mat'l	-376	-37	252	735	-218	-17.42%	612
24		23 Raw Mat'l	85	734	52	520	5	-51.40%	625
25		24 Raw Mat'l	-312	-22	127	715	-159	10.41%	458
26		25 Raw Mat'l	-168	-51	-37	527	-172	35.09%	455
27		26 WIP	-1	10	145	44	145	-35.91%	569
28		27 WIP	247	-323	-422	783	-291	9.64%	478
29		28 WIP	621	246	-305	624	266	-2.26%	383
30		29 WIP	-120	-58	-167	356	-835	25.90%	412
31		30 WIP	270	-231	-195	570	-288	45.65%	362
32		31 WIP	136	114	588	178	7	9.70%	529

H ◀ ▸ H / Shipment History / Backorder / Forecasting History \ Forecasting Error / Lead-Time / Ir ◀ |

Figure 10.5 Forecasting Error Worksheet

Operational costs are also adversely impacted due to product expediting, overnight shipments, lost sales on gross margin, and increases in direct labor caused by extra material handling. Lean Six Sigma improvement projects can be deployed to improve forecasting accuracy based on the problems identified by the improvement team using root cause analysis.

Lead-Time Worksheet

The Lead-Time worksheet is shown in Figure 10.6. Lead-time information is very important because it is the major driver of inventory investment, assuming that forecasting error rates are at an average level. In this analysis it is preferable to use actual rather than system estimated lead-times to calculate optimum inventory quantities. However, as mentioned earlier, actual lead-time information is seldom available in practice. In the simulated case study shown in this chapter, average lead-time as well as its standard deviation is incorporated into the model. Using the model, changes in inventory investment

	A	B	K	L	M	N	O	P
1	Part Number	Type	September	October	November	December	Lead Time Average	Lead Time StdDev
2	1	Raw Mat'l	1	6	10	12	6	3
3	2	Raw Mat'l	6	11	15	17	11	3
4	3	Raw Mat'l	11	16	20	22	16	3
5	4	Raw Mat'l	16	21	25	27	21	3
6	5	Raw Mat'l	21	26	30	32	26	3
7	6	Raw Mat'l	26	31	35	37	31	3
8	7	Raw Mat'l	31	36	40	42	36	3
9	8	Raw Mat'l	36	41	45	47	41	3
10	9	Raw Mat'l	41	46	50	52	46	3
11	10	WIP	46	51	55	57	51	3
12	11	WIP	51	56	60	62	56	3
13	12	WIP	1	6	10	12	6	3
14	13	WIP	6	11	15	17	11	3
15	14	WIP	11	16	20	22	16	3
16	15	WIP	16	21	25	27	21	3
17	16	WIP	21	26	30	32	26	3
18	17	Finished Goods	26	31	35	37	31	3
19	18	Finished Goods	31	36	40	42	36	3
20	19	Finished Goods	36	41	45	47	41	3
21	20	Finished Goods	41	46	50	52	46	3
22	21	Finished Goods	46	51	55	57	51	3
23	22	Raw Mat'l	51	56	60	62	56	3
24	23	Raw Mat'l	1	6	10	12	6	3
25	24	Raw Mat'l	6	11	15	17	11	3
26	25	Raw Mat'l	11	16	20	22	16	3
27	26	WIP	16	21	25	27	21	3
28	27	WIP	21	26	30	32	26	3
29	28	WIP	26	31	35	37	31	3
30	29	WIP	31	36	40	42	36	3
31	30	WIP	36	41	45	47	41	3
32	31	WIP	41	46	50	52	46	3

◄ ◄ ► ►◄ Shipment History / Backorder / Forecasting History / Forecasting Error \ Lead-Time / Ir ◄ ◄

Figure 10.6 Lead-Time Worksheet

can be simulated to show where lead-time improvements will reduce investment by the greatest amount. This approach is also a good way to deploy Kaizen events since they will be directly linked to major areas of inventory investment. Long or highly variable lead-times will necessitate deployment of one or more Lean Six Sigma improvement projects to eliminate their root causes.

Inventory History Worksheet

Figure 10.7 shows the Inventory History worksheet. This worksheet is important because calculated optimum inventory levels are compared for each item and location against estimated average on-hand inventory. The month-to-month inventory quantities are used to calculate the item's average unit inventory (sample mean) and its unit standard deviation. These statistics are also inputs into the make-to-stock model. The "actual inventory" quantity must be estimated

	A	B	K	L	M	N	O	P
1	Part Number	Type	September	October	November	December	Inventory Average	Inventory StdDev
2	1	Raw Mat'l	3123	3616	3945	2301	3397	596
3	2	Raw Mat'l	3846	4453	4858	2834	4183	734
4	3	Raw Mat'l	63	73	80	47	69	12
5	4	Raw Mat'l	1964	2274	2480	1447	2136	375
6	5	Raw Mat'l	413	478	521	304	449	79
7	6	Raw Mat'l	2083	2412	2632	1535	2266	398
8	7	Raw Mat'l	2671	3093	3374	1968	2906	510
9	8	Raw Mat'l	3332	3858	4209	2455	3624	636
10	9	Raw Mat'l	3204	3710	4047	2361	3485	612
11	10	WIP	1756	2033	2218	1294	1910	335
12	11	WIP	781	905	987	576	850	149
13	12	WIP	2228	2579	2814	1641	2423	425
14	13	WIP	1172	1356	1480	863	1274	224
15	14	WIP	1852	2145	2340	1365	2015	354
16	15	WIP	2902	3360	3666	2138	3157	554
17	16	WIP	1710	1980	2160	1260	1860	326
18	17	Finished Goods	486	563	614	358	529	93
19	18	Finished Goods	824	954	1041	607	897	157
20	19	Finished Goods	2594	3004	3277	1911	2822	495
21	20	Finished Goods	2781	3220	3513	2049	3025	531
22	21	Finished Goods	1823	2110	2302	1343	1983	348
23	22	Raw Mat'l	3556	4118	4492	2620	3868	679
24	23	Raw Mat'l	198	229	250	146	215	38
25	24	Raw Mat'l	2126	2461	2685	1566	2312	406
26	25	Raw Mat'l	388	449	490	286	422	74
27	26	WIP	1357	1572	1714	1000	1476	259
28	27	WIP	2503	2898	3162	1844	2723	478
29	28	WIP	3105	3595	3922	2288	3378	593
30	29	WIP	1498	1735	1893	1104	1630	286
31	30	WIP	381	441	481	281	415	73
32	31	WIP	573	663	724	422	623	109

Backorder / Forecasting History / Forecasting Error / Lead-Time \ Inventory History / S

Figure 10.7 Inventory History Worksheet

carefully because the difference between "actual" and "optimum" inventory quantities, for each item and location, will become the "reduction/addition targets" calculated by the model. Lean Six Sigma improvement projects will be easily identified as inventory turns and investment levels are analyzed using the model. As an example, a business case can be developed to create projects to eliminate areas of high investment or where operational breakdowns occur within the process.

Standard Data Worksheet

The Standard Data worksheets shown in Figures 10.8, 10.9, and 10.10 consist of summarized data from the other worksheets as well as descriptive information such as standard cost, gross margin percentage, lot size, order quantity, market segment, and product group, to name just a few. Other relevant information is associated with the team's original questions (see Table 1.4), which the team must answer for senior management. These columns of descriptive information will eventually be analyzed to reveal relationships between the input and output variables as represented by their data fields.

	A	B	C	D	E	F	G	H	I	J
1	Part Number	Type	Standard Cost	Margin	Lot Size	Order Quantity	Lead-Time	Service Target	Market Segment	Product Class
2	1	Raw Mat'l	$ 5.00	10%	4103	4103	5	95.00	1	1
3	2	Raw Mat'l	$ 6.00	20%	4339	4339	10	96.00	2	2
4	3	Raw Mat'l	$ 7.00	30%	2872	2872	15	97.00	3	3
5	4	Raw Mat'l	$ 8.00	40%	2308	2308	20	98.00	4	4
6	5	Raw Mat'l	$ 9.00	50%	4035	4035	25	99.00	5	5
7	6	Raw Mat'l	$ 10.00	10%	4780	4780	30	95.00	1	6
8	7	Raw Mat'l	$ 11.00	20%	2869	2869	35	96.00	2	7
9	8	Raw Mat'l	$ 12.00	30%	3608	3608	40	97.00	3	8
10	9	Raw Mat'l	$ 13.00	40%	4802	4802	45	98.00	4	9
11	10	WIP	$ 14.00	50%	1647	1647	50	99.00	5	10
12	11	WIP	$ 15.00	10%	669	669	55	95.00	1	1
13	12	WIP	$ 16.00	20%	2574	2574	5	96.00	2	2
14	13	WIP	$ 17.00	30%	1775	1775	10	97.00	3	3
15	14	WIP	$ 18.00	40%	4309	4309	15	98.00	4	4
16	15	WIP	$ 19.00	50%	4677	4677	20	99.00	5	5
17	16	WIP	$ 20.00	10%	1775	1775	25	95.00	1	6
18	17	Finished Goods	$ 21.00	20%	4425	4425	30	96.00	2	7
19	18	Finished Goods	$ 22.00	30%	2328	2328	35	97.00	3	8
20	19	Finished Goods	$ 23.00	40%	1325	1325	40	98.00	4	9
21	20	Finished Goods	$ 24.00	50%	457	457	45	99.00	5	10
22	21	Finished Goods	$ 25.00	10%	2372	2372	50	95.00	1	1
23	22	Raw Mat'l	$ 26.00	20%	3530	3530	55	96.00	2	2
24	23	Raw Mat'l	$ 27.00	30%	719	719	5	97.00	3	3
25	24	Raw Mat'l	$ 28.00	40%	2695	2695	10	98.00	4	4
26	25	Raw Mat'l	$ 29.00	50%	4946	4946	15	99.00	5	5
27	26	WIP	$ 30.00	10%	2507	2507	20	95.00	1	6
28	27	WIP	$ 31.00	20%	3231	3231	25	96.00	2	7
29	28	WIP	$ 32.00	30%	2789	2789	30	97.00	3	8
30	29	WIP	$ 33.00	40%	4335	4335	35	98.00	4	9
31	30	WIP	$ 34.00	50%	381	381	40	99.00	5	10
32	31	WIP	$ 35.00	10%	4522	4522	45	95.00	1	1

Lead-Time / Inventory History \ **Standard Data** / Make-To-Stock Model /

Figure I0.8 Standard Data Worksheet

	A	B	K	L	M	N	O	P
1	Part Number	Type	Inventory Average	Inventory StdDev	Demand Average	Demand StdDev	Shipment Average	Shipment StdDev L
2	1	Raw Mat'l	3397	596	994	363	797	396
3	2	Raw Mat'l	4183	734	999	136	710	341
4	3	Raw Mat'l	69	12	854	316	824	386
5	4	Raw Mat'l	2136	375	911	187	687	327
6	5	Raw Mat'l	449	79	978	324	759	288
7	6	Raw Mat'l	2266	398	1034	172	939	253
8	7	Raw Mat'l	2906	510	1065	217	866	315
9	8	Raw Mat'l	3624	636	1001	313	796	500
10	9	Raw Mat'l	3485	612	963	106	820	354
11	10	WIP	1910	335	967	269	711	220
12	11	WIP	850	149	971	317	857	340
13	12	WIP	2423	425	915	239	995	494
14	13	WIP	1274	224	868	334	773	320
15	14	WIP	2015	354	967	213	814	253
16	15	WIP	3157	554	987	159	811	272
17	16	WIP	1860	326	906	131	782	385
18	17	Finished Goods	529	93	965	248	895	190
19	18	Finished Goods	897	157	1006	191	909	306
20	19	Finished Goods	2822	495	951	276	787	356
21	20	Finished Goods	3025	531	1117	333	980	415
22	21	Finished Goods	1983	348	1131	234	881	246
23	22	Raw Mat'l	3868	679	1025	171	886	331
24	23	Raw Mat'l	215	38	956	462	773	392
25	24	Raw Mat'l	2312	406	938	273	787	317
26	25	Raw Mat'l	422	74	1071	158	1051	309
27	26	WIP	1476	259	907	177	872	395
28	27	WIP	2723	478	1015	132	867	315
29	28	WIP	3378	593	1067	328	963	278
30	29	WIP	1630	286	951	235	739	287
31	30	WIP	415	73	1099	527	955	500
32	31	WIP	622	109	1069	520	1047	880

`I◄ ◄ ► ►I \ Lead Time / Inventory History \ Standard Data \ Make-To-Stock Model /`

Figure 10.9 Standard Data Worksheet (continued)

Make-to-Stock Model Worksheet

The worksheets making up the Make-to-Stock model are shown in Figures 10.11, 10.12, 10.13, and 10.14. The make-to-stock model calculates optimum inventory quantities for each item and location

	A	B	Q	R	S	T	U	V
1	Part Number	Type	Lead Time Average	Lead Time StdDev	Backorder Average	Backorder StdDev	Percent Error	RMSD
2	1	Raw Mat'l	6	3	-197	385	-33.54%	480
3	2	Raw Mat'l	11	3	-289	355	-56.81%	497
4	3	Raw Mat'l	16	3	-30	389	-76.66%	533
5	4	Raw Mat'l	21	3	-224	318	-61.03%	522
6	5	Raw Mat'l	26	3	-219	255	-13.22%	388
7	6	Raw Mat'l	31	3	-95	127	-19.88%	324
8	7	Raw Mat'l	36	3	-199	292	-4.33%	423
9	8	Raw Mat'l	41	3	-205	325	90.81%	517
10	9	Raw Mat'l	46	3	-144	349	-75.40%	487
11	10	WIP	51	3	-256	331	-14.14%	425
12	11	WIP	56	3	-113	393	14.98%	489
13	12	WIP	6	3	80	517	66.36%	890
14	13	WIP	11	3	-95	383	11.57%	512
15	14	WIP	16	3	-142	318	64.68%	481
16	15	WIP	21	3	-176	256	55.57%	523
17	16	WIP	26	3	-124	446	-43.35%	558
18	17	Finished Goods	31	3	-70	151	18.39%	224
19	18	Finished Goods	36	3	-97	301	11.13%	462
20	19	Finished Goods	41	3	-164	320	19.48%	405
21	20	Finished Goods	46	3	-137	300	44.33%	597
22	21	Finished Goods	51	3	-260	316	80.57%	552
23	22	Raw Mat'l	56	3	-139	371	-17.42%	612
24	23	Raw Mat'l	6	3	-183	502	-51.40%	625
25	24	Raw Mat'l	11	3	-152	322	10.41%	458
26	25	Raw Mat'l	16	3	-20	249	35.09%	455
27	26	WIP	21	3	-34	546	-35.91%	569
28	27	WIP	26	3	-148	365	9.64%	478
29	28	WIP	31	3	-104	197	-2.26%	383
30	29	WIP	36	3	-212	310	25.90%	412
31	30	WIP	41	3	-144	167	45.65%	362
32	31	WIP	46	3	38	466	0.79%	528

`I◄ ◄ ► ►I \ Lead Time / Inventory History \ Standard Data / Make-To-Stock Model /`

Figure 10.10 Standard Data Worksheet (continued)

	A	B	W	X	Y	Z	AA	AB	AC
1	Part Number	Type	LdTm Rdctn	Efct LdTm	LdTmVarRedct	EffLdTmVar	Dmnd Var Rdctn	Efct Dmnd Var	Service Factor
3120	3119	WIP	0	31	0	3	0	424	1.65
3121	3120	WIP	0	36	0	3	0	353	2.33
3122	3121	WIP	0	41	0	3	0	262	1.65
3123	3122	Finished Goods	0	46	0	3	0	231	1.65
3124	3123	Finished Goods	0	51	0	3	0	346	1.65
3125	3124	Finished Goods	0	56	0	3	0	179	1.65
3126	3125	Finished Goods	0	6	0	3	0	271	2.33
3127	3126	Finished Goods	0	11	0	3	0	233	1.65
3128	3127	Raw Mat'l	0	16	0	3	0	259	1.65
3129	3128	Raw Mat'l	0	21	0	3	0	283	1.65
3130	3129	Raw Mat'l	0	26	0	3	0	304	1.65
3131	3130	Raw Mat'l	0	31	0	3	0	401	2.33
3132	3131	WIP	0	36	0	3	0	217	1.65
3133	3132	WIP	0	41	0	3	0	328	1.65
3134	3133	WIP	0	46	0	3	0	260	1.65
3135	3134	WIP	0	51	0	3	0	273	1.65
3136	3135	WIP	0	56	0	3	0	250	2.33
3137	3136	WIP	0	6	0	3	0	291	1.65
3138	3137	WIP	0	11	0	3	0	255	1.65
3139	3138	WIP	0	16	0	3	0	118	1.65
3140	3139	WIP	0	21	0	3	0	111	1.65
3141	3140	WIP	0	26	0	3	0	255	2.33
3142	3141	Finished Goods	0	31	0	3	0	305	1.65
3143	3142	Finished Goods	0	36	0	3	0	249	1.65
3144	3143	Finished Goods	0	41	0	3	0	482	1.65
3145	3144	Finished Goods	0	46	0	3	0	249	1.65
3146	3145	Finished Goods	0	51	0	3	0	260	2.33
3147	3146	Finished Goods	0	56	0	3	0	237	1.65
3148									
3149									

H ◀ ▶ H / Lead-Time / Inventory History / Standard Data \ Make-To-Stock Model /

Figure 10.11 Make-To-Stock Model Worksheet

based on the average inventory required over the order cycle T, that is, $\frac{1}{2} Q_1$ and required safety stock. Specific inputs to the model include unit service level, average lead-time LT, lead-time variation σ_L^2, the

	A	B	AD	AE	AF	AG	AH	AI
1	Part Num	Type	Avrg Dmnd Ovr Ld Time	Cmbnd Variance	Cmbnd StdDev	Safety Stock	Optimum Inventory	Optimum Investment
3124	3123	Finished Goods	1211	332325	576	951	2163	$209,765.08
3125	3124	Finished Goods	1436	120641	347	673	2009	$196,888.95
3126	3125	Finished Goods	162	65488	236	549	711	$70,366.28
3127	3126	Finished Goods	273	58489	242	399	672	$67,190.57
3128	3127	Raw Mat'l	374	79148	281	464	838	$84,636.03
3129	3128	Raw Mat'l	577	119728	346	571	1148	$117,081.12
3130	3129	Raw Mat'l	588	144390	380	627	1215	$125,161.40
3131	3130	Raw Mat'l	862	285455	534	1245	2107	$219,134.69
3132	3131	WIP	1016	121891	349	576	1592	$167,147.09
3133	3132	WIP	924	244280	494	816	1740	$8,699.86
3134	3133	WIP	1322	193909	440	727	2048	$12,289.88
3135	3134	WIP	1473	229572	479	791	2264	$15,047.72
3136	3135	WIP	1254	198157	445	1037	2291	$18,328.09
3137	3136	WIP	136	49381	222	367	503	$4,527.04
3138	3137	WIP	262	62220	249	412	673	$6,731.80
3139	3138	WIP	382	37662	194	320	702	$7,726.84
3140	3139	WIP	555	45477	213	352	906	$10,876.92
3141	3140	WIP	692	117834	343	800	1491	$19,388.26
3142	3141	Finished Goods	664	166674	408	674	1337	$18,721.59
3143	3142	Finished Goods	912	141606	376	621	1533	$22,997.64
3144	3143	Finished Goods	1182	515454	718	1185	2366	$37,062.98
3145	3144	Finished Goods	1197	174805	418	690	1887	$32,070.82
3146	3145	Finished Goods	1316	204256	452	1063	2369	$42,649.13
3147	3146	Finished Goods	1428	187717	433	715	2142	$40,706.62
3148								
3149								$248,147,207.93
3150								
3151								
3152								
3153								

H ◀ ▶ H / Lead-Time / Inventory History / Standard Data \ Make-To-Stock Model /

Figure 10.12 Make-To-Stock Model Worksheet (continued)

	A	B	AI	AJ	AK	AL	AM	AN
1	Part Num	Type	Optimum Investment	Add(Reduce) Units	Add(Reduce) Cost	Current Investment	Annual COGS	Optimized COGS
3124	3123	Finished Goods	$209,769.08	-38	$ (3,695.80)	$ 213,464.88	$ 989,929.94	$ 1,070,993.28
3125	3124	Finished Goods	$196,888.96	315	$ 30,882.27	$ 166,006.68	$ 938,864.35	$ 1,180,332.73
3126	3125	Finished Goods	$70,365.28	45	$ 4,449.02	$ 65,916.25	$ 1,012,611.45	$ 1,252,058.03
3127	3126	Finished Goods	$67,190.57	-3612	$ (361,246.25)	$ 428,406.82	$ 1,210,639.92	$ 1,122,631.16
3128	3127	Raw Mat'l	$84,636.03	-1921	$ (194,026.35)	$ 278,662.38	$ 1,007,197.70	$ 1,081,622.32
3129	3128	Raw Mat'l	$117,061.12	-2412	$ (246,030.59)	$ 363,111.71	$ 1,197,139.39	$ 1,299,551.81
3130	3129	Raw Mat'l	$125,161.40	995	$ 102,495.56	$ 22,665.84	$ 1,193,050.00	$ 1,092,578.55
3131	3130	Raw Mat'l	$219,134.69	-495	$ (51,482.66)	$ 270,617.35	$ 1,301,868.17	$ 1,370,838.91
3132	3131	WIP	$167,147.09	-291	$ (30,602.04)	$ 197,749.13	$ 1,219,646.69	$ 1,347,912.69
3133	3132	WIP	$3,699.86	1180	$ 5,897.71	$ 2,802.15	$ 45,091.56	$ 51,845.05
3134	3133	WIP	$12,289.88	-630	$ (3,180.77)	$ 15,470.64	$ 68,977.31	$ 80,124.35
3135	3134	WIP	$15,847.72	139	$ 970.57	$ 14,877.15	$ 76,014.45	$ 94,973.42
3136	3135	WIP	$18,328.09	1890	$ 15,116.22	$ 3,211.87	$ 78,017.04	$ 84,989.66
3137	3136	WIP	$4,527.04	-1391	$ (12,516.03)	$ 17,043.07	$ 80,489.86	$ 91,980.82
3138	3137	WIP	$6,731.80	-3362	$ (33,618.83)	$ 40,350.63	$ 99,691.25	$ 108,764.48
3139	3138	WIP	$7,726.84	-411	$ (4,524.90)	$ 12,251.75	$ 98,532.67	$ 121,718.16
3140	3139	WIP	$10,876.92	-936	$ (11,231.68)	$ 22,108.61	$ 133,791.69	$ 148,471.70
3141	3140	WIP	$19,388.26	14	$ 182.40	$ 19,205.86	$ 147,521.63	$ 163,796.16
3142	3141	Finished Goods	$18,721.59	-1044	$ (14,620.52)	$ 33,342.11	$ 120,535.55	$ 136,299.02
3143	3142	Finished Goods	$22,997.64	448	$ 6,725.39	$ 16,272.25	$ 166,660.58	$ 174,751.47
3144	3143	Finished Goods	$37,862.98	954	$ 15,267.87	$ 22,595.11	$ 180,835.12	$ 214,297.57
3145	3144	Finished Goods	$32,070.82	-1033	$ (17,563.06)	$ 49,633.88	$ 181,363.07	$ 207,655.31
3146	3145	Finished Goods	$42,649.13	-2674	$ (48,136.36)	$ 90,785.49	$ 183,369.21	$ 220,416.66
3147	3146	Finished Goods	$40,706.62	313	$ 5,937.51	$ 34,769.11	$ 198,454.24	$ 220,537.33
3148						$ -		
3149			$248,147,207.93	(3,455,980)	$ (193,756,756.78)	$ 441,903,966.71	$ 1,793,027,637.30	$ 2,014,637,802.89
3150								
3151								
3152								
3153								

Figure 10.13 Make-To-Stock Model Worksheet (continued)

expected demand, D, over the order cycle T, and demand variation σ_D^2. In Figure 10.11, columns W–AA show simulated reductions in lead-time and demand in a manner similar to Figure 6.1. The first step

	A	B	H	AO	AP	AQ	AR	AS
1	Part Num	Type	Service Target	Current Turns	Optimum Turns	Current Service	Current Profit	Proj Profit
3125	3124	Finished Gd	98.00	5.66	5.99	77.95%	$ 375,545.74	$ 481,768.46
3126	3125	Finished Gd	99.00	15.36	17.79	80.06%	$ 506,305.72	$ 632,372.74
3127	3126	Finished Gd	95.00	2.83	16.71	102.45%	$ 121,063.99	$ 118,171.70
3128	3127	Raw Mat'l	96.00	3.61	12.78	89.39%	$ 201,439.54	$ 225,337.98
3129	3128	Raw Mat'l	97.00	3.30	11.10	89.36%	$ 359,141.82	$ 401,923.24
3130	3129	Raw Mat'l	98.00	52.64	8.73	107.01%	$ 477,220.00	$ 445,950.43
3131	3130	Raw Mat'l	99.00	4.81	6.26	94.02%	$ 650,934.09	$ 692,342.89
3132	3131	WIP	95.00	6.17	8.06	85.96%	$ 121,964.67	$ 141,885.55
3133	3132	WIP	96.00	16.09	5.96	83.49%	$ 9,018.31	$ 10,801.05
3134	3133	WIP	97.00	4.46	6.52	83.51%	$ 20,693.19	$ 24,780.73
3135	3134	WIP	98.00	5.11	5.99	78.44%	$ 30,405.78	$ 38,764.66
3136	3135	WIP	99.00	24.29	4.64	90.88%	$ 39,008.52	$ 42,924.07
3137	3136	WIP	95.00	4.72	20.32	83.13%	$ 8,048.99	$ 9,682.19
3138	3137	WIP	96.00	2.47	16.16	87.99%	$ 19,938.25	$ 22,659.27
3139	3138	WIP	97.00	8.04	15.75	78.52%	$ 29,559.77	$ 37,644.79
3140	3139	WIP	98.00	6.05	13.65	88.31%	$ 53,516.68	$ 60,600.70
3141	3140	WIP	99.00	7.68	8.45	89.16%	$ 73,760.81	$ 82,725.33
3142	3141	Finished Gd	95.00	3.62	7.28	84.01%	$ 12,053.55	$ 14,347.27
3143	3142	Finished Gd	96.00	10.24	7.60	91.55%	$ 33,330.12	$ 36,406.56
3144	3143	Finished Gd	97.00	8.00	5.66	81.85%	$ 54,250.54	$ 66,277.60
3145	3144	Finished Gd	98.00	3.65	6.47	85.60%	$ 72,553.23	$ 84,757.27
3146	3145	Finished Gd	99.00	2.02	5.17	82.36%	$ 91,679.60	$ 111,321.55
3147	3146	Finished Gd	95.00	5.71	5.42	85.49%	$ 19,845.42	$ 23,214.46
3148								
3149			97.00	4.06	8.12	86.37%	$ 536,753,395.12	$ 622,707,597.12
3150								
3151								
3152								
3153								
3154								

Figure 10.14 Make-To-Stock Model Worksheet (continued)

	Before		Optimized	
Metric	Column	$Million	Column	$Million
Annual COGS	AM	1,793	AN	2,015
Inventory	AL	442	AI	248
Average Turns	AO	4	AP	8
Unit Service%	AQ	86	H	97
Gross Margin	AR	537	AS	623

Table 10.1 Initial Inventory Balance

of the analysis is to calculate the "optimum" inventory quantity by item and location and compare it to the estimated on-hand inventory quantity. This balance analysis, which is shown in Figure 6.2, will show if inventory investment should be increased or decreased (balancing the inventory) for each item and location. As an example, in the current scenario, Figure 10.12, column AI and the second line of Table 10.1 (the "optimized inventory" metric) show that the optimum inventory investment is calculated to be $248,147,207.93 (or 248 in Table 10.1). Table 10.1 shows the changes in key metrics when the inventory is optimized and operational problems are assumed to be eliminated using Lean Six Sigma improvement projects. As an example, in Table 10.1 the cost-of-goods-sold (COGS) increases due to elimination of backorders (additional shipments); average inventory investment decreases when the inventory is balanced (average turns increases); unit service level increases to the average target of 97%; and gross margin increases (simply based on incremental sales due to the hypothetical elimination of backorders). Additional data for Table 10.1 is shown in Figures 10.13 and 10.14. These figures are only approximate since they do not consider incremental costs and the assumptions do not capture all aspects of the actual inventory system. In practice, making the model more realistic is not very difficult. Also, the results could be very different for inventories characterized by low customer service levels and long lead-times or poor demand estimation. In these situations Lean Six Sigma projects must be deployed to significantly reduce lead-times prior to reducing inventory investment levels.

Obtaining the Inventory Balance

The improvements shown in Table 10.1 were due to simply balancing the inventory to achieve the unit service target for each item. This concept is shown graphically in Figure 6.2. The Excel model represents a simulated population of 3,146 items (part numbers). Some of the data for particular items, in particular inventory turns appearing in the Figure 10.14, may appear strange; but it can be explained if we consider all model assumptions. Remember, the initial goal of the analysis was to minimize inventory investment to achieve customer unit service targets without improvements in lead-time or demand management. In the model, shipments were set equal to external monthly customer demand through elimination of backorders. However, several Lean Six Sigma improvement projects would be required to actually reduce these backorders. Second, if on-hand inventory exceeded the calculated optimum quantity, it was reduced to the optimum level. Again, improvement projects would be required to reduce excess or obsolete inventory levels. But there might also be a constraint on how far the improvement team could actually go in this direction. On the other hand, if on-hand inventory levels were less than the calculated optimum quantity, they were increased to meet targeted per-unit service levels. This latter action had the effect of reducing overall inventory turns to achieve the item's customer service level target.

To show how the analysis is done in practice, we will analyze the first five items or part numbers using the example shown in Table 10.2. The assumptions are shown for each part number in the comments section of Table 10.2. Referring to the first part number in Table 10.2, the model calculates a reduction in inventory investment of ($14,098) while increasing turns from 3 to 20 and unit service level from 80.1% to 95.0%. Two process improvements were assumed by the model. The first was a reduction in inventory from 3,397 units to 578 (optimum level) through excess inventory elimination. Also, the backorders were eliminated by matching shipments to customer demand. As previously mentioned, achievement of these improvements depends on model assumptions being met in practice as well as effective deployment of Lean Six Sigma projects. But one advantage of using a model is that the business benefits can be aggregated and categorized by type of problem to estimate the potential business benefits gained by elimination of the problems to create project charters. This approach will greatly accelerate your Lean Six Sigma deployment process because the analysis can be used as financial justification to deploy improvement projects in operational areas having significant return-on-investment (ROI) and other business benefits. Returning again to Table 10.2 discussion, inventory investment can be decreased for the fourth part number because the average

P/N	$Add (Reduce)	Current Turns	Optimum Turns	Current Service%	Optimum Service%	Lead-Time Days (Months)
1	($14,098)	3	20	80.1	95.0	6(.3)
2	($21,471)	2	19	71.1	96.0	11(.55)
3	$5,609	143	11	96.5	97.0	16(.8)
4	($9,981)	4	12	75.4	98.0	21(1.05)
5	$10,220	20	7	77.6	99.0	26(1.3)
P/N	Average Inventory	Optimum Inventory	Average Demand	Average Shipments	Average Backorders	
1	3,397	578	994	797	−197	
2	4,183	605	999	710	−289	
3	69	870	854	824	−30	
4	2,136	888	911	687	−224	
5	449	1,584	978	759	−219	
P/N	Comments					
1	Excess inventory is reduced, backorders are reduced, service increased.					
2	Excess inventory is reduced, backorders are reduced, service increased.					
3	Not enough inventory on-hand, inventory increased, backorders reduced.					
4	Excess inventory is reduced, backorders are reduced, service increased.					
5	Excess inventory is reduced, backorders are reduced, service increased.					

Table 10.2 Part Number Detail

on-hand quantity is greater than the calculated optimum quantity. The customer service target increased because the backorders were assumed to have been eliminated through process improvements.

Admittedly, in the current inventory model, the assumptions are somewhat simplistic. But, depending on requirements, more sophisticated assumptions can be built into these types of models, which would more accurately reflect your specific operations. In fact, we have successfully used this analytical methodology over the past several years. Although the model is simplistic, in aggregate across many items and locations, sensitivity analyses provide very good estimates of business benefits necessary to justify Lean Six Sigma improvement projects. Consideration of the myriad of system constraints is also an important part of the supply chain modeling process since every organization will have its specific modeling constraints based on its unique supply chain design.

Root Cause Analysis of the Model Using Lean and Six Sigma Methods

Using the analytical information derived from the model, the improvement team can develop a problem statement, define the overall project, and calculate the project's initial business benefits. Also, the analytical results derived from the model, if correctly summarized, will act as an effective communication vehicle throughout the organization by identifying the root causes of major operational issues driving inventory investment or other supply chain problems. The analytical results will also ensure the root cause analysis is quantitatively linked to an organization's strategic goals and objectives. Using this information, focused Lean Six Sigma improvement projects can be developed to identify and eliminate the operational problem and standardize the optimized system using control strategies that are specific to the improvement team's recommended solutions. This concept was shown in Table 1.2. Creating a supply chain model will guarantee metric linkage will in fact occur.

If the inventory model had shown that inventory investment was caused, in part, by "failure to meet schedule," Lean Six Sigma improvement projects could be focused on this problem area. The "enabler"

Lead-Time Reduction	Lead-Time Days	Current GM Profit	Current COGS	Current Investment	Current Turns
0%	31	$536	$1,793	$442	4
Lead-Time Reduction	Lead-Time Days	Optimized GM Profit	Optimized COGS	Optimized Inventory	Optimized Turns
0%	31	$663	$2,014	$248	8
−20%	25	$663	$2,014	$212	9
−40%	19	$663	$2,014	$175	11
−60%	12	$663	$2,014	$137	15
−80%	6	$663	$2,014	$97	21
Lead-Time Reduction	Lead-Time Days	Add(Reduce) Inventory	Effective Interest(10%)		
0%	31	−$193	$19		
−20%	25	−$230	$23		
−40%	19	−$266	$27		
−60%	12	−$304	$30		
−80%	6	−$335	$34		

Table 10.3 Impact of Balancing and Lead-Time Reduction

▪ Customer billing adjustments	▪ Inventory holding costs
▪ Sales policy errors	▪ Inventory obsolescence
▪ Product destroyed in field	▪ Overtime expense
▪ Returned product	▪ Product transfer
▪ Warranty expense	▪ Rework expense
▪ Shipping errors	▪ Scrap expense
▪ Premium freight	▪ Past due receivables
▪ Margin improvement	▪ Sales success rate

Table 10.4 Common Problems

toolkit would be specific tools associated with the "Lean supply chain" initiative, as discussed in Chapters 2, 3, and 4. Alternatively, if the root causes of the problem were due to poor manufacturing yields, tools associated with the Six Sigma initiative as discussed in Chapter 8 would be applied to eliminate the root causes of the problem. Table 10.4 shows other operational problems, which supply chain models will help identify. If any of these problems are identified by the analysis, they could become the focus of Lean Six Sigma improvement projects. These projects can take many forms depending on the specifics of the model. More than 25 of these types of projects will be discussed at the end of this chapter.

Recent Trends in Distribution Inventory and Supply Chain Modeling

Over the past several years there has been an active interest in bringing together into one place all the data an organization collects within its MPS, MRPII, and distribution systems to create analytical models to more effectively manage supply chain work streams. However, these off-the-shelf models may not have the flexibility your organization requires to analyze operational problems. As an example, most of these models show material or information flow through a system and identify system bottlenecks and constraints or non-value-adding operations within the process. An analogy would be a computerized value stream map (VSM), which can be simulated to show how the process responds to changing levels of the key process input variables (KPIVs), such as lead-time, demand variation, and other system inputs. However, the Lean Six Sigma team will require information sufficient to drive to root causes at third, fourth, and fifth levels of analysis. In the case of inventory analysis, the required information will be at the item and location level and include specific descriptive information to help the team identify the most likely root causes for the process breakdowns. In many cases, simple Excel models may be more efficient to build and use since they capture the operational detail required to analyze major work streams. However, modifications to off-the-self software or customized Excel models can be developed by your organization if analyses of your supply chain's operations will be made on an ongoing basis.

Lean Six Sigma Applications

In the past 20 years, Lean Six Sigma methods have been successfully applied throughout many industries, organizations, and functions. The following supply chain examples show how the concepts in this book can be applied to real-world supply chain problems including distribution centers and major work streams including inventory investment. Some of the examples depend on the key Lean concepts of process simplification, waste reduction, work standardization, and mistake proofing. Other examples use Six Sigma methods and employ its phased problem-solving methodology called Define, Measure, Analyze, Improve, and Control (DMAIC). Also, many of the examples require simultaneous use of both methodologies to make the necessary process improvements, hence the term "Lean Six Sigma."

Improving Line-Item Availability

Problems with line-item availability are found through customer complaints or a baseline analysis of the line-item error rates using dock audits within the distribution center. Building the business case for this type of Lean Six Sigma improvement project is not usually difficult in most organizations because this is often a well-documented problem.

Line-item availability issues occur when inventory is not available for order picking or if manual operations within the distribution center impacting the inventory replenishment process have not been sufficiently standardized or mistake-proofed. In this sense, poor line-item availability is a high-level symptom of lower-level root causes. Poor line-item availability increases operational costs and degrades customer service levels. Increases in operational costs are seen as incremental freight expenses and higher direct labor costs due to handling the back-order shipments necessary to fill the original order.

To investigate the specific reasons why orders have missing or incorrect line items, the Lean Six Sigma improvement team uses dock audits and statistical sampling or orders waiting to be shipped to the customer. These reasons are categorized and analyzed by the improvement team using statistical tools to identify the root causes for the problem. The root causes may be the result of process break-downs in internal distribution center operations or due to external factors. Examples of internal process breakdowns associated with breakdowns in the ordering process include not correctly capturing order information from the customer and inefficient scheduling of the work relative to order demand, resulting in expedited orders. Additional examples of internal operational breakdowns causing inventory not to be available for order picking include not ordering correct inventory items or quantities from suppliers, not placing the inventory in the correct stocking location, or losing and damaging inventory. Internal breakdowns are also caused by lack of standard-ization, poor training, and lack of mistake proofing resulting in incor-rect line items or quantities. External causes include poor forecasts, poor on-time delivery by suppliers, orders that exceed the distribu-tion center's system capacity, and placing the wrong demand on the distribution center.

Lean Six Sigma tools and methods can be used to analyze poor line-item availability and the backorder and other operational prob-lems this situation creates. As an example, control charts can be used to measure the line-item error rates over time to set the line-item accuracy performance baseline. Capability analysis can help determine improvement targets for the team. In the dock auditing process, measurement system analysis (MSA) can help ensure that the auditing process is accurate and precise for subsequent data col-lection and analysis. After the MSA evaluation, Pareto analysis can help break the collected data into categories by their frequency of occurrence. This will help focus the project's problem statement on the major causes for line-item errors. Subsequent statistical analysis can help the improvement team compare error rates between work-ers and shifts and other demographic factors to continue the root cause analysis.

In the improve and control phases of the project, Lean tools and methods will become important as the team begins to identify the countermeasures necessary to eliminate the root causes of poor line-item availability. The improvement team should make the necessary process changes to sustain the process improvements over time. As an example, as the Lean Six Sigma team begins to transition the improved process back to the process owner and local work team, process improvements including process simplification, application of 5-S techniques, and mistake-proofing methods will help ensure permanent control of the modified process.

Reducing Distribution Center Schedule Changes

Schedule changes occur in distribution centers for a myriad of reasons. But, regardless of the reason, the existence of schedule changes increases the operational costs of a distribution center and may lower customer service levels. There are many possible reasons for schedule changes. Internal reasons include poor supervision of employees, especially over different working shifts; an inadequate range of employee skills due to inadequate training; and the failure of employees to report to work due to sickness or other reasons. Some external reasons may include demand that exceeds the capacity of the distribution center to meet its original schedule commitment, unexpected customer orders being placed on the distribution center within agreed-upon order lead-time, and poor on-time supplier delivery.

Reducing unexpected schedule changes begins by documenting the incident rates by line-item type, customer, and other related information. This requires verification of the accuracy of the historical record using basic measurement system analysis (MSA) techniques. Once the MSA is completed, control charts will be helpful in establishing the process baseline, and capability analyses can be used to identify performance gaps relative to improvement goals and targets. The collected data is analyzed using basic quality tools such as Pareto analysis, histograms, and other simple tools to further focus the project on one or more major categories associated with scheduling issues. Depending on the high-level issues that are identified in the baseline analysis, either Lean or Six Sigma tools may be used to identify and eliminate the root causes of the problem. Given that schedule changes may also be caused by organizational functions external to the distribution center, it is important to use a fact-based problem-solving methodology to gain support from the organization to identify and eliminate root causes external to the distribution center.

Reducing Distribution Center Overtime Expense

Overtime expense is a significant cause of margin erosion in a distribution center. The business benefits obtained through reductions of overtime expense are direct cost reductions as well as other less quantifiable benefits. There are times when overtime expense may actually be planned into the system in lieu of hiring full-time employees or because of limited capacity; however, if overtime has not been planned into the system, then the root causes for its existence must be discovered and eliminated by the Lean Six Sigma improvement team. As overtime is broken into its major categories using Pareto analysis or other analytical tools, Lean Six Sigma projects for one or more of the categories will be identified by the improvement team. Some internal causes of overtime expense may be poor employee training, schedule changes, poor on-time supplier delivery, issues with inventory availability, poor forecasting, damaged product, and inefficient operational procedures. External causes for overtime expense may include work that was not originally planned into the schedule. In summary, almost anything that lowers operational efficiency can increase overtime expense. As a result, Lean Six Sigma projects are usually applied with great success to the reduction of overtime expense.

To initiate the improvement project, the Lean Six Sigma team works backward from the overtime transaction summary to determine the high-level reasons for the overtime expense. As part of this improvement process, the team verifies the accuracy of the overtime expense data by major category. One or more of the major categories are then chosen for detailed analysis. The second major task of the improvement team is to verify the accuracy of the baseline overtime expense data. After verifying its accuracy, the team can begin data collection and analysis. Once the data analysis has been completed, the project is focused on one or more of the major reasons for the overtime expense until their root causes are eliminated and the improved process is transitioned back to the process owner and local work team.

Reducing Logistic Transfer Costs

In logistic systems there are usually several interacting distribution centers. Some of these centers may act as central depositories for low-volume, specialized, or expensive items. In other systems, distribution centers may be located next to unique manufacturing sources or major suppliers.

In any of these situations, the distribution center may stock items and supply to other distribution centers within the network. The reason for stocking inventory items centrally is to reduce overall inventory investment as shown in Figures 7.4 and 7.5. These centrally inventoried items are periodically shipped to other distribution centers based on system forecasts of expected demand.

Given the complexity of these systems, there are many ways in which this process can break down. Typically, process breakdowns are caused by poor system forecasts, inefficient loading of trailers, and damaged and lost materials and products, as well as other factors. To eliminate these operational problems, the Lean Six Sigma improvement team must completely analyze the processes associated with the product transfer process. The analysis begins with verification of the process baseline using measurement system analysis. Potential measurement problems may exist because transfer costs are aggregated at many steps within the system. Part of this analysis work includes collection and analysis of the transaction history for every trailer shipped during the time period of interest. In addition to statistical analysis of the available data, creating a value stream map (VSM) and reviewing information technology (IT) logic flows will also be useful in identification of root causes of the problem. The analysis of this type of problem normally requires simple analytical tools ranging from graphical and statistical analyses to evaluation of the waste elements found in the VSM. In the control phase of the project, the application of 5-S techniques within the distribution center and minor modifications to the IT system logic flow will normally result in major operational improvements and business benefits.

Reducing Premium Freight Costs

Although premium freight costs are caused by several situations, three in particular have been found to significantly drive premium freight expense. The first situation is the routine shipment of packages overnight or second-day delivery without any justification from the customer. In other words, people ship packages overnight through personal habit without regard to the associated costs. The second situation occurs when products or services are promised to a customer without regard for the system's capability to deliver according to that promise. As an example, salespeople may promise an order will be delivered in three business days, but current transportation systems cannot deliver orders in less than five business days. The third situation is the failure of the organization to have the correct inventory available to meet customer demand at a particular location within its distribution system. In each case, the Lean Six Sigma team must begin its data collection efforts at the source of the problem, for example, the customer complaints, and analyze the shipment transaction history.

The Lean Six Sigma improvement team begins the analysis by identifying the reasons for the overnight shipment of packages. This is done by separating shipments that the customer asked to have expedited from those not requiring the expedited shipping service. The incident rate of the latter group can usually be significantly decreased to lower expedited transportation expense. If customers have been promised service levels that are beyond system capability to provide on an ongoing basis, the Lean Six Sigma team can analyze the system to determine where additional system capability is required to provide the higher service levels. Also, on a short-term basis the team could temporarily stage inventory closer to the customer to mitigate the problem. While inventory staging is not an optimum solution, the holding costs of the inventory typically represent between 10% to 30% of the premium freight expense, so this short-term solution has justification from both a customer and business viewpoint. However, the proper course of action is for the Lean Six Sigma improvement team to eliminate the operational breakdowns associated with building inaccurate or incomplete orders, poor inventory availability, and related issues, which may be the causes of expedited shipments and their associated premium freight expense.

Optimizing Forecasting Parameters to Reduce Safety-Stock Levels

Forecasting models use the historical demand for an item by location to estimate demand out into future time periods. Most organizations use time-series models to make their monthly sales forecasts. However, the accuracy of forecasting models can degrade for a variety of reasons. Here are some examples: customer demand is incorrectly assigned to an inventory location; incorrect specification of modeling parameters; using the historical demand histories, which are too long or short; or basing an item forecast on its shipment rather than demand history. To compound the problem, the forecasting analysts may not have been properly trained to build accurate forecasting models or may not have the time to build the best model for the particular item and location. Lean Six Sigma tools and methods can be very helpful in these situations.

As an example, experimental design methods can be used to optimize modeling parameters to improve the accuracy of the forecasting model.

Experimental designs can be thought of as a very efficient way to conduct a sensitivity analysis of the impact of changing parameter levels on the forecasting error metric, that is, mean absolute percentage error (MAPE) or root mean standard deviation (RMSD), that is, $Y = f(X)$. As an example, Winter's method is a commonly used time-series forecasting model using three parameters to build the seasonal forecasting model. The first parameter is used to estimate the average level of the time series, the second parameter estimates the time series trend (if any), and the third parameter is used to estimate the seasonal pattern of the time series. The continuous range of the three parameters is 0 to 1. In the absence of efficient experimental designs, the forecasting analyst would evaluate different levels of the each parameter. Eventually, the best combination of parameter levels would be selected using the minimum forecast error (highest accuracy) as measured by one or more error statistics. These error statistics were defined in Chapter 3 and Figure 3.15. However, evaluating parameter levels at random is an inefficient way to obtain the best combination of levels because there are many possible combinations. As an example, if each parameter were evaluated at 10 levels, the total number of parameter sets would be $10 \times 10 \times 10 = 1,000$. However, a Lean Six Sigma black belt would know, given three continuous and linear variables, that just eight experiments would be required to complete the analysis if the relationship was linear or a slightly larger number of evaluations if the relationship was curvilinear. Although most modern forecasting systems can automatically optimize the parameter sets, forecasting analysts may interfere with the system optimization by manually setting parameter levels to nonoptimum levels.

In addition to experimental design methods, the Lean Six Sigma black belt has a powerful tool set to fully evaluate the best way to build the forecasting model. It should be understood there are other considerations in building a correct forecasting model. These include the length of the historical time series that will be incorporated into the model as well as the general form of the model relative to its position within its life cycle. From a Lean perspective, value-stream-mapping the forecasting process relative to the flow of customer demand information and applying 5-S techniques will ensure standardization and higher forecasting accuracy.

Reducing Excess and Obsolete Inventory

Excess and obsolete inventory occur for many reasons. Some of the reasons may be outside the control of the organization or due to past operational practices, that is, legacy issues. In many Lean Six Sigma projects, the improvement team makes major reductions in excess and obsolete inventory through selling the inventory back to suppliers or customers; but then it has to move on to other pressing problems (driven by cost-savings goals and objectives) without eliminating all the root causes of the excess and obsolete problem. Not following through to eliminate the root causes for the problem is a mistake since excess and obsolete inventory will quickly return to their former levels due to the chronic nature of the root causes of the problem. These problems are also exacerbated by system complexities and politics within an organization. To the greatest extent possible, the improvement team should attack and eliminate the root causes for the excess and obsolete inventory problem.

Some of the major causes of excess and obsolete inventory are shown in Figure 6.7. These include new products that do not sell due to poor marketing research; forecasting errors; the purchase of large lot sizes, which in some situations, represent multiple years of supply; and design changes that obsolete items before their inventories have been sold. In the latter case, poor organizational communication and engineering change control systems are responsible for the process breakdowns causing excess and obsolete inventory. Excess and obsolete inventory can build up very quickly when lot sizes represent multiples of demand quantities. As an example, in some organizations, lot sizes represent 10 or more years of estimated use requirements. If large lot sizes or similar supplier problems exist, the Lean Six Sigma team should bring the supplier onto their team to help identify and eliminate the root causes for the large incoming lots. The other major contributor to excess and obsolete inventory is poor new-product planning and design. In these situations, marketing, sales and design engineering must be involved in the project.

Regardless of the high-level reasons for the excess and obsolete inventory problem, the Lean Six Sigma team must start with a transaction summary of inventoried items. This information should be verified by an onsite physical count of the inventoried items to ensure that the management reports are accurate. After verification of the measurement system, the improvement team begins data collection. The purpose of data collection is to capture information relative to the identification of the inventoried items. This information includes suppliers of the item, where it is located, when it was purchased, customers who bought the item, and other relevant facts that will help the improvement team in its root cause investigation of the problem.

A value stream map should also be constructed by the team since Lean tools and methods are particularly important in the execution of these projects. Application of 5-S techniques will become particularly important in the improve and control phases of the project. The project direction might change if the root cause analysis points to problems in estimating demand of new products. In these situations, Six Sigma tools and methods will be also be very useful in improving the organization's marketing research capabilities since the quantitative modeling focus of the Six Sigma tool kit is particularly useful in planning, collecting, and analyzing marketing research data. Excess and obsolete inventory problems are good examples of why Lean Six Sigma "belts" must understand Lean and Six Sigma toolsets as well as the tools, techniques, and methods of proper supply chain design, management and control.

Improving Inventory Turns for Slow-Moving Products

Low inventory turns is a chronic problem in many organizations. There are many reasons for the low turns, but the first step in problem resolution is separating the high-level turns metric into current inventory turns and legacy inventory turns. At a high level the causes can be attributed to lead-time and demand issues. The Lean Six Sigma improvement team delayers these high-level causes into lower-level root causes such as lot size, poor forecasting, and other operational breakdowns, as discussed earlier in this chapter. Countermeasures and improvement strategies are developed based on the root cause analyses.

In the case of legacy inventory, Figure 6.8 provides several commonly accepted strategies to improve legacy turns, including selling the inventory into distribution channels that will not impact current sales, donating the inventory for a tax deduction, or just disposing of it and incurring a financial loss. There are also other forms of the turn's metric. As an example, in addition to current and legacy turns ratios, another ratio is important. If inventory is purchased in advance due to external or internal factors, which impact available capacity, the inventory turns ratio could be modified as a "forward-looking turns ratio." This turn's ratio matches current inventory investment to future cost-of-goods-sold (COGS). It is important to understand that the inventory investment in materials or products purchased in advance has been planned by management for specific contingencies which impact capacity.

Reducing Damaged Product Incident Rates

Given the number of steps and operations that exist even in the simplest of supply chains, the probability of product damage is high. Product can become damaged at many steps in the order-handling process. As an example, product could arrive damaged from a supplier, but the damage might not be found at incoming receiving. Damage could also occur to a product while storing or removing it from an inventory storage location. Product could also be damaged while loading products onto trucks for shipment to the customer. Finally, the carrier could damage the product during transit to the customer.

The types of product damage also vary by operation within the supply chain. Products can be crushed, torn apart, or pilfered, or become contaminated or damaged due to variety of environmental conditions including light, heat, and moisture. Because of the myriad of possible damage causes, the Lean Six Sigma improvement team must work backward from the high-level damage incident reports down to lower operational levels to understand the root causes for the damage problem. Lower-level root causes may include poor packaging design, poor stacking of product for storage or during transit, failure to protect the product from warehouse equipment such as forklifts and failure to protect the product from environmental damage either within the distribution center or where it is loaded in trailers for shipment. Since the possible root causes of damage are so different due to the complexity of the distribution system, the Lean Six Sigma team must be very careful that in addition to reviewing management reports, they also "walk the process" to really understand the conditions under which damage occurs. Product damage projects normally require application of 5-S techniques in the improve and control phases of the project unless the root cause is found to be due to poor packaging design or factors external to the distribution center.

Improving BOM Accuracy

Organizations rely on accurate bills of material (BOMs) to build their products. Design engineering creates BOMs when they specify the components and materials necessary to build a product including their hierarchical build sequence. The specifics of BOMs were discussed in Chapter 5 and Figure 5.9. Incorrect BOMs result in incorrect parts and materials being ordered and higher-level assemblies and products being incorrectly built. However, given the complexity of the design process with its many manual operations and design changes, it is very probable BOM errors will occur over time. BOM problems are noticed when excess or obsolete inventory is built or schedule changes occur due to missing or incorrect materials and components. The Lean Six Sigma team should analyze the specific transaction history associated with process breakdowns and work backward into the process to identify and eliminate the root cause of BOM inaccuracies.

Improving Cycle-Counting Accuracy

Ensuring the accuracy of the cycle-counting system is very important for an organization. This is because cycle counts are an important measurement system for almost every inventory improvement project relative to inventory quantity and location. For this reason, the improvement team must analyze the cycle-counting system to ensure that an accurate baseline can be established for their improvement project. In some cases, critical elements of the cycle-counting system must be fixed prior to beginning the inventory project. The best way to begin an analysis of the cycle counting system is to "walk the process" and create a value stream map detailing the flow of materials, people and information through the system. A transaction summary should also be created for every item and location to determine their baseline accuracy levels. This will show where the process breakdowns are occurring.

In Chapter 7, the major steps in creating an effective cycle-counting system were shown in Table 7.3. Although many organizations must cycle-count due to breakdowns within their complex processes, the best cycle count is no cycle count. As an example, in a "Leaned-out" system, inventory is very easy to count since it is controlled using Kanban quantities and assigned to visibly marked locations. This situation effectively minimizes or eliminates the need to cycle-count the work-in-process (WIP) materials. However, in distribution system environments, inventory may exist over several thousand products and locations. In these large and complicated inventory systems, transactional errors occur over time due to shrinkage (damage, loss, and theft of product), misplacement of the inventory within the distribution, or other reasons.

There are many efficient cycle-counting strategies that can be employed within complicated inventory systems. If the distribution center has a warehouse management system, then the cycle-counting system can automatically schedule inventory counts when a picker is near a location with zero or very low inventory levels. This strategy decreases travel time to the inventory location as well as the actual time to cycle count the inventory. In addition, statistical sampling techniques can be employed to count a subset of inventory locations over time to measure inventory valuation and location accuracies. The advantage of statistically based periodic cycle counting is that problems can be continuously investigated as they are encountered by the auditor. This allows the root causes for inventory discrepancies to be systematically eliminated from the process to improve inventory accuracy. In the improve and control phases of the project, application of Lean, 5-S, and mistake-proofing techniques will help improve the cycle-counting system and sustain accuracy improvements.

Reducing Returned-Product Incident Rates

Product is returned to distribution centers for many reasons. These can include reasonable reasons such as sales policies guaranteeing the customer the right to return a percentage of product purchases each year or product that is defective or damaged. Lean Six Sigma projects can be applied to reduce the incident rate of these return reasons based on root cause analysis of the problem. However, customers may return products for other reasons that may be contrary to the original sales agreement or agreed-upon return policies. As an example, a customer may over-order products and then attempt to return them to the supplier when they do not sell. Also, customers have been known to return competitive product or product that is obsolete. Analysis of these types of returns can also form the basis of Lean Six Sigma projects.

Reducing the incident rates of returned product will also lower operational costs. Additional benefits include less required floor space and simplified return operations. Because the root causes of product returns are so varied, the Lean Six Sigma improvement team must carefully measure the extent of the product return problem through their data collection and analysis. Sales must also be included in these projects since customers are an integral part of the problem statement, analysis and final resolution. This is especially true when the final solution directly impacts the customer or the organization's sales policies and procedures.

Reducing Kitting Inventory

Some distribution centers build and package service kits for external customers. Kit building is usually a very complex set of tasks depending on the product design as reflected in its BOM structure. In particular, in distribution centers without pull systems or having standardized and stable operations, work-in-process (WIP) inventory can build up very rapidly within the distribution center. In worst-case situations, the kitting area becomes littered with half-assembled kits. The most common reasons for kitting problems include poor forecasts, scheduling changes, non-availability of materials and components, or complicating factors such as missing or incorrect components due to assembly errors.

There are many potential process breakdowns in kitting operations due to the complexity of the systems relative to the number of materials and components as well as the extensive use of manual operations. The best way to attack inventory and process breakdowns in kitting operations is through application of Lean tools and methods. These include establishment of a pull demand system and application of 5-S and mistake-proofing methods throughout the operation. At a higher level, to the extent the organization can modularize and simplify the kit BOM, kitting quality and on-time delivery will be higher and per-unit costs will be lower. Application of design for manufacturing (DFM) principles can also have a very positive impact in kitting environments.

Reducing Work Batching

Chapter 4 showed that batching of work increases lead-time, which impacts the total order cycle time. Also, as lead-time lengthens, system variation increases causing additional operational problems and defects. One good way to reduce or eliminate the batching of work is to move the process toward a transfer batch system. A transfer batch system moves items through the process as soon as they are ready to be worked on by downstream operations. Transfer batches were discussed in Chapter 4 and shown in Figure 4.7. However, moving to a transfer batch system requires making changes to process designs. Design changes may be easy or difficult to implement depending on the specific industry and its technological base. However, there are some process changes that the Lean Six Sigma team can make almost immediately to improve the process flow. These improvements include changing the workflow to a U-shaped work-cell design, implementing at least a local pull system, and applying basic 5-S and mistake-proofing strategies to the local work area. Reducing work batching will reduce lead-time, improve quality, and reduce the costs associated with rework and scrap. These types of projects rely more on Lean rather than Six Sigma tools and methods.

Reducing Shipping Errors

Shipping errors arise for a number of reasons including customer issues, process breakdowns within the distribution center, and problems with carriers transporting products between facilities and to the external customer. Internal process breakdowns include generating incorrect picking lists, incorrect picking of products resulting in line-item shortages or overages, and damage done to the product because of poor handling by the distribution center. External process issues occur when carriers pick up the order at the supplier's outbound loading dock and transport it to their central "break bulk" operations. In these centralized operations product could be damaged or lost as it is taken off the trailer and loaded onto other trailers for transport to the freight terminal closest to the external customer. At the customer, product could be incorrectly counted, damaged, or lost.

Verification of the shipping error baseline by error category will help provide a focus to the project to reduce the shipping error incident rate. This is done using both Lean and Six Sigma tools to design data collection strategies and eventually analyze the collected data. Once the data has been analyzed and the high-level root causes of the problem categorized and quantified, the improvement project can be focused on one major root cause for the shipping errors. Since shipping errors can occur almost anywhere within the distribution system, it is important that the Lean Six Sigma improvement team eventually focus the project on one or more closely related areas. These could include employee training, incorrect work standards, and poor environmental conditions such as excessive noise or low lighting conditions, or perhaps the people cannot read or write proficiently. Additional problems can occur when products having similar numbers are stocked near each other. As an example, the product numbers "78" and "87" might be confused by the order pickers. Additional errors may be caused by external factors. External factors might include mistakes in customer orders made by the customer or incorrect translation of customer order requirements by the call center agent. Application of 5-S techniques and mistake proofing will help sustain the process improvements over time.

Improving MRPII System Accuracy

Lean Six Sigma projects may be identified during the development of the supply chain or inventory model. The MRPII system in particular may have a large impact on inventory investment and other supply chain metrics. Chapter 5 described how the MRPII system drives the material requirements planning system and the purchase and material and components required in manufacturing. Inaccuracies in the MRPII system relative to parameter settings such as lot size, lead-time, and others will result in excess inventory and other operational breakdowns within the supply chain. In some Lean Six Sigma projects, data inaccuracies within the MRPII system may become Lean Six Sigma projects.

Data problems may occur if data was incorrectly entered into the system by analysts or if changes are not updated in the system. Since manufacturing schedules, capacity levels, and inventory targets are off-set by these parameters, their accuracy or lack thereof can have a major impact on your supply chain.

Reducing Account Receivable Aging

An efficient account receivable process exhibits low aging of its account receivables. *Aging* refers to how long the invoice has been open. Typical aging categories include 0–30 days, 31–60 days, 61–90 days, and greater than 90 days outstanding. The goal of Lean Six Sigma projects is to lower the aging of the number of accounts and their monetary amounts within each aging category. As an example, if the number of account receivable accounts is shifted toward the 0–30 day category cash flow will be increased and the interest expense necessary to fund the account receivables will decrease. Also, the higher the aging of an account, the lower the probability it will be eventually collected by the organization. In these projects, the business benefits could actually be very significant since highly aged accounts must be written off completely.

In a manner similar to other Lean Six Sigma projects, the improvement team must work backward from account receivable management reports to verify the extent and root causes for the aging problem. After measurement verification of the account receivable aging baseline, the improvement team begins data collection and analysis to focus the project on one or more major root causes for the highly aged accounts. In the improve and control phases of the project, the team will likely implement a combination of standardized work procedures and controls using 5-S techniques and mistake proofing.

Reducing Billing Errors

Because the billing process has numerous manual operations and interfaces between the customer and organization, many organizations have problem with billing errors. Billing errors may slow cash flow and increase direct labor costs because of the rework necessary to sort out the reasons for the errors. Customer satisfaction is also lowered during these complicated investigations. However, the true financial and operational impacts of billing errors on the organization may not appear on management reports. This is because these reports include documented customers complaints, but not errors that may be in the customer's favor or errors received by customers who do not notice or choose to ignore them. Eventually, the organization may lose a portion of its customers due to dissatisfaction with billing errors.

To begin its billing error investigation, the Lean Six Sigma improvement team should conduct a sampling audit on invoices created by the billing department. This audit should baseline the error percentages by type of each billing error, the impacted customers, and other important demographic factors as well as their financial impact on the customer and organization. In parallel, the Lean Six Sigma team should construct a value stream map (VSM) of the billing department to identify rework loops and cross-reference the errors found in their samples to critical parts of the VSM to identify areas of improvement. The improvement team should focus the project's problem statement and its objective on either similar error types or specific errors that have the largest impact on the organization. The Six Sigma tools required to collect and analyze the data should be relatively simple and should be graphical and statistical in nature. Elimination of billing errors normally requires process simplification, 5-S including work standardization, and mistake-proofing manual operations within the billing department.

Reducing Call Average Handling Time (AHT)

Most modern call centers use advanced IT technology to ensure that customers are serviced according to marketing and operational policies. Call centers consist of many operations including management of incoming and outgoing calls, complaint resolution, providing technical advice, and perhaps even customer billing. Also, each customer market segment may have different operational standards relative to average call duration and service level targets, such as answering the call within a specified time period and answering all pertinent questions for a particular market segment. The different operational areas within the call center may also have different work standards reflecting the specific work being done in that area of the call center. As an example, the work standards for complaint resolution versus standardized calls will be different since complaint resolution usually takes a longer average time to resolve than a standardized call.

The interesting fact about call centers is that although they are very sophisticated due to applied technology, they are also very manually intensive. For this reason there are chronic process breakdowns in these systems resulting in lower productivity (as specified by average call duration) and quality (as specified by time to answer the call and providing accurate information based on the customer expectation). The other interesting fact about call centers is they have significant amounts of operational data that can be analyzed to create Lean Six Sigma improvement projects. By using information available from operational reports, the Lean Six Sigma team can verify the project's data has been accurately collected by the automated measurement systems. After the measurement systems have been verified as accurate, Lean Six Sigma projects can be deployed to close performance gaps. As an example, if the average call duration, by agent, exceeds the specific work standard, then an entitlement analysis can be conducted to compare the best agents with those not meeting the standard. This will allow the improvement team to estimate by how much the process can be improved over its current baseline given the current system design. Of course quality metrics must also be analyzed to ensure that agents having high productivity also have high quality levels (service levels).

In addition to statistical analysis, the Lean Six Sigma improvement team must also walk the process. As they walk the process, the team should analyze all aspects of the call center agent and customer interaction by major functional group within the call center. This measurement activity includes quantifying time, cost, and quality elements associated with the agent and customer interaction. Achieving an entitlement level of performance is accomplished through application of 5-S and mistake-proofing techniques to the work area. These process improvements may include training agents to use standardized scripts for each customer market segment and ensuring that agents have the information and tools necessary to maintain their productivity and quality standards over time. This is a classic 5-S and mistake-proofing application.

Reducing Scrap and Rework

Almost every process creates scrap and rework and to the extent the supply chain wastes labor and material, especially in its manufacturing, packaging, and kitting operations, there will be documented expenses associated with scrap and rework. Unlike manufacturing operations, which are heavily dependent on machines for assembly operations, distribution centers rely to a great extent on manual operations. For this reason, the data collection, analysis, and eventual solutions to each problem will require application of Lean tools and techniques rather than Six Sigma statistical methods. Like most processes, not all the scrap and rework costs may be captured by the measurement systems, or these systems may be misallocating the scrap and rework costs. For these reasons, the improvement team must walk the process and implement data collection systems to ensure that scrap and rework baselines are accurate by defect category. These actions should enable the team to identify improvement projects based on defects having a significant operational or financial impact on the organization. As the improvement team works backward from the defect data, it will be able to apply the appropriate Lean and Six Sigma tools necessary to identify the root causes of the problem. Improvement and control will require application of 5-S and related techniques in direct proportion to the number of manual operations within the process.

Improving On-Time Delivery

On-time delivery is a chronic problem in most organizations. The reasons vary, but all require accurate measurement of the on-time delivery metric to create an accurate metric for the supplier and customer. Since the on-time delivery metric is calculated using information gathered from several places throughout the supply chain, the team must first verify its accuracy prior to beginning the project. There may be instances in which the supplier disputes its on-time delivery record. To resolve the discrepancy, the improvement team must create a value stream map of the customer and supplier interaction. This value stream map would include metrics related to the initial release of the order to the supplier and all operations up to receipt of the order by the customer.

As an example, an order cycle depends on the initial-lead-time definitions agreed to by the customer and supplier during purchasing negotiations. It is important for both sides to have the same lead-time definition. Also, in complex supply chains, it may be difficult to keep the lead-time promises if the process has not been broken down into its time elements to allow people to understand what would be a realistic lead-time promise in the first place. As an example, in a typical order cycle, the customer's MPS sends an order to the supplier via electronic data interchange (EDI) with an expected delivery date based on the agreed-upon lead-time. The supplier builds the product but may place it on the customer's truck or designated carrier for delivery to the customer's distribution center rather than its own truck. In some supply chains, the customer's system may not officially receive the product until it is scanned in at the item location within its distribution center.

It is at this point in time that the on-time delivery metric is calculated by the customer's system. There is obvious ambiguity in this on-time delivery measurement system. However, if the Lean Six Sigma improvement team breaks the process into its time elements, it will be able to realistically set the supplier on-time delivery metric based on the actual design of the supply chain. This will increase the baseline accuracy of the on-time delivery metric and allow the improvement team to move the project through its measure phase. Later on during the pilot study, the team will be able to measure changes in the on-time delivery metric as the levels of the KPIVs change. Improvement and control will require application of 5-S and related techniques in direct proportion to the number of manual operations within the process. in conclusion, on-time delivery requires a firm grasp of the Six Sigma concept of measurement system accuracy (MSA) but relies for its solution on Lean tools and techniques.

Reducing Employee Hiring and Retention Costs

High employee turnover can significantly increase hiring and retention costs for a large organization. Annualized hiring and retention costs can easily exceed millions of dollars per year for some organizations. In most organizations, baseline hiring and retention costs are usually well estimated by human resources, but the data collection analysis is not designed toward proactive elimination of the problem. In this type of project, it is very important to verify that the hiring standards are accurate for the job description and that employees are hired based on proven success criteria. However, more is required to reduce this chronic problem. The normal approach in this type of project is to measure each *cohort* of employees (employees hired at the same time) and follow their retention rates over time, while accurately recording their reasons for leaving the organization. These interviews must be very carefully designed to ensure that accurate information is obtained from the exiting employees. These projects normally have significant financial and personal benefits for the organization.

Another application of the employee retention problem is applying Lean Six Sigma methods up-front in the hiring process to ensure that the organization hires the right employee in the first place. The key concept in this situation is to ensure that the organization has a clear and standardized approach to the identification of potential employees and the right combination of advertising and incentives to attract these employees. Voice-of-the-customer (VOC) survey techniques are very useful in identifying ideal employees as well as the effectiveness of advertising for them. Lean Six Sigma improvement projects become particularly important in situations in which the potential number of employees is small due to either the geographical location of the facility or competing employers.

An extension of the employee hiring and retention cost problem is the entire concept of employee benefits, management, and promotion over time. Standardized Lean Six Sigma tools and methods are used in situations where the current process has broken down. However, in cases where the process must be redesigned to meet new HR require-ments, DFSS tools and methods might be more useful. Complicating deployment of Lean Six Sigma projects into the HR area is the legal and regulatory environment in which these problems are identified and analyzed and the fact that process improvements must be fully integrated into the HR systems.

Reducing Lost-Time Accidents

Reducing lost-time accidents is a major strategic goal of every organiza-tion. In these systems very accurate data is usually maintained on every lost time incident. This includes the causes of the accident, including all relevant demographic information, as well as the individual and organiza-tional impact. To the extent the Lean Six Sigma team can identify chronic problems within their process as opposed to isolated incidents. Lean Six Sigma projects can be created to reduce lost time and accident incident rates. However, in situations in which incident rates are very low and isolated, it becomes very difficult to directly apply Lean Six Sigma tools and methods. In these low incident environments it is normally better to bring in *health, safety, and environmental (HS&E)* experts to redesign the HS&E system to attain new operational baselines. A design-for-Six-Sigma (DFSS) approach may also yield significant operational and business ben-efits for the organization.

However, there are many situations in which the application of Lean Six Sigma methods to these HS&E problems such as lost-time accidents can yield significant improvements in safety as well as employee satis-faction. As an example, Six Sigma tools can be applied up-front in the project definition phase to identify those areas having the most impact on the organization relative to incident rate and costs. As the team col-lects data for analysis, it will usually find numerous opportunities to apply Lean tools and methods. These methods include 5-S applications within the local work area as well as application of rigorous mistake-proofing strategies to prevent further accidents. Interestingly, in some systems such as retail stores, the customer may also be injured while on the premises. Deployment of Lean Six Sigma projects takes on extreme urgency in these situations.

Reducing Facility Maintenance and Energy Costs

Because most organizations have similar maintenance, facility, and energy systems, optimized solutions to facility, maintenance, and energy management have been implemented across many organizations. However, to the extent these systems are poorly designed or are significantly impacted by manual operations, there will be opportunities to develop Lean Six Sigma projects. The advantage of these types of projects is that solutions may be readily available since the problems are commonplace and general solutions have probably been developed for a variety of similar situations, even though the root causes of your particular situation may be unique.

As an example, in buildings having automated lighting controls that shut off the lights as people leave an area, there may still be opportunity to improve the system if situations occur in which the automatic lights can be overridden by people leaving an area, or if many lights are connected together on the same sensor and must be turned on simultaneously, or if areas have too much or too little light. The improvement team should start the investigation by measuring the light intensity across the buildings to determine whether the lighting systems meet or exceed OSHA standards. Or perhaps special causes of high energy usage will be identified, like a janitor who turns on an entire floor of lights to read a newspaper. After the team identifies where the energy usage is highest; but, not being utilized efficiently, plans can be developed to optimize the lighting system.

Organizations may or may not have a preventive maintenance (PM) program. Lack of a maintenance program will certainly guarantee unexpected process breakdowns; but, even if a PM program exists, there may be issues relative to high operational costs and unexpected equipment breakdowns that may be unique to your system. PM systems depend on accurate information regarding hours of machine or equipment usage; but, the maintenance logs used to record this information may be in error due to poor training or inadequate work procedures. Also, environmental conditions may contribute to maintenance problems. As an example, perhaps forklift batteries are not being sufficiently charged by the workers resulting in their unavailability to move materials. Or perhaps maintenance jobs were incorrectly prioritized or the maintenance staffing levels have not been optimized, resulting in long lead-times for equipment service. These process breakdowns as well as others can form the basis of very good Lean Six Sigma improvement projects.

Energy costs also form the basis for good Lean Six Sigma projects. Analyzing where the greatest energy usage is occurring within the facility using Six Sigma tools and methods can quickly identify improvement opportunities. As an example, correlating energy usage to billing rates by power companies can help to shift energy demand within the facility across time to reduce overall costs for the organization. There will usually be many other opportunities to improve energy usage. The key basis of these improvement projects is the establishment of operational baselines as well as effective data collection and analysis resulting in sustainable process improvements.

Reducing Emergency Replenishment

Emergency replenishment is a rework operation caused when inventory is unexpectedly depleted at an item's inventory stocking location. There could be many reasons for emergency replenishment including inventory not being available to place at the stocking location, not anticipating customer demand for the item in the first place, the arrival of unexpected customer orders within established lead-times, or incorrect sizing of the stocking location, that is, storage racking. There could be other issues. Emergency replenishment is easy to track since people are usually assigned full-time to this task. The Lean Six Sigma team should have a good performance baseline (hours required for emergency replenishment) against which to measure operational improvements. The baseline information can be obtained by analyzing the replenishment transaction history.

However, like every Lean Six Sigma project, the replenishment data must be verified prior to finalizing the project charter to build a financial justification for the project. Once the operational baseline has been established, the team should develop a data collection and analysis plan. Since the data is analyzed, the root causes of the problem may be complex and range from external customer issues to inadequate storage racking within the distribution center. Inadequate storage racking requires a high restocking frequency. The team should work through the project's root cause analysis and place countermeasures against the specific causes of the emergency replenishment problem.

Improving Line-Item Picking Productivity

Many variables impact line-item productivity. These include the warehouse layout, worker training as well as the tools and methods workers are provided with to pick product. Warehouse layout has an important impact on picking productivity since any order picking route other than the optimum will take a longer cycle time. Also, if the worker must travel to pick up information, forms and other things their productivity will decrease. Also, poorly trained or distracted workers will make mistakes in filling orders that must be corrected later, further lowering productivity. Automating portions of the picking process such as information gathering, retrieval, and recording will increase worker productivity and quality since manual work tasks will be eliminated from the process.

The best way to initiate a Lean Six Sigma project to increase productivity or quality is to value-stream-map the picking process for each picking zone to determine the time elements associated with key work tasks in each zone. Analyzing the non-value-adding (NVA) time elements in the value stream map will provide opportunities to improve operational efficiencies and quality through elimination of unessential tasks. Applying 5-S and mistake-proofing strategies to those tasks which remain, will help standardize and control the process.

Summary

In addition to the information gained from management reports and value stream mapping, building simple Excel models will help the Lean Six Sigma team identify Lean Six Sigma improvement projects. This will be especially true if the distribution system has a warehouse management system or a similarly advanced technological infrastructure. Accessing operational and financial data using these automated systems is usually not difficult and may allow the improvement team to immediately drive down to major root causes of the high-level problems. However, the accuracy of the data must be verified since it is collected across many areas of the supply chain using a combination of manual and automated systems. Having the data in one place (after merging disparate files) shows relationships between input and output variables. The analysis also identifies operational performance gaps and financial benefits that are the justification for Lean Six Sigma improvement projects which close the performance gaps. The summarized results from the model also act as a communication vehicle throughout the supply chain since everyone can see the project's status and root cause analysis. While the inventory example presented in this chapter is a very simplistic model, more realistic applications to other work streams can be easily developed by the improvement team. In fact, off-the shelf software can be purchased and modified to build work stream models. This chapter also discussed more than 25 Lean Six Sigma supply chain applications. The specific root causes of your organization's process breakdowns may differ from those mentioned in this chapter; but, the general solution methodologies will more likely than not be similar.

Appendix I

Important Supply Chain Metrics

Twelve Key Metrics

I. Inventory Investment

An organization is required to invest in inventory to maintain customer service levels because of limited manufacturing capacity. In other words, in these systems, inventory is built in advance since it cannot be built on demand. However, high cash investment in inventory does not guarantee that an organization will achieve its customer per-unit service levels. Inaccurate demand estimation may result in some items having too much inventory and others too little. Inventory investment should be calculated by item and location based on the item's lead-time and expected demand. The major theme of this book is that organizations should manage inventory scientifically to set optimum inventory and safety-stock levels by item and location. Also, it is important to identify slower-moving items to reduce excess and obsolete inventory. These concepts are also discussed in Chapter 10 using an applied example.

Another important consideration when calculating the optimum inventory investment level for an item is inventory type or classification As an example, corporations that have deployed Lean systems may have low raw material and WIP inventories, but still have high finished goods inventories. Also, the service level targets will vary by inventory classification with raw material and WIP inventory typically having a higher service target than finished goods inventory. It should be also noted that high inventory investment by itself does not indicate there is an investment problem. This is because inventory investment is correlated to sales of an item. In other words, the higher the sales of an item, the higher will be the required inventory investment. For this reason inventory investment must be evaluated in conjunction with an inventory turns ratio which shows the efficiency of inventory utilization relative to the sales level it must support.

2. Profit and Loss (P/L) Expense

A profit and loss (P/L) statement summarizes an organization's revenues and expenses for a specific time period. Expenses on the P/L are costs incurred in the course of business operations. Important P/L expenses, relative to inventory investment, are excess and obsolete inventory write-offs and the interest expense for carrying inventory as well as the expenses necessary to maintain inventory at various stocking locations. There are also P/L expenses caused by poor supply chain practices. These include premium freight, direct labor, overtime, shipping costs, and related expenses. Analysis of P/L expenses provides an excellent financial justification for Lean Six Sigma projects including expense reductions or revenue and gross margin increases.

3. Inventory Efficiency (Turns Ratio)

The number of times that an inventory "turns over" during a year is calculated as the ratio of "annualized cost-of-goods sold (COGS)" divided by monthly "average inventory investment." If customer service levels are being met, high inventory turn ratios indicates inventory management practices are effective at least relative to the current turn's ratio target. Companies which must build inventory in advance, because of anticipated sales promotions, labor issues, or other factors, may adjust the standard inventory turns ratio to provide a "forward-looking turns ratio." Forward-looking turns ratios match on-hand inventory to future time periods in which inventory will be converted for sale. A third version of the inventory turns ratio is obtained by subtracting out excess and obsolete inventory. This adjusted turns ratio allows the organization to track its inventory improvement activities without the distraction caused by its legacy investment. In summary, organizations should use more than one turn's ratio to provide focus for its Lean Six Sigma process improvement efforts.

Although inventory turns ratio benchmark data often exists, the most effective way to estimate inventory targets is developing an internal benchmark based on lead-time, demand variation, and target unit service level. This information is used to estimate the optimum inventory investment by item and location. This benchmark calculation provides a reasonable inventory turns target based on current process performance. Potential improvements in the inventory turns ratio can be estimated by calculating reasonable reductions in lead-time or demand variation, based on current system performance and constraints relative to the products process design. Lead-time and demand variation can also be reduced using Lean Six Sigma methods.

4. On-Time Supplier Delivery

Supplier on-time delivery performance is calculated based on an agreed-upon versus actual delivery time. There could be several variations of the metric. The metric could be stated in terms of on-time percentage (pass or fail). It could also be stated in terms of hours early or late (continuous). The latter performance metric is more useful for process improvement activities. The on-time delivery metric is important because if incoming supplier orders are late, manufacturing schedules may be missed. This results in higher operational costs, but also longer system lead-times. Longer lead-times increase higher inventory investment.

The first step in evaluating on-time delivery issues is to verify that it is being measured correctly. To accurately assess "on-time delivery" performance, organizations must consider capacity allocation, demand quantity, lot size, and lead-time as well as other specific on-time delivery metrics across the system. As an example, a supplier cannot be expected to maintain on-time delivery schedules if its customer demand exceeds allocated capacity levels or the product mix changes within agreed-upon lead-times. On the other hand, the supplier should ensure its operations are standardized so that products are manufactured on schedule to the agreed-upon quality levels. In other words, both sides must keep their contractual promises.

To keep promises, the delivery targets must be correctly specified based on actual system capability. Unfortunately, this is seldom set correctly since the people setting the targets (purchasing and sales) have not been properly trained in value-stream-mapping (VSM) techniques. Recall that VSM breaks a work stream down into its time components. A target delivery date should be developed based on the cumulative lead-time through the system including order generation, manufacturing, transportation to the customer, and receipt by the customer into their distribution center. After calculating the cumulative lead-time, the on-time delivery metric can be calculated and the supplier's performance evaluated against this calculated and realistic target. Inventory investment levels should be set based on the lead-time calculated using actual supplier on-time delivery performance.

5. Forecasting Accuracy

Forecasting accuracy is an important supply chain metric since it estimates future demand and capacity requirements. Also, other decisions are made using these demand estimates. There are many ways to measure forecasting accuracy including several common methods discussed in Chapter 3. Although, many organizations have automated systems that measure forecasting accuracy, they usually do not attempt to understand the reasons for their system's poor accuracy. In contrast, Lean Six Sigma improvement teams use forecasting accuracy metrics to systematically improve supply chain performance. Also, if the root causes of other projects are found to be due to poor forecasting accuracy, then the Lean Six Sigma team takes actions to improve the forecasting process for products with low accuracy.

Accurate forecasts are important since they drive manufacturing schedules and other organizational planning activities. Poor forecasts mean that either not enough or too much product has been scheduled for manufacturing. In either situation, capacity is wasted and operational costs are higher than optimum. Inventory turns will also be adversely impacted by poor forecasting accuracy. However, in low-volume manufacturing environments such as job shops where products are manufactured to order, Lean methods can be effectively utilized to match customer demand to manufacturing schedules. In fact, if Lean methods have been well deployed, visual control systems, such as Kanbans, can be used to manage material flow through the system. However, in make-to-stock environments, where capacity is prorated across hundreds or thousands of products, demand must be estimated using forecasting models. In these make-to-stock systems, forecast accuracy is analyzed at the product-line level and then broken down to an item and location level. Using this information, the Lean Six Sigma improvement team will continuously improve forecasting accuracy over time through an integrated deployment of projects tied to root cause analysis. Improved forecasting accuracy will result in lower operational costs and inventory investment.

6. Lead-Time

Chapter 4 discusses lead-time in detail including how it is defined and strategies to reduce it. Lead-time is the time required to perform a single operation (or series of operations in a network). Individual components of lead-time can include order preparation time, queue time (waiting), processing time, move or transportation time, and receiving and inspection time. Important parameters used to describe lead-time are the average completion time and standard deviation of each activity. Lead-time must be calculated along a system's critical path since this critical path represents the longest cumulative lead-time through the system. In fact, many improvement activities will take place on the critical path to reduce the system's cumulative lead-time. This is because lead-time drives a large portion of supply chain cost. Inventory investment is the most obvious example. It should be noted, as the cumulative lead-time along the critical path is reduced, another critical path may emerge within the system's network.

7. Unplanned Orders

Unplanned orders degrade supply chain operational efficiency. This is because additional machine setups and other orders may be delayed if schedule modifications occur. The two basic characteristics of unplanned orders are that they are scheduled within standard lead-time (time fence) or they exceed their original capacity allocation. Productivity decreases in either situation. The problem with unplanned orders is that they force changes to the original manufacturing schedule, causing the number of manufacturing setups to increase. This situation tends to increase lead-times across the entire product mix since other products must be rescheduled to a later date. A compounding factor occurs when raw materials or WIP is diverted to the unplanned order, resulting in additional delays for the orders currently in the schedule.

Isolated unplanned orders, due to unforeseen circumstances, can usually be integrated into the manufacturing schedule, but poor management practices often allow unplanned orders to become a chronic problem. There are many reasons for unplanned orders. These include poor demand forecasts by customers, internal process breakdowns, or failure to adhere to standard operating procedures regarding lead-time and capacity requirements. In these situations, it is important for the Lean Six Sigma improvement team to determine the extent of the problem and prioritize projects based on their business benefits. By properly measuring the incidence rate of unplanned orders and discovering their root causes, effective action plans can be developed by the improvement team to eliminate unplanned orders. These actions will improve productivity and customer satisfaction while reducing inventory investment.

8. Schedule Changes

Schedule changes are different from unplanned orders in the sense that they are usually caused by process changes or unforeseen circumstances. Of course one reason for a schedule change could be an unplanned order placed into the current schedule. But measuring schedule changes separately is useful for process improvement efforts because the root causes may be different than unplanned orders. Schedule changes also have an adverse impact on operational efficiency if they are within the product's time fence or exceed its allocated capacity. Some of the major reasons for schedule changes include unavailable materials, equipment, and personnel and quality problems. The Lean Six Sigma improvement team should measure the reasons for the schedule changes and estimate their business impact to identify and deploy Lean Six Sigma projects.

9. Overdue Backlogs

Overdue order backlogs occur for many reasons. In most cases, available capacity may not exist (for a variety of reasons) or the backlog represents industry practice based on technology constraints, that is, make-to-order systems. Overdue backlogs cause an organization to lose the time value of money relative to delayed sales. But whether the backlog is detrimental to the organization to a great extent depends on industry practice. In industries characterized by contract manufacturing or high capital cost, backlogs may be created to ensure that system capacity is effectively utilized over time. In these situations, a certain amount of overdue backlogs would be normal. But an organization should periodically re-evaluate assumptions regarding overdue backlogs to confirm that the current policy remains optimum. There are many reasons for overdue backlogs. These range from poor scheduling, quality problems and machine breakdowns to obsolete assumptions regarding the optimum backlog level. Where there are process breakdowns, especially those which adversely impact system capacity, the Lean Six Sigma team should create projects to improve process performance.

10. Data Accuracy

The Lean Six Sigma improvement process is designed to identify measurement system problems. In fact, measurement systems analysis (MSA) is one of its major strengths. Chapter 8 shows MSA methods commonly used in manufacturing environments must be modified for supply chain applications. Also, each of the main components of measurement system error in manufacturing has an analogue in supply chain applications. As an example, many of the analyses performed in supply chains involve automatic calculation of quantities by the information technology (IT) system. In these applications, it is important to ensure there is no bias in the system. Also, MSA checks of the incoming data quality are very important. In many Lean Six Sigma projects, fixing the MSA will eliminate the higher level problem.

Problems with data accuracy are found in almost every Lean Six Sigma project. This is true whether it is a manufacturing or non-manufacturing project application. In most cases, the organization is measuring either the wrong metric or the right metric, but measuring it in the wrong way. But, unlike manufacturing, where data accuracy problems may impact one machine, data inaccuracies existing within a supply chain often impact one or more functions within an organization as well as several organizations. Since decisions regarding how much product to make and which subcomponents to order, as well as their quantities, depend on accurate data within supply chain systems, it is important to understand the degree of measurement error in supply chain information technology (IT) systems. In supply chains, data is used in the form of historical patterns such as demand and inventory histories as well as parameter settings within IT systems. This system data is used to calculate manufacturing schedules and what and how much to order through the MRPII system, to set inventory and safety-stock levels, and control supply chain functions. Poor data accuracy increases operational costs and inventory investment throughout the supply chain.

11. Material Availability

All resources required to manufacture a product must be available at the scheduled time of manufacture. These required resources include materials, labor, and equipment as well as the information necessary to assemble the customer order. Poor resource availability causes work stoppage and rescheduling of work. These process breakdowns result in late customer orders and increased operational costs. Depending on the root causes of the problem, there may be several reasons for poor material availability. These include incorrect inventory data, poor supplier quality machine breakdowns, lack of skilled workers, lack of information, and similar issues as well as a failure to effectively coordinate resources. Lack of available material also drives up inventory investment since raw material and WIP inventory remain unutilized because of missed manufacturing schedules.

12. Excess/Obsolete Inventory

Excess inventory is calculated based on multiples of lead-time. A correlating metric is days-of-supply (DOS). If DOS either exceeds the demand quantity expected over the order cycle (or lead-time in some situations) or the required lot size, there may be excess inventory in the system. In an inventory population some items and locations may have excess inventory while others do not have enough available inventory. Chapters 6 and 10 show how to determine excess inventory quantities using an inventory balance analysis. In this balance analysis, the optimum inventory is calculated using the item's lead-time, expected demand, and target service level to calculate an optimum quantity. Excess is defined as average on-hand inventory minus the optimum inventory quantity. The reasons for excess inventory were discussed in Chapter 6 and shown in Figures 6.1 and 6.2 and Table 6.1. Excess inventory for items having a high demand can usually be rapidly decreased, but excess inventory associated with low demand items may become obsolete over time. Lean Six Sigma projects will help lower excess inventory by attacking the root causes for the problem.

Obsolete inventory may have started as "excess inventory", but now it can no longer be used to generate sales. This results in a loss of stated book value. Obsolete inventory may be caused by replacement of the current product by newer versions because of improved design, more advanced/economical manufacturing methods, changes in consumer preferences, or unrealized forecasted demand (normally for new products). Obsolete inventory is difficult to eliminate without a balance sheet write-off. However, there are strategies that can be employed by the Lean Six Sigma improvement team to reduce current levels of obsolete inventory. But it is very important the team attack the root causes for the obsolete inventory since obsolete inventory is recurring and has a cumulative negative impact on an organization's balance sheet. As an example, if obsolete inventory is accumulating due to several failed new product introductions, then the new product design process, consisting of sales, marketing, purchasing, and engineering, must be analyzed and improved to ensure that new product forecasts are accurate and the product design meets all intended requirements. On the other hand, if obsolete inventory is accumulating because purchasing routinely purchased large lot sizes to obtain price discounts, then this will be a different Lean Six Sigma project. In other words, there can be many reasons for the creation of obsolete inventory and as a result there will be different Lean Six Sigma projects applications.

Additional Lean Supply Chain Metrics

13. Customer Service Target

Customer service targets are one of the foundations for a Lean supply chain since the system is designed to deliver products and services based on customer promises and requirements. There are several definitions of customer service targets defined relative to inventory analysis in Chapter 6 and shown in Figure 6.3. These service-level targets can be used to set optimum inventory level targets in conjunction with lead-time and expected demand for the time period under consideration. As an example, service-level targets can be expressed in per-unit fill, line fill, order fill, and financial terms. However, depending on the Lean Six Sigma project application, service-level targets can be defined from several perspectives such as on-time delivery (delivery to promise) and manufacturing schedule attainment as well as other processes which touch the customer.

14. Net Operating Profit after Taxes (NOPAT)

Net operating profit after taxes is calculated by dividing income after taxes by total revenue. Higher NOPAT levels are better than lower ones. But NOPAT tracks industry averages, so it is difficult to compare performance across different supply chains. As a result NOPAT should be evaluated against direct competitors and an organization's competitive strategy. As an example, an organization might have a strategy to forego certain NOPAT targets in the short term in order to attain greater market share and higher NOPAT in future time periods. For this reason Lean Six Sigma projects normally impact NOPAT at a tactical level. These projects either increase revenue or gross profit margin. Increases in revenue can be obtained by increasing sales over current sales projections.

Gross margin improvements are normally obtained through expense reductions. However, at a supply chain level, NOPAT is one of several financial metrics that indicate the efficiency of the supply chain in converting material, labor, and capital inputs into product or service outputs that are sold to customers.

15. Asset Efficiency (Turns) = Sales/Assets

Assets include fixed investments in building, plants and equipment, account receivables, inventory investment, goodwill, long-term investments, and other miscellaneous assets. From a financial perspective, these asset categories are also defined as invested capital and short-term and long-term debt. A Lean supply chain is measured relative to its ability to efficiently utilize its assets to generate sales revenue and meet customer requirements. Asset efficiency (or turnover) is calculated by dividing total sales revenue by the total asset investment necessary to obtain these sales for the time under analysis. Asset efficiency is an important metric to measure the degree of supply chain "Leanness." Lean supply chains have high asset efficiencies relative to their competitors. Asset efficiency can be broken into three subcategories. These are discussed below from an operational perspective since these are the asset categories the Lean Six Sigma team can improve. These categories include fixed asset efficiency, account receivables efficiency, and inventory efficiency.

16. Fixed Asset Efficiency (Turns) = Sales/ (Average Property + Plant + Equipment)

Effective management of fixed assets is critical to developing a Lean supply chain since supply chains meeting financial and operational goals and objectives using fewer assets are by definition Leaner than their competitors. Assets can be leased rather than owned by an organization. But operating expenses will be higher for leasing scenarios due to the expenses associated with leasing. So although asset investment levels can be reduced by leasing, other financial metrics may be adversely impacted. At any one point in time there will be an optimum balance between asset ownership versus leasing as well as how efficiently an organization utilizes their owned and leased assets. Lean Six Sigma improvement teams are very useful in these types of projects since they can build total cost models to understand the trade-offs between all system constraints. The analysis will also be useful when implementing the solutions.

17. Receivables Efficiency (Turns) = Net Credit Sales/Average Accounts Receivables

The efficiency of account receivables (receivables turnover) is calculated as credit sales divided by net receivables for the time period under analysis. A correlating metric is account receivables aging, which is calculated as the number of days in the time period divided by the account receivables turnover metric. Cash flow into the organization increases as the efficiency of account receivables increases. This has a positive impact on other financial metrics. Lean Six Sigma improvement projects have been deployed to reduce the aging of accounts receivables and in some cases to reduce the incident rate of account receivable write-offs.

18. Gross Profit Margin = Gross Profit/Sales

Gross profit margin is a measure of an organization's operational efficiency in converting inputs into outputs. It is calculated as revenue minus cost-of-goods sold *(COGS)*. Lean Six Sigma projects can impact both revenue and COGS. Lean Six Sigma applications to increase revenue and reduce expenses have been discussed above under NOPAT and in other sections of the appendix. Also, gross profit margin does not consider adjustments to revenue (customer returns and allowances).

19. Return-On-Assets (ROA) = Net Profit Margin x Asset Efficiency

Lean supply chains should have high ROA levels relative to their competitors. Or to put it another way, their ROA levels should improve year-over-year if their Lean Six Sigma deployment and other improvement initiatives are effective. ROA is calculated by multiplying the net profit margin by asset efficiency. Net profit margin is calculated by dividing net income by total revenue. Net income is defined as total revenue minus all expenses including taxes. Asset efficiency is defined above in Section 15. Return-on-assets (ROA) shows how much income has been generated for every $1 of assets utilized to create the sales revenue. The term "assets" as used in this context are defined as shareholder's invested capital as well as short- and long-term borrowed cash. ROA is industry-specific since some capital-intensive industries such as those in telecommunications, heavy equipment, and transportation tend to have much lower ROA levels than other industries. For this reason, a supply chain's ROA should be compared against competitive supply chains within the same industry. Lean Six Sigma projects can help improve ROA by increasing profit margins or improving asset efficiencies.

20. Gross-Margin-Return-On-Investment (GMROI= Gross Margin/Average Inventory Investment at Cost)

GMROI is a good operational metric that shows how much money should be invested in inventory to generate increased return-on-investment (ROI). GMROI shows the percentage return (or realized gross margin) for every $1 invested in inventory. As an example a GMROI of 200% means that for every $1 of inventory invested, the percentage return was 200% and the gross margin per item was $2.00. GMROI when used in conjunction with other financial metrics can be a very effective metric in guiding the Lean Six Sigma improvement team to effectively utilize inventory investment to increase an organization's ROI.

Appendix II

Key Lean
Six Sigma Concepts

Chapter I

1. You should understand there are many tools and methods that are specific to supply chain improvement. In particular, your team should be familiar with the topics shown in Figure 1.1. These topics include a basic understanding of supply chain concepts including materials planning, forecasting, and inventory analysis as well as Lean Six Sigma tools and methods. Remember to balance your team based on your project charter's problem statement to include functional experts from relevant areas of your process.

2. Use metrics to measure how your process is performing. A list of 20 important supply chain metrics was shown in Figures 1.1 and 1.2. Clearly define these metrics. Make sure your team can accurately and precisely measure the metric levels to detect process improvements. Use more than one metric to clearly show performance. As an example, show that customer service has not deteriorated as inventory turns increase.

3. Inventory investment and turns are major barometers of supply chain performance. The inventory improvement actions shown in Table 1.1 ensure that the team collects relevant data to answer the required project questions and drives to the root cause of the problem using Lean Six Sigma methods to meet the project's improvement objectives.

4. Process improvements require development of a quantified system model map. This map could be a higher-level SIPOC, a functional map, or a value stream map.
5. Project identification and selection follows a top-down deployment strategy to ensure alignment with senior management's high-level goals and objectives.
6. Operational linkage is critical to process improvement. Working backward from the high-level measurable improvement, the correct Lean Six Sigma "tool set" is selected to drive to the root causes of the problem to improve lower-level operational metrics that tie into higher-level financial metrics. These financial metrics are used to measure supply chain improvement.
7. There are different project types as shown in Figure 1.5. Some projects are "just-do-it" types, and others may require capital expenditure to enable the process improvement. The Lean Six Sigma team should understand the differences between each project and which set of improvement tools to apply to execute the project.
8. Every Lean Six Sigma improvement team should use a project charter to scope out the project and communicate its status to the organization. A formal project plan should also be used to control the project until its completion.
9. The 10-Step Solution Process, which has been slightly modified for supply chain applications, parallels the Lean Six Sigma DMAIC methodology and ensures business alignment, relevant data collection, use of correct supply chain analysis tools, and identification of root causes and effective control of the process improvements over time.
10. Tables 1.5, 1.6, and 1.7 provide many ideas for the creation of Lean Six Sigma supply chain projects.

Chapter 2

1. Misalignment of functional metrics creates process breakdowns within a supply chain. Lean Six Sigma projects are created around these misalignments to improve process performance.
2. It is important to use the correct tool kit to analyze the problem. As an example, Lean tools are used to simplify a process and eliminate obvious sources of waste. Six Sigma tools are used to analyze root causes and build optimized process models to

explain key process output variables (KPOVs) in terms of key process input variables (KPIVs). There are many other types of supply chain analysis tools, including forecasting models, scheduling algorithms, and optimization techniques. Many of these latter tools fall within the disciplines of operations management and operations research.

3. The proper execution of the sales and operating plan requires that all supply chain participants perform their functions well and meet their operational and financial objectives. If operational performance is below target, Lean Six Sigma improvement projects can be deployed to improve operational metrics.

4. Lead-time reduction is one of the major ways to reduce inventory investment. Important tools to reduce lead-time include elimination of process steps and operations, bottleneck management, application of mixed-model scheduling, application of transfer batch methods, and deployment of Lean, JIT and QRM techniques.

Chapter 3

1. A sales and operations planning (S&OP) team is essential to the identification of Lean Six Sigma improvement projects that will have a significant business impact.

2. Lean Six Sigma tools can be used to build more accurate forecasting models to reduce forecasting error.

3. Many problems at an operational level arise because forecasts are reconciled in monetary units at a product group level, and then demand is linearly estimated, without regard for historical demand patterns, at an item and location level.

4. Violation of the organization's stated lead-times, that is, time fences, causes significant operational problems. Lean Six Sigma projects need to work through cross-functional teams and carefully measure the negative impacts of schedule changes within lead-time on the organization.

5. Lean Six Sigma belts can help forecasting analysts develop advanced models to more effectively manage products with unusual demand patterns.

6. Lean Six Sigma tools and methods will benefit marketing to prevent poor forecast of new product sales, which is a major contributor of excess and obsolete inventory.

Chapter 4

1. Creating a value stream map (quantified process map) and breaking operations down into their time elements is a powerful way to identify areas within the process where lead-time can be reduced and waste eliminated to improve quality.
2. Identify the system's critical path and understand the lead-time variation components as well as the average time to effectively reduce the lead-time of the system.
3. In most Lean and Six Sigma projects, many of the root causes of long lead-times can be found in Table 4.1.
4. Lean Six Sigma belts can use Tables 4.2, 4.4, and 4.6 to reduce lead-time.
5. Using the system's takt time as a target will show the team much of the process waste (non-value-adding resources necessary to maintain the system takt time).
6. Capacity analysis using queuing and simulation models is an important tool to understand why lead-times are too long in a system. Lean Six Sigma belts should learn operation research tools and methods to properly understand how a system dynamically changes.

Chapter 5

1. Understanding how your organization's MPS and MRPII systems work will allow your Lean Six Sigma improvement team to identify many areas for improvement projects. Also, working backward from chronic problems at receiving, production activity control, purchasing and inventory investment will also direct the improvement team to issues related to MPS and MRPII systems.
2. Comparing advantages and disadvantages between make-to-stock, assemble-to-order, and make-to-order systems will allow your improvement team to identify opportunities to improve your process and reduce inventory investment.
3. Analyzing system capacity especially at bottlenecks and system-constrained resources will allow your team to increase system throughput to increase revenue or reduce operational cost.
4. Understanding how to move your operations from push to pull scheduling systems will reduce inventory, improve quality, and reduce operational costs. MRPII-driven Kanban will allow your organization to implement key elements of a pull system in an MPS/MRPII scheduling environment.

Chapter 6

1. Calculating the optimum investment level for every item by its stocking location will not only allow a quick balancing of the item's inventory quantity, but also show if the item has an excess quantity. Aggregating inventory investment up to product levels identifies where Lean Six Sigma projects should be strategically deployed to reduce inventory investment.
2. Figures 6.7 and 6.8 provide information useful to reduce excess and obsolete inventory.
3. Look for ways to outsource noncore activities to more efficient organizations.
4. There are many different types of cycle-counting systems. The best cycle-counting system is one with very little inventory that is controlled using visual system display, that is, Kanban systems, which specify the exact allowed inventory quantities.

Chapter 7

1. Analysis of measurement system error is an important part of every Six Sigma project. The basic categories of measurement error, that is, resolution, accuracy, reproducibility, repeatability, linearity and stability, have direct analogues to supply chain systems.
2. Prior to collecting and analyzing data, the Lean Six Sigma improvement team should have recorded the critical questions that needed to be answered by the analysis prior to data collection.
3. The specific analytical tools that are used by the Lean Six Sigma improvement team reflect the information required to drive to root cause. Although there are many Lean Six Sigma statistical tools available, there are other analytical tools including queuing analysis, linear programming, scheduling algorithms, financial models, and others depending on the industry and function within the industry.
4. Creating a quantified system map allows the team to look for areas of obvious process waste. These areas of waste are called "non-value-adding" operations and are differentiated from "value-adding" or "business-value-adding operations."
5. Creating a quantified model of your process will allow "what-if" analyses to be performed to show the business benefits for changing portions of your process. This methodology will identify improvement projects.

Chapter 8

1. Evaluation of alternative solutions, using a cost-benefit analysis, is critical to ensuring that an organization realizes the maximum business benefit from the Lean Six Sigma project with minimum risk.
2. Using a pilot study to confirm the final solution minimizes the project implementation risk and gains support from process owners since they can see the impact of the process changes under controlled conditions. This allows modifications to the original implementation plan prior to complete conversion to the modified process.
3. The process changes must be integrated into the organization's other process improvement and control systems to ensure that they are incorporated into daily work.
4. An important part of the project transitioning process is showing changes to key project metrics following up with rigorous statistical analysis.

Chapter 9

1. There are different control tools that vary in the effectiveness of control as well as the resources necessary to implement and maintain the countermeasures over time.
2. Process maps can be used throughout the Lean Six Sigma project depending on the analytical and communication requirements of the project.
3. Failure mode effect and analysis *(FMEA)* tools are useful in ensuring there are no breakdowns relative to implementing the project's countermeasures.
4. Before and after graphs are useful in communicating the project's status to the process owner and local work team.
5. All Lean Six Sigma project improvements must be integrated back into the organization's standard control systems including ISO systems.

Chapter 10

1. Inventory models can be very useful in analyzing an organization's current inventory population to identify excess and obsolete inventory.
2. The analysis of inventory models can also help identify the Lean Six Sigma projects.

3. All the projects identified using the inventory model will be strategically linked and business benefits can be immediately estimated to create project charters.
4. There are many applications of Lean Six Sigma projects within the supply chain. Wherever there are manual interventions, variation will be increased and Lean Six Sigma improvement project will be useful in eliminating process breakdowns.

—

Conclusion

Using Lean Six Sigma tools and methods to improve supply chain performance is a complicated and long-term set of inter-related tasks requiring cross-functional teams to identify Lean Six Sigma improvement projects, investigate the reasons for process breakdowns, and systematically eliminate their root causes. Many of the resultant process improvements will be linked to the demand management or lead-time issues. As an example, it was shown that demand management had a critical impact on supply chain performance including inventory investment. If a product's forecast accuracy was poor, manufacturing schedules would break down, causing deterioration in customer service levels and reducing operational efficiencies. At a lower operational level, low forecasting accuracy can be due to many factors. These include situations in which sales incentives distort future demand patterns or inaccurate forecasts of new products are caused by erratic demand patterns. In addition to Lean Six Sigma tools and methods, which were based on statistical analyses, the importance of implementing a Lean supply chain was discussed in detail. The goal of a Lean supply chain system is expansion of operational competence to systematically eliminate waste from the supply chain. A key measure of Lean supply chain performance is the efficiency of asset utilization.

Lean Six Sigma belts should understand how supply chain systems work, at least on a basic level, in order to more effectively identify and deploy Lean Six Sigma projects. As an example, given the complex information technology (IT) infrastructure existing in modern supply chains in the form of forecasting systems and models, master production scheduling (MPS), master requirements planning (MRPII), and warehouse management systems (WMS), it makes sense to use tools and methods from operations management as well as operations research to analyze supply chain performance given the complex nature of the input and output relationships of key process variables. Also, rather than throwing generalists or internal consultants who must first learn how the process works before they can intelligently study the reasons for process breakdowns, it may be wiser to train people familiar with these complex systems as "belts"; but, with relevant tools and methods rather than complicated statistical methods alone. However, if your belts are new to the field of supply chain management, then several chapters in this book will serve as a good introduction to Lean supply chain structure, distribution, and inventory management as well as practical tools and methods to demand management, lead-time management, and related activities. Every chapter also had practical ideas to identify and deploy Lean Six Sigma projects within supply chains. An important part of understanding the performance of major work streams within a supply chain is build models of these systems. A simple model of an inventory system was created to help identify Lean Six Sigma projects which were the high level root causes for higher level financial metrics.

The 10-Step Solution Process was characterized by several sequential and inter-related steps that directly correlated to the DMAIC problem-solving methodology. The difference between the two problem-solving methods is an emphasis by the 10-Step Solution Process on analytical tools and methods more applicable to supply chain analysis, given their complexity and unique systems, rather than to manufacturing applications. However, both methods emphasize the importance of strategic business alignment, measurement system improvement, and an emphasis on analytical and fact-based methods, as well as detailed plans to sustain process improvements. Using the 10-Step Solution Process, Lean Six Sigma teams deployed Lean Six Sigma projects to strategically improve supply chain performance.

In Step 1 of the 10-Step Solution Process, the project was aligned with business goals and objectives. This strategic alignment ensured that the improvement team obtained the necessary resources and organizational support to deploy the Lean Six Sigma project within the organization. The project charter was an important tool to create the project team as well as communicate the project's status over time. In this step the team also measured baselines of operational and financial metrics as well as their performance gaps relative to target. These performance gaps served as the project justification. Using the performance gaps as well as a long list of relevant questions from senior management, the team brainstormed a long list of questions that the project was required to answer. These questions related to the 20 metrics listed in Figure 1-2 including 12 key operational metrics. Understanding current metric performance levels and their required improvement targets as well as their inter-relationships is an important part of supply chain analysis. These projects should improve both internal operational as well as external customer metrics. Along this line of thought, it is also important to ensure that Lean Six Sigma projects link to the voice-of-customer (VOC). To achieve these goals, the team was encouraged to build a balanced metric scorecard, similar to Figure 1-2, prior to deploying Lean Six Sigma projects to ensure their linkage across the supply chain. A second key element of Step 1 was creating an overall project plan to execute each work activity. The project plan, timeline, and resource estimates may be modified during the analysis portion of the project depending on the root causes found responsible for metric deterioration.

In Step 2, the Lean Six Sigma team worked with key organizational functions, including the process owner, to identify team members. An important part of this meeting was facilitating a review of the team's project charter, including the problem statement and objective, to ensure that everyone was aligned behind the process owner's goals and objectives. At this point in the project, it may be wise to split the project into several smaller projects to more clearly focus the team on the operational issues most relevant to the process owner. Of course, going to a lower level in the problem statement could only be done if high-level data analysis clearly showed this to be the correct course of action. Once everyone agreed on the project's problem statement, the final project plan was developed by the Lean Six Sigma improvement team. This final project plan listed all anticipated activities required to collect and analyze relevant data to identify and eliminate the root causes for the problem. This plan also showed how every project activity was cross-referenced to the people responsible for its completion as well as its required resources. If the project was complicated, scheduling software was used to manage it.

In an environment of poor process quantification, it is difficult to collect the relevant data necessary to analyze and eliminate operational breakdowns such as high inventory investment, poor on-time delivery, long lead-times, and poor demand estimation as well as other supply chain issues. On the other hand, if your organization can build a simple model of key work streams within its supply chain, it becomes possible to begin to understand the complex inter-relationships between system variables. Without a quantified model, the root causes of supply chain problems including high inventory investment are more difficult to systematically identify and eliminate by the Lean Six Sigma improvement team. Development of input/output matrices similar to Figure 1-4 and Table 1-2 was used to translate senior management's goals and objectives into tactical improvement projects. This is critical to building the work stream models. As the Lean Six Sigma improvement team worked though the ten steps, it began to build the quantitative skills and competencies necessary to provide the fundamental knowledge of how their supply chain operates. Gaining this competence ensures the systematic elimination of major supply chain performance problems work stream by work stream according to a strategically designed deployment plan.

In Step 3, an effective project communication plan was designed (in the context of an overall deployment plan) to maintain support for the project and ensure that it remained on schedule. An important aspect of communication is involvement of all interested parties up-front before the project gains momentum. This allows modifications to the project scope and planning that everyone can support. Effective communication is critical in supply chain improvement projects because of the numerous organizational functions and competing priorities. The project charter was an integral part of the project's communication process. In addition to the written project charter, there are many ways to communicate the team's progress over time including face-to-face meetings, conference calls, and e-mails as well as other communication vehicles. However, the type of communication depends on the complexity and emotional content of the message. As an example, if the project has a large impact on the process owner's process, an e-mail would be the wrong initial communication vehicle. But a face-to-face meeting would be appropriate. On the other hand, to convey simple and routine information, e-mails would be an acceptable form of project communication.

In Step 4, the Lean Six Sigma team collected and analyzed data to identify the root causes for the process breakdowns and identify countermeasures to eliminate these root causes. The optimum solutions were based on proving the causal effect between process input variables (the "X's") and the process output variables (the "Y's"), that is, $Y = f(X)$. Although the specific analytical tools varied depending on the complexity of the problem as well as the specific project objective, one of the most important analytical tools was the process map. A process map is an important analysis tool for every project. Various types and applications for process maps throughout the project were shown in Figure 9-4. Process maps are developed in one or more formats depending on the team's requirements. These formats include a high-level SIPOC (supplier-inputs-process-outputs-customer), a value stream map (VSM), or functional process maps (quantified work streams). These process maps were used to quantify metric linkages at process input and output boundaries. These metric linkages were associated with functional interfaces across the work stream under analysis. The resultant work stream models were based on the simple analytical concept $Y = f(x)$. The initial business benefits of the project were estimated using the identified relationships between the inputs and outputs versus the anticipated performance improvements. Other important considerations in the root cause analysis and development of countermeasures included measurement system analysis (MSA), data collection and analysis tied to the project's problem statement and objective, and the use of analytical tools such as queuing analysis, linear programming, scheduling algorithms, simulation, and financial modeling to identify the root causes for the process problem.

Accurate and precise measurement are critical to the project throughout the entire 10-Step Solution Process. In the early steps of the project, accurate measurements of key metrics (KPOV variables) ensured that the team had correctly estimated performance gaps using capability analysis. In addition, the potential business benefits were estimated for the project. Later in the project, in Step 5, information technology (IT) measurement systems were analyzed to ensure that they could be used to control the modified process over time. The changes to the modified process typically involved changes to the information technology (IT) software systems and the system parameters supporting major supply chain work streams. As an example, the improvement team may have found lead-time was incorrectly specified in the MRPII system resulting in incorrect inventory levels, missed supplier deliveries, or large lot sizes. Forecasting parameters may have been incorrectly specified, reducing forecasting accuracy resulting in higher inventory investment. In the work stream analyses, there are many other situations where measurement problems are associated with the root causes of major process breakdowns and need to be fixed prior to continuing with the problem analysis.

Strong project management skills were shown to be necessary in the early stage of the project to break down major project activities into detailed work tasks. This work breakdown included identification of project milestones, major work activities, and work tasks. Work tasks were further analyzed relative to task duration, their inter-relationships, and resource requirements. These critical project management tasks were necessary to ensure that improvement projects met cost, schedule, and business objectives. However, effective project management is also required during implementation of the solution in Steps 6, 7, 8, 9, and 10. The basis of Steps 6 and 7 was development of effective improvement plans based on countermeasures tied to the root causes of the problem under investigation. In Steps 8, 9, and 10 of the 10-Step Solution Process, process changes are integrated into the process using standardized work methods and other Lean tools and methods to ensure that the process improvements were incorporated into the local work team's daily work. An important output of the transition phase is communication of the process changes to the organization using fact-based methods. Control strategies were developed by the Lean Six Sigma improvement team based on the pilot study results. In Step 4, the pilot was run on the optimum solution to verify the Lean Six Sigma improvement team's analysis. The control strategy was also based on knowledge gained from data collection and analysis throughout the project. Also associated with each countermeasure were the anticipated business benefits along with an implementation date. These actions ensured that the project remained on schedule and there was an effective project transition to the process owner and local work team.

Inventory is a key barometer of Lean supply chain effectiveness. In other words, just how "Lean" the Lean supply chain really is can be analyzed relative to its asset utilization efficiencies. A subcomponent of asset utilization is inventory utilization efficiency. Inventory utilization efficiency or its turns ratio along with other metrics shows the effectiveness of an organization's Lean Six Sigma deployment as well as other supply chain improvement initiatives. In Chapter 10, inventory was analyzed as one major work stream within a supply chain to show the advantages of building and analyzing simple inventory models. Building models allowed the improvement team to understand the complex relationships between key process input variables (KPIVs) and their associated key process output variables (KPOVs). Understanding these relationships as they impacted inventory investment and its turns ratio helped the Lean Six Sigma improvement team to answer senior management questions such as, "Why is inventory investment so high?"; "What are the most feasible and beneficial Lean Six Sigma improvement projects that should be deployed and are they strategically linked to senior management's high-level goals and objectives?"; and "When will we know if the process breakdown is fixed?"

The major drivers of inventory investment (or its root causes) can be more easily understood if an inventory model is built, analyzed, and verified by the improvement team. However, development of a formal inventory or other work stream model may not be required to successfully execute the project's objective and the questions the team must answer for senior management through its analytical work. In other words, some projects can be completed using simple analyses. It was shown that the improvement team can reduce inventory investment in two ways. First, inventory quantities can be balanced by item and location to identify areas in which to reduce investment. But, if customer service levels are already low, then additional investment may be required, to ensure that the item's customer target is met. In balancing an item's inventory quantity, the optimum quantity is calculated using target service levels, actual lead-time, and an estimate of demand variation (root mean squared deviation or error from the forecasting model as shown in Figure 3-15). Differences between the optimum versus average on-hand inventory quantity show if a reduction in inventory quantity is possible. Once the inventory balance is achieved, further inventory reductions may be possible by reducing the item's lead-time or more effectively managing its demand. This was shown to depend on deployment of one or more improvement projects based on specific contributing factors or cause for high investment including large lot sizes, excessive order quantities, and poor schedule adherence. The root cause analysis was taken to a point where the vital few causes impacting the problem became apparent to the improvement team. At this point in the project, the team began to identify countermeasures and solutions to eliminate the root causes for low inventory turns to improve process performance. However, in some instances, implementing the solutions was complicated by the existence of high levels of excess and obsolete inventory. In summary, it was shown in Chapter 10 that the proper use of Lean Six Sigma tools and analytical methods in the context of effective project management, that is, the 10-Step Solution Process, will allow an organization to intelligently analyze its inventory work stream. This analysis can then serve as the basis to select and deploy Lean Six Sigma projects to improve the efficiency of inventory utilization across its supply chain. Other work streams can be analyzed in a similar manner using process models.

Design-for-Six-Sigma (DFSS)

Design-for-Six-Sigma (DFSS) is a methodology applied to create new product or service systems. DFSS moves performance past its entitlement (entitlement was discussed in Chapter 8) up to a Six Sigma capability level. At a Six Sigma performance level, defects are created at a rate of 2 per billion (short-term variation analysis with the process on target). In some Lean Six Sigma discussions this performance level is reduced to 3.4 parts per million based on an assumption of higher variation caused by exposure to many sources of process variation over extended time periods. A Six Sigma process will seldom fail from a customer perspective, although few processes ever attain this performance level without creating product and process designs based on DFSS principles. DFSS design principles enable systematic identification of the voice-of-the-customer (VOC) as well as the specific system elements that will ensure that customer requirements relative to quality, cost, and time are met at Six Sigma performance levels.

The DFSS process begins with identification of customer needs, expectations, and requirements (CTQs) using marketing research, quality function deployment (QFD), and related tools and methods (these were discussed in Chapter 1). The major objective of understanding the VOC is to develop performance specifications for the new product or process design. Using these customer specifications as an initial guide, systems are designed to ensure that performance targets are achieved with minimum variation under a variety of expected usage environments. In this DFSS development process, the performance of numerous key process output variables (KPOVs) is jointly analyzed relative to key process input variables (KPIVs) using advanced analytical tools. In supply chain system design, these include tools and methods including those from operations management and research as well as the basic DFSS tools and methods. Interactions between one or more independent variables are also analyzed to ensure that the correct level for each input variable is chosen in such a way as to minimize overall system variation when the KPOVs are at their target levels.

The DFSS team evaluates various design alternatives using FMEAs, alternative ranking methods, focus groups, and QFD, as well as other relevant methods based on necessary system design trade-offs. After the preliminary system design has been chosen, the DFSS team selects important system elements using experimental design methods, reliability analyses, system simulations, tolerance design, and related optimization tools and methods to establish the $Y = f(X)$ relationships. Pilot studies and prototypes are also developed according to initial design specifications to match predictions relative to performance and cost. After prototyping, the DFSS team uses FMEAs and related tools to mistake-proof the design. Working concurrently, the process engineering team designs a Six Sigma–capable process using basic Lean Six Sigma tools including capability analysis, mistake proofing, control plans, statistical process control, and other relevant tools and methods. Finally, the ability of the new product design to meet manufacturability and customer requirements over time is verified by the process owner and local work team, and a Lean Six Sigma balanced metric scorecard is developed for the new product.

Using DFSS for Supply Chain Design

Applying DFSS principles to supply chain systems and their major work streams requires modifications of the standard manufacturing DFSS tools and methods, but the basic concepts are analogous. In any design process, customer requirements must be translated into system specifications and performance targets. The basics of this translation process remain the same regardless of the system, industry, or organizational function being analyzed. As the VOC information is firmed up into system requirements, the information is collected and analyzed by the DFSS team using design reviews, concurrent engineering methods, and specific tools and templates that aid the communication process. These tools and methods include process maps of the supply chain system including the major work streams. These work streams are the material and information flows through the system. Prototypes of the system might also have been created including simulations of how the system dynamically responds to changing levels of the input variables. These process simulations may have been built on previous system models, queuing theory, or linear programming models or developed using other operations management and research methods. In the case of distribution and inventory systems, specialized methods related to operations management and operations research may have been used to design distribution facilities to model the movement and storage of materials. In these applications, basic industrial engineering techniques are extremely useful to efficiently design the work streams.

Standard project management techniques such as those discussed in Chapter 1 are used to control and manage the project's work tasks over time. These project management tools and methods discussed in Chapter 1 included Gantt charts and network analysis models. A Gantt chart of the 10-Step Solution Process was shown in Figure 1-7. In the world in which we live today, communication of the system design status is relatively easy.

The DFSS team can be located anywhere in the world to work on the modifications to the supply chain system design 24 hours a day and 7 days a week until the project is completed by the improvement team. Analytical evaluations of the system's inputs and outputs can be conducted to ensure that the new system design is fully understood and optimized to achieve Six Sigma performance levels and satisfy VOC requirements.

DFFS applications to supply chain systems and work streams require satisfying external customer demand utilizing a minimum amount of resources, that is, a Lean supply chain. To achieve this objective, to the greatest extent possible, the system design should use the minimum number of operations and work tasks. Manual interventions should be minimized and even eliminated to prevent operational failures. Information should be captured at its source to be available in real time across the supply chain to allow users to correctly and easily interpret system status. This will facilitate coordination of work tasks across the supply chain. In addition to capturing raw data from system transactions, value-adding and intellectual tasks should be automated and controlled by the system to the greatest degree possible. Also, intermediaries should be eliminated from the supply chain system to minimize organizational and functional interfaces. Common metrics and performance measurements should also be established for all participants of the supply chain to ensure strategic and tactical alignment of business goals and objectives throughout the system. The DFSS team should design their supply chain system using these high-level design concepts as a guide.

Over time, as the DFSS team becomes experienced in using DFSS techniques (many of which must be customized for their organization), the development of new systems will become a core competency. This design DFSS methodology will eliminate the rework caused by implementation failures due to poorly designed work streams. The secret to DFFS is to execute every step of the process flawlessly starting with VOC analysis, including optimization of the system using the correct supply chain tools and methods. This will allow an organization to design its supply chain from a global perspective and achieve real competitive dominance over its competitors.

Managing Organizational Change for Lean Six Sigma Deployments

Is your organization ready to change? Is it ready to systematically improve its operational performance using Lean Six Sigma methods? Organizational change is very difficult. Even under the best of circumstances it may take

5 to 20 years to convert an organization to a new behavioral system. This process will only occur if the new system and its methods have been useful to the organization. Because Lean Six Sigma operational improvements support strategic initiatives and organizational change, operational competencies are developed over time throughout the organization. This occurs as Lean Six Sigma belts analyze and improve key work streams, project by project, throughout the supply chain. These process improvements are accomplished as much by effective strategic deployment as through process modifications. As an organization begins to effectively change its work processes, customer service levels increase and its core competencies are improved over time. Integral to these improvement efforts are significant business benefits including cost and cycle-time reductions as well as quality improvements. Over time a synergy develops within an organization that accelerates process improvements, resulting in a high sense of urgency and a bias for change. The result is an accelerated learning rate for the organization. On the business side there is a growth in market share and operating margin.

However, to become a learning organization that can effectively formulate and execute strategy, an organization must develop core competencies that give it a competitive edge relative to its competition. As an example, some organizations compete based on price while others compete on operational performance or convenience. Integral to these competitive strategies is the concept of value (this concept was discussed in Chapter 1). Perceived customer value, by market segment, can be categorized relative to purchase price, time to acquire the product or service, the functionality of the product or service, the usefulness to the customer of the products and service, and the relative importance to the customer. To compete across these value dimensions, organizations must ask themselves where their core competencies are and where they need to partner with other organizations to cover their weak areas. Competencies can be developed in product and process design, supply chain design and operational efficiency, quality, and customer service as well as other areas of competence. Lean Six Sigma methods, if properly modified for supply chain applications, can be an extremely useful addition to aid operational improvement and the development of core competencies throughout the organization.

Whereas an organization wants to be agile, proactive, growth-oriented, and profitable, it must avoid the cultural degeneracy many organizations fall victim to over time. If an organization just stays level, without generating real growth, an environment is created in which change cannot occur. In these environments, there is often high bureaucracy, an internal focus rather than customer focus, and an inability to develop people and leverage resources. Many organizations fall into this category, but they are very successful. Why are they successful? Any successful organization has a competitive advantage due to one or more factors. Some of these factors may be internal to the organization while others are external. As an example, some organizations have a monopoly on the marketplace due to product development. Having a core competency in this area effectively protects them from having to change their culture or the day-to-day way they work. Other organizations are the lowest-cost producers in their industry. Customers may not even like them, but they cannot switch to other sources of supply. In either situation, if a competitor finally creates higher perceived value for the customer, then old competitive advantages may be lost. In summary, just because an organization is doing well today does not mean it will maintain its competitive advantage over time. What should an organization do? They should understand customer value and other relevant customer needs and requirements and then design simple and efficient systems to deliver this value flawlessly to their customers. Also, they should be able to dynamically adapt to changing customer value expectations over time.

How does an organization embrace change in an organized and systematic way to generate significant benefits for its customers and business? There are some basic principles which, if followed, will drive an organization onto the path of cultural change. This will allow it to more effectively compete over time across the world. Effective change begins with strategy. It must be part of senior management's strategic direction. Also, change must be planned across the organization with firm improvement targets for important operational and financial metrics. Change must be created and supported by key stakeholders throughout the business and an infrastructure created to identify what needs to be changed at a tactical level and how to correctly change the underlying systems to execute strategy. People throughout the organization must become change agents and have the ability to systematically make process changes based on fact within and between their local work groups. These changes must be based on process modifications and improvements that are sustainable over time and aligned with senior management's goals and objectives. Finally, over time as the organization executes change, project by project, the lessons learned relative to what works and does not work must be integrated into the organizational culture.

Lean Six Sigma Maturity

The Lean Six Sigma initiative has most of the characteristics of an effective change program. This why is has been so successful over the past several years. The goal of the Lean Six Sigma initiative is to proactively enable the transformation of an organization from a functionally oriented, reactionary set of isolated operations to a cross-functional, process-focused, continuously improving learning organization. As the organization systematically analyzes its current performance using Lean Six Sigma and other relevant methods, it will be able to proactively predict product and process performance. This will ensure high levels of customer satisfaction as the inherent process variation of major work streams is reduced by Lean Six Sigma improvement teams. These organizations will become the benchmarks within their industries. But successfully completing this journey will require management commitment at all levels of the organization. Enabling organizational success will be aggressive goals and objectives, common metrics used throughout the organization (all over the world), as well as accountability for effective project execution. Finally, honest communication and dialogue will accelerate organizational change and performance.

Final Thoughts

Effective deployment of a Lean Six Sigma initiative saves an organization time and money and improves customer satisfaction. Lean Six Sigma concepts apply to every industry, organization, and function throughout the world. The result of using the Lean Six Sigma change initiative is that your organizational culture will be more agile, flexible, leadership-oriented, employee-empowered, externally oriented toward its customers, and risk-tolerant. There are no secrets to effective Lean Six Sigma deployments. However, in supply chain environments the tools and methods must be correctly applied to answer the questions necessary to drive operational improvements. Many of the required tools and methods currently lie in the areas of operations management and operations research. Supply chains across the world have a lot to gain from the effective application of Lean Six Sigma principles.

Glossary

5-S techniques   Lean techniques that encourage local work teams to sort, set in order, shine, standardize, and sustain operations.

10-Step Solution Process   The 10-Step Solution Process is the basis for improving the 12 key metrics and is used to systematically improve process performance in conjunction with the team's balanced scorecard using the five-phase Lean Six Sigma methodology i.e. to define, measure, analyze, improve, and control (DMAIC) .

A

ABC inventory classification   Classification of inventory accounts based on dollar valuation or other criteria.

account receivable collection period   Three hundred sixty-five days divided by accounts receivable turnover.

account receivable turnover   Credit sales divided by average account receivables.

Advanced Shipping Notification (ASN)   System in which items are electronically received using bar code scanning.

affinity diagram   Used to organize large amounts of data into natural groups based on a common theme.

analysis of collected data   The process of applying analytical tools to understand relationships in data to obtain information.

analyze phase   The third DMAIC phase, in which data is analyzed to obtain relevant information.

Andon   A signal that shows abnormal processing condition.

assemble-to-order A system of manufacturing products from components and subassemblies based on firm customer orders.

B

backorder worksheet A Microsoft Excel worksheet that records backorder information by item and location.

Black belt A process improvement specialist who works projects between business functions.

Bill-of-Material (BOM) A document showing the hierarchical relationship of a product's components and build sequence.

bottleneck management Management of the key process constraint to ensure that it is fully utilized to meet the takt time requirements.

box plot A graphical analysis tool showing the first, and third, quartiles as well as the median (2nd quartile) of the sample data.

breakthrough improvement Changing process performance by a discontinuous amount, that is, a step change improvement.

business metrics A measure of operational performance used to provide information or to control a process.

business-value-added operations Operations valued by the external customer and required to meet their requirements.

C

capability ratio The ratio of the "range of customer specification limits" divided by the "process variation measured in standard deviations."

capacity The amount of material or information that can be put through a system in a unit amount of time.

capacity requirements The amount of capacity required to produce a unit of material or information.

cash investment The amount of cash required to maintain inventory in a system.

cause-and-effect diagram (C&E) A graphical tool that shows qualitative relationships between causes (X's) and their effect (Y).

Champion A person assigned to assist the Lean Six Sigma improvement team identify and execute projects.

Cluster sampling Drawing a sample based on natural grouping of the population, that is, by a stratification variable.

common sampling methods The four basic ways in which a random sample can be drawn from a population.

consignment inventory Supplier inventory that has been assigned to a customer usually without payment to the supplier until used by the customer.

continuous improvement Process improvement activities that are deployed over an extended time.

control phase   The fifth DMAIC phase, in which process improvements are transitioned back to the process owner.

control strategies   Combining various control tools to ensure that process improvements are sustained over time.

Cost-of-Goods-Sold (COGS)   The standard cost of goods and services sold.

Cp   The process capability when the process is on target and estimated using short-term variation.

Cpk   The process capability when the process is off target and estimated using short-term variation.

current ratio   Current assets divided by current liabilities.

cycle counting   The process of conducting physical audits to estimate inventory value.

cycle-counting system design   The process of designing inventory audits using labor, information, and materials in combination to provide information regarding inventory quantities and values.

D

data accuracy   The requirement data is not biased (or off-set) from its true average level.

data analysis   Various tools and methods designed to analysis data to create information to answer questions.

data collection   The process of bringing people and systems together to obtain data samples from a process for analysis

defect   A process nonconformance relative to a specific customer requirement for a particular characteristic.

Defects-Per-Million-Opportunities (DPMO)   A normalized quality metric calculated by multiplying Defects-per-Unit (DPU) by 1,000,000 and dividing by the opportunity count.

Defects-Per-Unit (DPU)   The ratio of total defects in a sample to total units sampled.

define phase   The first DMAIC phase, in which a project is defined relative to business benefits and customer requirements.

demand aggregation   The process of combining unit demand for several items and locations by time period (usually one month).

demand forecast   The process of estimating demand in the future (Usually 12 months) by discrete time period (usually one month).

demand history worksheet   An Excel worksheet containing actual customer demand history.

demand pull   The scheduling process in which actual customer demand is used to schedule upstream operations.

demand push   The scheduling process in which forecasted demand is used to schedule operations.

dependent variable   A variable that is explained by one or more other variables (which are independent).

descriptive statistics   Summarized information relative to the central location and dispersion of a sample.

Design-for-Manufacturing (DFM)   A design methodology using several rules and techniques, which attempts to simplify and modularize designs or easy assembly.

display board   A board placed in work areas to show key operational metrics and other information useful to improve and control processes.

Distribution Requirements Planning (DRP)   Inventory controlled in a distribution network based on demand placed on each distribution center.

DMAIC   The five-phase Six Sigma problem-solving methodology, in which the phases are define, measure, analyze, improve, and control.

E

Economic Order Quantity (EOQ)   An optimally calculated quantity that balances inventory ordering cost versus inventory holding cost.

economy of scale   A situation in which the variable cost of a product decreases as its manufacturing volume increases.

Electronic Data Interchange (EDI)   An electronic system that allows information technology (IT) systems to exchange information between organizations.

enabler initiative   An initiative that is used to execute improvement projects.

entitlement analysis   An analysis showing the best possible process performance, that is, higher capability level due to the process being on target with short-term variation.

equal variance test   A statistical test that compares the equivalence of two variances.

e-supply   Electronic supply chain processes between supplier and customers.

e-supply applications   Various organizational functions using e-supply technology and methods.

excess inventory   Inventory quantities exceeding expected demand during the order cycle.

execution of the plan   The ability to make a plan and ensure that it is completed on time, stays within budget, and meets all goals and objectives.

experimental design   An analysis of the impact on a dependent variable of changing levels of one or more independent variables in a controlled and planned manner.

exponential smoothing   A type of forecasting method in which one or more parameters are used to build a forecasting model using historical data of a time series to forecast future values of the same time series.

F

Failure Modes and Effects Analysis (FMEA)   An analytical tool showing the relationship of failure modes to causes and evaluating their risk levels.

FIFO inventory valuation   First-in-first-out inventory management.

financial justification   The process of developing financial estimates of improvement project benefits.

finished goods inventory   Inventory that is ready for shipment to an external customer.

finite loading   Scheduling a product with consideration for its subcomponent demand on the internal work centers (available capacity).

first-pass yield   The number of units meeting requirements at a single operation excluding rework and scrap.

forecasting accuracy   A metric that measures agreement between actual quantities demanded versus forecasts by time period (usually one month).

forecasting benchmarks   Forecasting accuracy metrics considered best-in-class.

forecasting error   A metric that measures variance between actual quantities demanded versus forecasts by time period (usually one month).

forecasting error worksheet   An Excel worksheet that records forecasting error information by item and location.

forecasting history worksheet   An Excel worksheet that records forecasting history information by item and location.

forecasting new products   The process of estimating demand for new products by time period (one month).

forecasting products   The process of estimating demand for current products by time period (one month).

forecasting reconciliation   The process of smoothing out a forecast based on factors external to the forecasting model to ensure "one number" for an organization.

forecasting roadmap   A map of how a forecast is developed within an organization.

forecasting stratification   The process of categorizing items by their demand variation and volume to focus forecast analysts' attention on the more critical time series.

forward scheduling   A scheduling method that uses a forecast to schedule manufacture of a product.

functional interface   The hand-off between organization functions.

functional silos   Different departments or work areas within an organization each having different responsibilities and work tasks.

G

Gantt chart A chart showing the interrelationships between the start and finish time of work tasks in a project.

Green belt Process improvement specialists who works projects within their function.

H

histogram A graphical tool that shows the central location and dispersion of an independent variable.

I

improve phase The fourth phase of the DMAIC improvement methodology.

infinite loading Scheduling a product without consideration of its subcomponent demand on internal work centers (available capacity).

input-output matrices Matrices that correlate input variables relative to output variables.

integrated supply chains A supply chain that is linked by information technology to provide system status in real time.

inter-relationship diagram Used to show the spatial relationships between many factors to qualitatively identify causal factors.

inventory age Three hundred sixty-five days divided by inventory turnover.

inventory analysis algorithm An algorithm that uses lead-time, demand variation, and service level to estimate the optimum inventory quantity.

inventory balance An analysis of optimum versus the actual inventory quantity to determine excess or insufficient inventory quantities.

inventory history worksheet An Excel worksheet that records inventory history information by item and location.

inventory improvement plan A plan to reduce inventory investment or improve customer service based on data analysis.

inventory investment The amount of money required to maintain inventory to meet target service levels.

inventory make-to-stock model A model that calculates required finished goods inventories by item and location.

inventory turnover Cost-of-goods-sold divided by average inventory investment.

J

Just-in-time (JIT) system An operational system which supplies materials as needed based on stable lead-times and demand obtained through operational standardization and high quality.

K

Kanban   A Kanban is a sign attached to inventory to show order status. There are transport, production, and signal Kanbans.

Kanban calculation   Sets the WIP inventory based on lead-time, demand and service level.

Kano   Kano needs consist of basic needs, performance needs, and excitement needs.

Key Process Input Variable (KPIV)   Independent variables that have a significant impact on key process output variables (KPOV).

Key Process Output Variable (KPOV)   Dependent variables that are directly correlated to voice-of-the-customer (VOC) requirements.

kitting   The process of bringing together components for use in service kits, replacement parts, and other applications.

L

lead-time   The time to complete all work tasks.

lead-time analysis   The process of decomposing lead-time into its time elements.

lead-time components   Lead-time can be broken into subcomponents related to transportation, setup, waiting, processing, inspection, and idle time.

lead-time reduction   The process of reducing non-value-adding time within a process.

lead-time variation   The dispersion of task completion times within an operation or work task.

lead-time worksheet   An Excel worksheet that records lead-time history information by item and location.

Lean performance measurements   A series of operational and financial metrics that can be used to evaluate the effectiveness of a Lean deployment.

Lean supply chain   A supply chain that uses a minimum amount of resources to satisfy customer demand.

LIFO inventory valuation   Last-in-first-out inventory valuation.

linear programming   A mathematical technique used in operations research to maximize or minimize an objective function subject to constraints.

linear regression   A statistical technique used to develop a linear relationship between several independent variables and a dependent variable.

long-term variation   The variation that impacts a process over a long period of time.

Lower Specification Limit (LSL)   Lower level of a customer specification.

M

make-to-order   A manufacturing system that makes products based on firm customer orders.

make-to-stock   A manufacturing system that makes product based on a forecast.

make-to-stock model worksheet   An Excel worksheet that calculates optimum inventory quantities by item and location.

Master black belt (MBB)   A process improvement specialist who works projects across business units and trains black belts and green belts.

Master Production Schedule (MPS)   System that aggregates demand for the MRPII system.

Master Requirements Planning (MRPII)   System that uses demand from the MPS and explodes it using the bill-of-materials (BOM) throughout the supply chain.

measure phase   The second phase of the DMAIC process.

measurement accuracy   The ability to measure a characteristic and be correct on average over many samples of the characteristic.

measurement linearity   The ability of a measuring device to measure a characteristic over its entire range with equal variation (error).

measurement repeatability   The variation between measurements of a characteristic when made by the same person (or machine) on the same thing.

measurement reproducibility   The ability of two or more people (or machines) to measure a characteristic with low variation between them.

measurement resolution   The ability of the measurement system to discriminate changes in the characteristic being measured (1/10 rule).

measurement stability   The ability of a measurement system to obtain the same measured value over time.

metric scorecard   A matrix that is used to record and track process metrics by metric type.

min/max inventory system   An inventory system that orders inventory when a predetermined amount is used based on the item's reorder point.

mistake proofing   A set of Lean techniques used to prevent or detect process errors and defects.

mixed-model operations   A scheduling system using the commonality of a product and its process design to more frequently schedule the manufacture of the product without increasing set-up cost.

MRPII-driven Kanban   A system in which the MRPII system is used as a Kanban system to control material or informational flow through the system.

multiple linear regression   A mathematical technique used to develop a linear relationship between several independent variables and one dependent variable.

N

non-value adding operations Operations that can be eliminated from a process because they have no value from an external customer viewpoint.

O

obsolete inventory Inventory quantities that can no longer be sold or if sold cannot be sold at standard prices.

One-sample t-test A statistical test that compares the equivalence of one sample mean to a test constant.

One Way Analysis-of-Variance (ANOVA) A statistical test that compares the equivalence of several means.

on-time delivery Delivery of supplier's shipment quantity at the agreed-upon lead-time without process defects.

operational balancing Every operation in a system contributes the material or information flow necessary to maintain the takt time.

operational linkage Ensuring that operational metrics are consistent across functional boundaries.

operational planning overview A group of cross-functional people who meet to determine the best strategy to ensure that available supply will meet expected customer demand.

operations management A field of study that includes systems, tools, methods and concepts useful in the analysis of the transformation of inputs into outputs.

operations research A field of study that includes mathematical tools, methods and concepts useful in the analysis of operations.

overdue backlog Quantity of product that could not be manufactured according to customer demand due to capacity constraints.

P

Pareto chart A graphical bar chart listing frequency or occurrence by classification category with the categories displayed in descending order.

Parts-per-Million (PPM) A metric used to measure very high-quality processes by multiplying the defect fraction by 1,000,000.

payback The number of years required to recover an initial investment.

percent error The number of defects found in the sample divided by the total sample and multiplied by 100.

performance gap The difference between target and actual performance level.

performance measurements Metrics used to measure process changes.

periodic review inventory system An inventory model that checks inventory at specific times and orders quantities to bring inventory up to a target.

Perpetual Inventory Model (PIM)   An inventory model that continuously tracks inventory quantities and orders an economic order quantity at their reorder point.

pilot-of-solution   A test of proposed process change within the actual process, but under controlled conditions.

Plan-Do-Check-Act   A process improvement cycle used for problem solving.

Pp   Process capability when the process is on target and estimated using long-term variation.

Ppk   Process capability when the process is off target and estimated using long-term variation.

prioritization matrices   Used to prioritize decisions based on various weighting methods.

problem statement   A verbal description of the problem that must be solved by the project.

process capability   A method used to compare process performance against customer specifications.

process improvement projects   Projects used to close metric performance gaps.

process mapping   A method used to show the movement of materials and information through the system.

product life cycle   The demand phases a product goes through over time including its introduction, growth, maturity, and decline.

product proliferation   Allowing products having little demand or margin contribution to remain active in the organization's systems.

Production Activity Control (PAC)   A manufacturing function on the shop floor used to schedule work through work centers.

profit and loss (P/L)   A key financial statement that compares income and costs to show whether profit has been made by the organization.

profitability index   The ratio of present value of cash inflows to present value of cash outflows.

project charter   A document in either electronic or paper format, which describes a project and provides financial justification for a project.

project management   A set of tools and techniques used to manage a project.

project objective   A section of the project charter that states the specific business benefits of the project as well as their timing.

project planning   The process of scheduling work tasks that are necessary to complete the project.

project resources   These are materials, labor, money, and information that are necessary to complete the project.

project selection A process of identifying improvement opportunities that will benefit the business.

pull scheduling A visual scheduling system in which the manufacturing system produces according to external customer demand.

push scheduling system A manufacturing scheduling system that uses a forecast through the master production schedule to schedule orders.

Q

quantitative forecasts Forecasts that rely on mathematical models to predict future demand.

Queuing analysis A mathematical technique used to analyze the probabilistic relationships between arrivals into a system relative to their service times.

quick ratio Cash plus marketable securities plus accounts receivables, divided by current liabilities.

R

random sampling A representative sample drawn from a population.

raw material inventory Inventory that has been received from an external supplier.

reclamation center A distribution system that is designed to receive returned product and process it according to organizational policies.

reorder point A point in time in an inventory system in which an EOQ is ordered to arrive just when current inventory becomes zero, that is, lead-time inventory quantity.

response variable A dependent variable that is also called a key.

Return-on-Investment (ROI) Net income divided by available total assets.

reverse logistics Logistical operations associated with receipt and processing of customer returns.

Rolled-Throughout-Yield (RTY) Calculated by multiplying first-pass yield of each operation (which are in serial relative to each other) together.

Root Mean Squared Error (RMSE) Forecasting error statistic comparing actual to forecast quantities to provide unit forecast error. It is also used to set inventory safety stocks.

run chart A time-series chart used to analyze time-dependent data in which observations are arranged by time order.

S

safety-stock calculation Inventory quantity calculated as service factor multiplied by the standard deviation of unit demand and the square root of its lead-time.

Sales and Operational Planning (S&OP)   The process of coordinating demand and supply based on system constraints.

sample mean   Calculated as the sum of the sample observations divided by their number.

sample standard deviation   The square root of the average (degrees-of-freedom minus one) squared deviations of sample observations from their calculated sample mean and expressed in units. It is useful in calculating safety-stock levels.

sample variance   The standard deviation squared.

sampling   The process of obtaining a sample from a larger population. There are many types of sampling methods.

scheduling algorithms   Mathematical models that help schedule resources.

shipment history worksheet   An Excel worksheet that records shipment history by item and location.

short-term variation   Variation acting on a process for a limited amount of time.

sigma level   The short-term capability of a key process output variable (KPOV).

simulation   A mathematical technique using a reference distribution and random numbers to re-create observations from the reference distribution.

Single-Minute Exchange of Dies (SMED)   A set of techniques to reduce the time of job changeovers.

SIPOC   A high-level process map. SIPOC is an acronym for supplier, input boundary, process, output boundary, and customer.

Six Sigma   A problem solving methodology consisting of five phases called Define, Measure, Analyze, Improve and Control. From a mathematical perspective, six sigma means plus or minus six standard deviations (short-term variation analysis) of the KPOV fitting within the customer's lower and upper specification limits.

Standard Data Worksheet   An Excel worksheet that records system standard data and constants by item and location.

Standard Operating Procedures (SOP)   Procedures that are determined to be the best way to do a job.

statistical inference   A set of mathematical techniques that allow population parameters to be estimated with predetermined confidence.

statistical sampling   A set of methods that specify how observations are to be drawn from a population.

stocking location   A storage location for an item within a distribution center.

strategic project selection   The process of ensuring that a project is selected to align with senior management's goals and objectives.

stratified sampling A type of sampling method in which a population is broken into subgroups each having minimum variation.

system model map A quantified process map of a process showing input and output metrics.

systematic sampling A sampling method in which a sample is taken from every *n*th ordered subgroup.

T

takt time calculation A calculation that determines how much time it takes to manufacture one unit.

third-party logistics Outsourcing one or more internal logistical functions to external organizations.

time fence The cumulative lead-time to build a product.

time series Data that is sequentially arranged by time order.

total productive maintenance (TPM) A set of methods to ensure that machines are maintained and available at a predetermined percentage of the time.

transfer batches Work that is transferred to downstream operations without being batched.

tree diagram Used to map higher-level to lower-level relationships.

Twelve key metrics Twelve metrics that a Lean Six Sigma team can analyze to identify supply chain improvement projects.

Two-sample t-test A statistical test comparing the equivalence of two means.

U

unidentified task Goals and objectives that do not currently have projects assigned to ensure their execution.

unplanned orders Orders put into the schedule without regard for the product's lead-time or time fence.

Upper Specification Limit (USL) The customer's upper requirement level for the product characteristic.

V

value-adding-operations Operations that the customer is willing to pay for.

value stream maps Process maps that help analyze operations the customer values from those not valued by the customer.

visual displays Graphics used to convey system status in the workplace.

visual workplace A workplace in which system status can be determined immediately by looking at visual metric displays.

Voice-of-the-Business (VOB)   The variation that is exhibited by a process.

Voice-of-the-Customer (VOC)   The customer requirements or specifications.

W

waiting-line models   Mathematical models showing relationships between available resources versus arriving customers.

working capital   Current assets minus current liabilities.

Work-in-Process Inventory (WIP)   Inventory within a manufacturing process acting as a buffer against disruptions in material flow.

Index